T0145085

Smart Innovation, Systems and Technologies

Volume 58

Series editors

Robert James Howlett, KES International, Shoreham-by-sea, UK
e-mail: rjhowlett@kesinternational.org

Lakhmi C. Jain, University of Canberra, Canberra, Australia;
Bournemouth University, UK;
KES International, UK
e-mails: jainlc2002@yahoo.co.uk; Lakhmi.Jain@canberra.edu.au

About this Series

The Smart Innovation, Systems and Technologies book series encompasses the topics of knowledge, intelligence, innovation and sustainability. The aim of the series is to make available a platform for the publication of books on all aspects of single and multi-disciplinary research on these themes in order to make the latest results available in a readily-accessible form. Volumes on interdisciplinary research combining two or more of these areas is particularly sought.

The series covers systems and paradigms that employ knowledge and intelligence in a broad sense. Its scope is systems having embedded knowledge and intelligence, which may be applied to the solution of world problems in industry, the environment and the community. It also focusses on the knowledge-transfer methodologies and innovation strategies employed to make this happen effectively. The combination of intelligent systems tools and a broad range of applications introduces a need for a synergy of disciplines from science, technology, business and the humanities. The series will include conference proceedings, edited collections, monographs, handbooks, reference books, and other relevant types of book in areas of science and technology where smart systems and technologies can offer innovative solutions.

High quality content is an essential feature for all book proposals accepted for the series. It is expected that editors of all accepted volumes will ensure that contributions are subjected to an appropriate level of reviewing process and adhere to KES quality principles.

More information about this series at http://www.springer.com/series/8767

Gordan Jezic · Yun-Heh Jessica Chen-Burger
Robert J. Howlett · Lakhmi C. Jain
Editors

Agent and Multi-Agent Systems: Technology and Applications

10th KES International Conference, KES-AMSTA 2016 Puerto de la Cruz, Tenerife, Spain, June 2016 Proceedings

Editors
Gordan Jezic
Faculty of Electrical Engineering and Computing
University of Zagreb
Zagreb
Croatia

Yun-Heh Jessica Chen-Burger
School of Mathematical and Computer Sciences
Heriot-Watt University
Edinburgh
UK

Robert J. Howlett
KES International
Shoreham-by-sea
UK

Lakhmi C. Jain
University of Canberra
Canberra
Australia

and

Bournemouth University
Poole
UK

and

KES International
Shoreham-by-sea
UK

ISSN 2190-3018 ISSN 2190-3026 (electronic)
Smart Innovation, Systems and Technologies
ISBN 978-3-319-81988-4 ISBN 978-3-319-39883-9 (eBook)
DOI 10.1007/978-3-319-39883-9

Printed on acid-free paper

This Springer imprint is published by Springer Nature
The registered company is Springer International Publishing AG Switzerland

Preface

This volume contains the proceedings of the 10th KES Conference on Agent and Multi-Agent Systems—Technologies and Applications (KES-AMSTA 2016) held in Puerto de la Cruz, Tenerife, Spain, between 15 and 17 June 2016. The conference was organized by KES International, its focus group on agent and multi-agent systems and University of Zagreb, Faculty of Electrical Engineering and Computing. The KES-AMSTA conference is a subseries of the KES conference series.

Following the successes of previous KES Conferences on Agent and Multi-Agent Systems—Technologies and Applications, held in Sorrento, Italy (KES-AMSTA 2015); Chania, Greece (KES-AMSTA 2014); Hue, Vietnam (KES-AMSTA 2013); Dubrovnik, Croatia (KES-AMSTA 2012); Manchester, UK (KES-AMSTA 2011); Gdynia, Poland (KES-AMSTA 2010); Uppsala, Sweden (KES-AMSTA 2009); Incheon, Korea (KES-AMSTA 2008); and Wrocław, Poland (KES-AMSTA 2007), the conference featured the usual keynote talks, oral presentations and invited sessions closely aligned to the established themes of the conference.

KES-AMSTA is an international scientific conference for discussing and publishing innovative research in the field of agent and multi-agent systems and technologies applicable in the digital and knowledge economy. The aim of the conference was to provide an internationally respected forum for both the research and industrial communities on their latest work on innovative technologies and applications that is potentially disruptive to industries. Current topics of research in the field include technologies in the area of mobile and cloud computing, big data analysis, business intelligence, artificial intelligence, social systems, computer embedded systems and nature inspired manufacturing. Special attention is paid on the feature topics: business process management, agent-based modelling and simulation, anthropic-oriented computing, learning paradigms, and business informatics and gaming.

The conference attracted a substantial number of researchers and practitioners from all over the world who submitted their papers for main track covering the methodologies of agent and multi-agent systems applicable in the digital and knowledge economy, and five invited sessions on specific topics within the field. Submissions came from 16 countries. Each paper was peer-reviewed by at least two members of the International Programme Committee and International Reviewer Board. 28 papers were selected for oral presentation and publication in the volume of the KES-AMSTA 2016 proceedings.

The Programme Committee defined the main track entitled Agent and Multi-Agent Systems and the following invited sessions: Agent-based Modeling and Simulation (ABMS), Business Process Management (BPM), Learning Paradigms and Applications: Agent-based Approach (LP:ABA), Anthropic-Oriented Computing (AOC), and Business Informatics and Gaming through Agent-based Modelling.

Accepted and presented papers highlight new trends and challenges in agent and multi-agent research. We hope that these results will be of value to the research community working in the fields of artificial intelligence, collective computational intelligence, robotics, dialogue systems and, in particular, agent and multi-agent systems, technologies, tools and applications.

The Chairs' special thanks go to the following special session organizers: Dr. Roman Šperka, Silesian University in Opava, Czech Republic; Prof. Mirjana Ivanović, University of Novi Sad, Serbia; Prof. Costin Badica, University of Craiova, Romania; Prof. Zoran Budimac, University of Novi Sad, Serbia; Prof. Manuel Mazzara, Innopolis University, Russia; Max Talanov, Kazan Federal University and Innopolis University, Russia; Prof. Jordi Vallverdú, Universitat Autònoma de Barcelona, Spain; Prof. Salvatore Distefano, University of Messina, Italy; Prof. Robert Lowe, University of Skövde/University of Gothenburg, Sweden; Prof. Joseph Alexander Brown, Innopolis University, Russia; Assoc. Prof. Setsuya Kurahashi, University of Tsukuba, Japan; Prof. Takao Terano, Tokyo Institute of Technology, Japan; Prof. Hiroshi Takahashi, Keio University, Japan; and España for their excellent work.

Thanks are due to the Programme Co-chairs, all Programme and Reviewer Committee members, and all the additional reviewers for their valuable efforts in the review process, which helped us to guarantee the highest quality of selected papers for the conference.

We cordially thank all authors for their valuable contributions and all of the other participants in this conference. The conference would not be possible without their support.

April 2016

Gordan Jezic
Yun-Heh Jessica Chen-Burger
Robert J. Howlett
Lakhmi C. Jain

KES-AMSTA 2016 Conference Organization

KES-AMSTA 2016 was organized by KES International—Innovation in Knowledge-Based and Intelligent Engineering Systems.

Honorary Chairs

I. Lovrek, University of Zagreb, Croatia
L.C. Jain, University of Canberra, Australia; and Bournemouth University, UK

Conference Co-chairs

G. Jezic, University of Zagreb, Croatia
J. Chen-Burger, The Heriot-Watt University, Scotland, UK

Executive Chair

R.J. Howlett, University of Bournemouth, UK

Programme Co-chairs

M. Kusek, University of Zagreb, Croatia
R. Sperka, Silesian University in Opava, Czech Republic

Publicity Chair

P. Skocir, University of Zagreb, Croatia

International Programme Committee

Dr. Dariusz Barbucha, Gdynia Maritime University, Poland
Prof. Costin Badica, University of Craiova, Romania
Dr. Marina Bagić Babac, Faculty of Electrical Engineering and Computing, University of Zagreb, Croatia
Dr. Iva Bojic, Singapore-MIT Alliance for Research and Technology, Singapore
Dr. Gloria Bordogna, CNR IREA, Italy
Joseph Alexander Brown, Innopolis University, Russia
Dr. Grażyna Brzykcy, Poznań University of Technology, Department of Control and Information Engineering, Poland
Prof. Zoran Budimac, University of Novi Sad, Serbia
Prof. Frantisek Capkovic, Slovak Academy of Sciences, Slovak Republic
Dr. Jessica Chen-Burger, The Heriot-Watt University, Scotland, UK
Dr. Angela Consoli, Defence Science and Technology Group, Australia
Prof. Ireneusz Czarnowski, Gdynia Maritime University, Poland
Prof. Radhakrishnan Delhibabu, Kazan Federal University, Russia
Salvatore Distefano, University of Messina, Italy; Kazan Federal University, Russia
Dr. Arnulfo Alanis Garza, Instituto Tecnologico de Tijuana, Mexico
Prof. Chihab Hanachi, University of Toulouse, France
Dr. Quang Hoang, Hue University, Vietnam
Prof. Zeljko Hocenski, Faculty of Electrical Engineering, University Josip Juraj Strossmayer in Osijek, Croatia
Prof. Tzung-pei Hong, National University of Kaohsiung, Taiwan
Dr. Adrianna Kozierkiewicz-Hetmańska, Wroclaw University of Technology, Poland
Prof. Mirjana Ivanovic, University of Novi Sad, Serbia
Prof. Piotr Jedrzejowicz, Gdynia Maritime University, Poland
Prof. Dragan Jevtic, University of Zagreb, Zagreb, Croatia
Dr. Arkadiusz Kawa, Poznan University of Economics, Poland
Prof. Petros Kefalas, The University of Sheffield International Faculty, Greece
Assoc. Prof. Setsuya Kurahashi, University of Tsukuba, Japan
Prof. Mario Kusek, University of Zagreb, Croatia
Prof. Kazuhiro Kuwabara, Ritsumeikan University, Japan
Dr. Konrad Kułakowski, AGH University of Science and Technology, Poland
Robert Lowe, University of Skövde/University of Gothenburg, Sweden
Dr. Marin Lujak, University Rey Juan Carlos, Spain
Dr. Manuel Mazzara, Innopolis University Russia
Dr. Daniel Moldt, University of Hamburg, Germany
Prof. Cezary Orłowski, Gdansk School of Banking, Poland
Assist. Prof. Vedran Podobnik, University of Zagreb, Croatia
Prof. Bhanu Prasad, Florida A&M University, USA
Prof. Radu-Emil Precup, Politehnica University of Timisoara, Romania
Rajesh Reghunadhan, Central University of South Bihar, India
Prof. Silvia Rossi, University of Naples “Federico II”, Italy

Mr. James O'Shea, Manchester Metropolitan University, UK
Dr. Roman Sperka, Silesian University in Opava, Czech Republic
Prof. Darko Stipanicev, University of Split, Croatia
Prof. Ryszard Tadeusiewicz, AGH University of Science and Technology, Kraków, Poland
Prof. Hiroshi Takahashi, Keio University, Japan
Prof. Yasufumi Takama, Tokyo Metropolitan University, Japan
Max Talanov, Kazan Federal University and Innopolis University, Russia
Prof. Takao Terano, Tokyo Institute of Technology, Japan
Dr. Wojciech Thomas, Wroclaw University of Technology, Poland
Dr. Krunoslav Trzec, Ericsson Nikola Tesla, Croatia
Prof. Taketoshi Ushiama, Kyushu University, Japan
Prof. Jordi Vallverdú, Universitat Autònoma de Barcelona, Spain
Prof. Bay Vo, Ho Chi Minh City University of Technology, Ho Chi Minh City, Vietnam
Prof. Toyohide Watanabe, Nagoya University, Japan
Mrs. Izabela Wierzbowska, Gdynia Maritime University, Poland
Prof. Mahdi Zargayouna, University of Paris-Est, IFSTTAR, France
Prof. Arkady Zaslavsky, Data61 at CSIRO, Australia

Workshop and Invited Session Chairs

Business Process Management

Dr. Roman Šperka, Silesian University in Opava, Czech Republic

Agent-Based Modelling and Simulation

Dr. Roman Šperka, Silesian University in Opava, Czech Republic

Anthropic-Oriented Computing

Prof. Manuel Mazzara, Innopolis University, Russia
Max Talanov, Kazan Federal University and Innopolis University, Russia
Prof. Jordi Vallverdu, Universitat Autonoma de Barcelona, Spain
Prof. Salvatore Distefano, University of Messina, Italy; Kazan Federal University, Russia
Prof. Robert Lowe, University of Skovde, University of Gothenburg, Sweden
Prof. Joseph Alexander Brown, Innopolis University, Russia

Learning Paradigms and Applications: Agent-Based Approach

Prof. Mirjana Ivanovic, University of Novi Sad, Serbia
Prof. Zoran Budimac, University of Novi Sad, Serbia
Prof. Costin Badica, University of Craiova, Romania
Prof. Lakhmi Jain, University of Canberra, Australia and Bournemouth University, UK

Business Informatics and Gaming Through Agent-Based Modelling

Assoc. Prof. Setsuya Kurahashi, University of Tsukuba, Japan
Prof. Takao Terano, Tokyo Institute of Technology, Japan
Prof. Hiroshi Takahashi, Keio University, Japan

Contents

Part I
Agent and Multi-agent Systems

Faceted Query Answering in a Multiagent System of Ontology-Enhanced Databases

Tadeusz Pankowski and Grażyna Brzykcy

Abstract We discuss an architecture and functionality of a multiagent system supporting data integration based on faceted query evaluation in a system of ontology-enhanced graph databases. We use an ontology (a subset of OWL 2 RL profile) to support a user in interactive query formulation. A set of server agents cooperate one with another in query answering using the ontological knowledge for query rewriting and for deciding about query propagations. We propose some algorithms ensuring tractable complexity of query rewriting and query evaluation.

Keywords Multiagent systems · Graph databases · Ontologies · Faceted queries

1 Introduction

In last decade, we observe the rise of a new category of database systems, where classical extensional databases are enhanced by ontologies allowing for deducing new intensional knowledge contributing query answering. In particular, such ontological enhancement is applied in the case of WWW datasets, where RDF data [14], constituting the extensional part of a database, is enriched with rules (axioms) in compliance with OWL 2 profiles [11].

Related work The semantic technologies involving RDF data and OWL 2 ontologies, are often used to build specialized repositories based on a global schema defined by the terminological part of the ontology. This class of database systems is often referred to as *Ontology-Based Data Access* (OBDA). In particular, it is used to achieve semantic integration of data [4–7]. Then a challenging issue is to answer queries in a multidatabase environment. A standard way for querying graph data-

T. Pankowski (✉) · G. Brzykcy
Institute of Control and Information Engineering, Poznań University of Technology, Poznań, Poland
e-mail: tadeusz.pankowski@put.poznan.pl

G. Brzykcy
e-mail: grazyna.brzykcy@put.poznan.pl

G. Jezic et al. (eds.), *Agent and Multi-Agent Systems: Technology and Applications*, Smart Innovation, Systems and Technologies 58,
DOI 10.1007/978-3-319-39883-9_1

bases (mainly RDF databases) is SPARQL [15]. However, it is not a suitable language for end-users. Thus, another user-friendly querying facilities have been proposed, such as keyword search [18], and faceted search [2].

Contributions In this paper, we assume that the ontology-enhanced multidatabase system constitutes a multiagent system, called DAFO (*Data Access based on Faceted queries over Ontologies*). There is a *manager agent* (responsible for interaction with users and for query reformulation based on the ontology), and a set of *server agents* cooperating with one another in query answering. Each server agent has its own local database in which evaluates queries, and can consult other agents while producing answers. We assume that the query language is based on a faceted search [2].

Novelties of this paper are as follows: (a) we propose an architecture of multiagent system, where agents cooperate in query answering against ontology enhanced databases; (b) we propose an algorithm for translating *faceted queries* (FQs) created in interaction with a user, into a class of *first order* (FO) queries, called FO FQs; (c) we propose a way of efficient evaluation of FO FQs, including cooperation between agents and query reformulation.

Preliminaries A *graph database*, $G = (V, E)$, is a finite, labeled and directed graph [3]. Nodes (V) of the graph are elements from a set Const of *constants*, and from a set of *unary predicates* UP. Edges (E) are labeled by *binary predicates* from a set BP, in which two predicates, type and $\approx$, are distinguished. Additionally, we assume that a set LabNull of *labeled nulls* is a subset of Const. For constants not in LabNull we assume the *unique name assumption* (UNA), while for LabNull the UNA is not required [8]. Thus, like in OWL, two different symbols from LabNull can denote the same individual. In FO notation, $C(v)$ stands for an edge (v, type, C); $v_1 \approx v_2$—for an edge $(v_1, \approx, v_2)$, and $P(v_1, v_2)$—for an edge (v_1, P, v_2). A *rule* is a FO implication of the form $\forall \mathbf{x} \forall \mathbf{y}(\varphi(\mathbf{x}, \mathbf{y}) \rightarrow \exists \mathbf{z} \psi(\mathbf{x}, \mathbf{z}))$. If the tuple $\mathbf{z}$ of existentially quantified variables is empty, the rule is a *Datalog rule*. An *ontology* is a triple $\mathcal{O} = (\Sigma, R, G)$, where Σ is a signature, R is a finite set of rules, and G is a database graph (a set of facts). Σ consists of all and only unary and binary predicates occurring in R and G. A FO *query* (a query, for short) is a FO formula. If in a query occur only symbols of conjunction ($\wedge$), disjunction ($\vee$), and existential quantification ($\exists$), then the query is a *positive existential query* (PEQ). The query is *monadic* if it has exactly one free variable. Aforementioned FO FQs form a subset of monadic PEQs. A constant a is an *answer* to a query $Q(x)$ against $\mathcal{O}$, if $Q(a)$ is satisfied in $\mathcal{O}$, i.e., $G \cup R \models Q(a)$, where $G \cup R$ denotes all facts belonging to G and deduced from G using rules in R.

Outline In Sect. 2, we introduce an example, which motivates our research and is used to illustrate the considerations. Architecture of the system is discussed in Sect. 3. In Sect. 4 we show how faceted queries can be interactively specified and how can be translated into FO FQs. Evaluating of FO FQs in a multiagent system is studied in Sect. 5. Section 6 concludes the paper.

2 Motivation Example

Let us assume that there are three graph databases, G_1, G_2, and G_3, presented in Fig. 1. Against them we want to evaluate the query: "*Find ACM authors who are from University of NY, and some of their papers were published in 2014 at ACMConf or at KESConf*". The FO form of the query is:

$$\begin{array}{c} Q(x) = ACMAuthor(x) \wedge \exists w(univ(x, w) \wedge w \approx NY) \wedge \\ \exists y(authorOf(x, y) \wedge Paper(y) \wedge \exists z(pyear(y, z) \wedge z \approx 2014) \wedge \\ \exists v(atConf(y, v) \wedge (ACMConf(v) \vee KESConf(v)))). \end{array} \quad (1)$$

We see that without additional information, the answer to this query is empty. In particular, some predicates (e.g., *ACMAuthor*) do not even exist in considered databases. Such additional information can be provided by means of the following rules (only some relevant rules are listed):

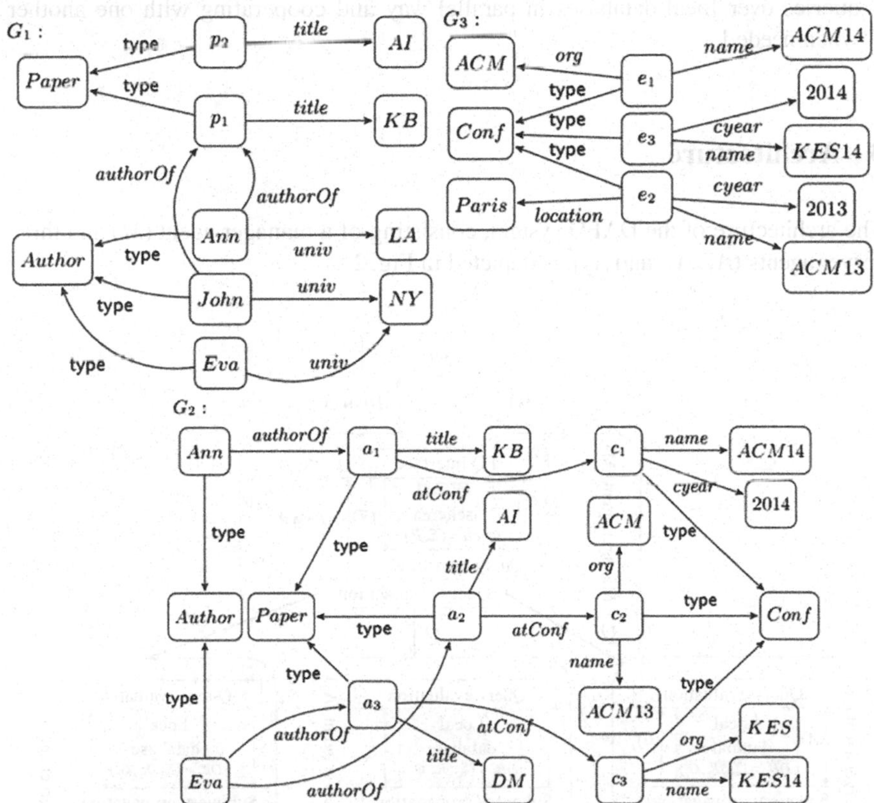

Fig. 1 Sample graph databases (lower case node identifiers are labeled nulls)

R :
$title(x,y) \to Paper(x) \wedge String(y)$, $authorOf(x,y) \to Author(x) \wedge Paper(y)$,
$title(x,y_1) \wedge title(x,y_2) \to y_1 \approx y_2$, $title(x_1,y) \wedge title(x_2,y) \to x_1 \approx x_2$,
$atConf(x,y) \wedge cyear(y,z) \to pyear(x,z)$, $Author(x) \to \exists y\, authorOf(x,y)$.

$atConf(x,y) \wedge ACMConf(y) \to ACMPaper(x)$, $org(x, ACM) \to ACMConf(x)$,
$authorOf(x,y) \wedge ACMPaper(y) \to ACMAuthor(x)$,
$ACMAuthor(x) \to Author(x)$.

A set R of rules, together with graph databases in Fig. 1, form an ontology $\mathcal{O} = (\Sigma, R, G_1 \cup G_2 \cup G_3)$. A crucial issue is then the organization of an ontology-enhanced database system, which integrates (virtually) a set of autonomous local databases while answering queries. In this paper, we assume the following:

- a multiagent architecture of the system,
- queries are *faceted queries* formulated by means of a graphical interface,
- query answering is based on an agent-oriented approach with agents evaluating queries over local databases in parallel way and cooperating with one another when needed.

3 Architecture

The architecture of the DAFO system, consisting of a manager agent (M) and three server agents (A_1, A_2, and A_3), is depicted in Fig. 2.

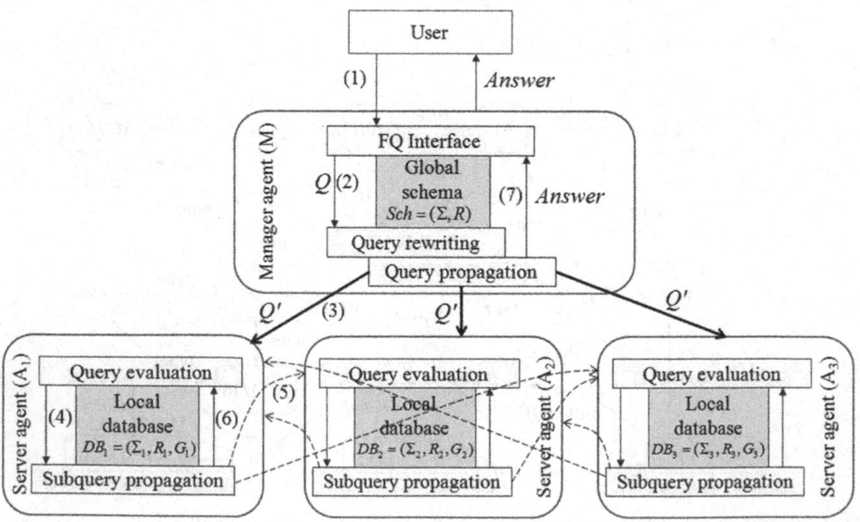

Fig. 2 Architecture of DAFO, a multiagent system for data access based on faceted queries over ontology-enhanced databases

The architecture follows so called *single ontology approach*, i.e., an approach when a single ontology is used as a global reference model in the system [9, 16]. Like OBDA systems, the architecture involves an ontology to effectively combine data from multiple heterogeneous sources [17].

It is essential in our approach, that the user is free from having to know the global schema. Instead, by interacting with the system using faceted query interface (FQ Interface), she or he builds a faceted query in an interactive manner (1). The global schema $Sch = (\Sigma, R)$ is the intentional (terminological) part of the ontology. The schema is used to support query creation (1). Next, the query is translated into FO FQ query Q, and rewritten into Q' (2). Then, query Q' is asynchronously sent to all server agents (3). An agent do some local database specific rewritings and evaluations (4). If there is a need for complementary queries, the agent propagates boolean queries to all partner agents (5), and gathers local answers (6). Finally, answers obtained from server agents are collected by manager agent and returned to the user (7).

Each server agent A_i has its local database G_i and a subset of the global ontology, i.e., $\Sigma_i \subseteq \Sigma, R_i \subseteq R$. Data in different local databases complement each other and can overlap. We assume, however, that databases are consistent and do not contradict one another. Local databases are created by local users and the global schema is used as the reference in introducing new facts.

4 Faceted Queries

4.1 Facets

A *facet* is a pair: $F = (X, \circ\Gamma)$, where: (a) $X \in \mathsf{BP} \setminus \{\approx\}$ is the *facet name*; (b) $\circ \in \{\vee, \wedge\}$, $\vee$ denotes *disjunctive* facets, and $\wedge$ *conjunctive* facets; (c) Γ is a subset of $X.2$ (range of X); if X is type, then $\Gamma \subseteq \mathsf{UP}$, otherwise $\Gamma \subseteq \{\mathsf{any}\} \cup \mathsf{Const}$, where any stands for *any* element in $X.2$ [2]. For example, the disjunctive type-facet $F_t = (\mathsf{type}, \vee\{ACMConf, KESConf\})$ can be used to select subsets of individuals satisfying at least one of the given two unary predicates, while the conjunctive *binary predicate*-facet $F_p = (univ, \wedge\{NY, LA\})$ to select subsets of persons who represent simultaneously both *NY* and *LA* universities.

Examples of some disjunctive facets corresponding to Fig. 1 and ontology $\mathcal{O}$, are listed in Fig. 3.

$F_1 = (\mathsf{type}, \vee\{Author, ACMAuthor, \dots\})$, $F_2 = (univ, \vee\{\mathsf{any}, NY, LA\})$,
$F_3 = (authorOf, \vee\{\mathsf{any}, p_1, p_2, a_1, a_2, a_3\})$, $F_4 = (\mathsf{type}, \vee\{Paper, ACMPaper, \dots\})$,
$F_5 = (pyear, \vee\{\mathsf{any}, 2013, 2014\})$, $F_6 = (atConf, \vee\{\mathsf{any}, c_1, c_2, c_3, e_1, e_2, e_3\})$,
$F_7 = (\mathsf{type}, \vee\{ACMConf, KESConf\})$.

Fig. 3 Facets corresponding to interactive FQ interface in Fig. 4

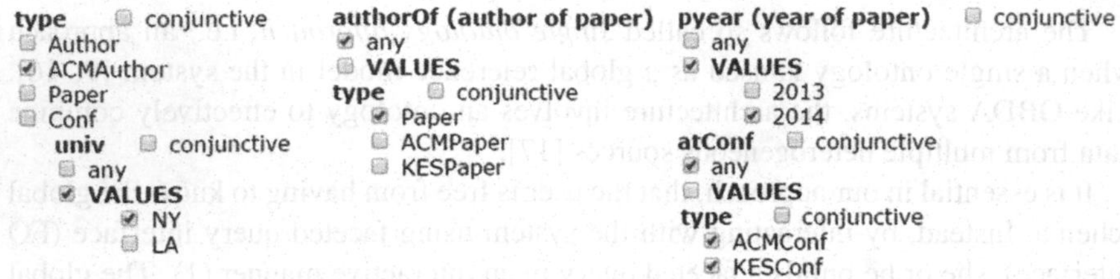

Fig. 4 Formulation of a faceted query using graphical form of interface (2)

From facets one can construct a complex expressions, called *faceted interfaces*, by means of *conjunction* ($\wedge$), *constraint* ([]) and *restriction* (/) operators:

$$F_1[F_2 \wedge F_3/F_4[F_5 \wedge F_6/F_7]]. \tag{2}$$

Then: (a) $F_t[F_{p_1} \wedge \cdots \wedge F_{p_k}]$ constraints the set of elements of types specified in $F_t.2$ to those which are in the intersection of domain sets of binary relations specified in $F_{p_1}.1, \ldots, F_{p_k}.1$, (b) F_p/F'_t restricts the set of codomain (range) values of the binary relation $F_p.1$ to types specified in $F'_t.2$.

A graphical form of the interface (2) is given in Fig. 4. Such form is used to support the user in defining a faceted query. The user selects some elements from presented sets of facet values. In this way a *faceted query* is being specified.

During the interaction with an user, the faceted query (3) can be created.

$$\begin{aligned} \alpha = \ & (F_1, \{ACMAuthor\})[(F_2, \{NY\}) \wedge (F_3, \{\mathsf{any}\})/(F_4, \{Paper\})[\\ & (F_5, \{2014\}) \wedge (F_6, \{\mathsf{any}\})/(F_7, \{ACMConf, KESConf\})]]. \end{aligned} \tag{3}$$

4.2 Syntax and Semantics

Definition 1 Let $F = (X, \circ\Gamma)$, $\circ \in \{\wedge, \vee\}$, be a facet. A basic faceted query determined by F is a pair (F, L), where $L \subseteq \Gamma$. A basic faceted query will be denoted by t, if $X = \mathsf{type}$, and by b when $X \in \mathsf{BP} \setminus \{\mathsf{type}\}$. A *faceted query* is an expression conforming to the syntax

$$\begin{aligned} \alpha & ::= t \mid t[\beta] \\ \beta & ::= b \mid b/\alpha \mid \beta \wedge \beta \end{aligned}$$

The semantic function $[\![\alpha]\!]_x$ assigns to each query α and a given variable x a FO FQ with one free variable x. Further on, we will restrict ourselves to disjunctive facets only.

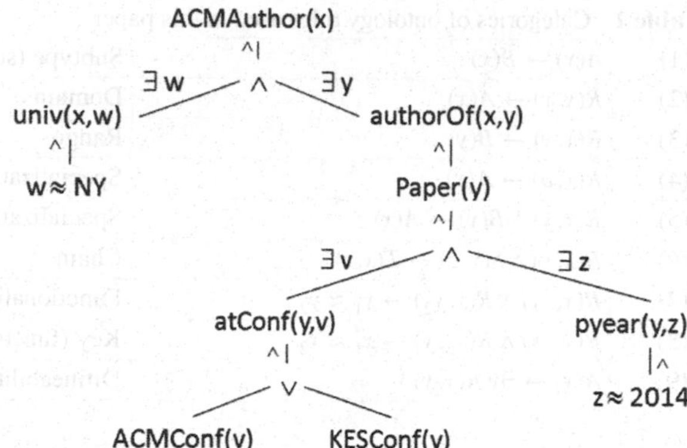

Fig. 5 Syntactic tree of FO FQ $[\![\alpha]\!]_x$, where α is given in (3)

Definition 2 Let t be a basic faceted query over $F_t = (\text{type}, \vee\Gamma)$, b be a basic faceted query over $F_P = (P, \vee\Gamma)$, $P \in \mathsf{BP} \setminus \{\text{type}\}$. Then for basic faceted queries:

1. if $t = (F_t, L)$, $L \subseteq \mathsf{UP}$, then $[\![t]\!]_x = \bigvee_{C \in L} C(x)$.
2. if $b = (F_P, \{\mathsf{any}\})$, then $[\![b]\!]_{x,y} = P(x, y)$,
3. if $b = (F_P, L)$, $L \subseteq \mathsf{Const}$, then $[\![b]\!]_{x,y} = P(x, y) \wedge \bigvee_{a \in L}(y \approx a)$

and for the general form of faceted queries (existentially quantified variables must be chosen in such a way that any variable is only once quantified):

$$\begin{aligned} [\![t[b]]\!]_x &= [\![t]\!]_x \wedge \exists y([\![b]\!]_{x,y}), \\ [\![t[b/\alpha]]\!]_x &= [\![t]\!]_x \wedge \exists y([\![b]\!]_{x,y} \wedge [\![\alpha]\!]_y), \\ [\![t[\beta_1 \wedge \beta_2]]\!]_x &= [\![t[\beta_1]]\!]_x \wedge [\![t[\beta_2]]\!]_x. \end{aligned}$$

Note that $[\![(\text{type}, \vee\{C\})[\beta]]\!]_x$ results in FO FQ of the form $C(x) \wedge \varphi(x)$. For α defined by (3), we obtain $[\![\alpha]\!]_x = Q(x)$, where $Q(x)$ is defined by (1). The syntactic tree for $Q(x)$ is presented in Fig. 5.

5 Evaluating Faceted Queries in a Multiagent System

5.1 *Complexity of Query Answering*

Faceted queries are monadic FO PEQs. Moreover, FQs are tree-shaped meaning that variables occurring in the query can be arranged in a tree (they are partially ordered). In addition, if the applied ontology is one of tractable OWL 2 profiles, i.e., RL, EL or QL, then query answering is tractable. In this paper, we assume that the ontology can contain rules of the categories listed in Table 1. In this paper, the last, (9), category of rules, where the existential quantifier occur in the rule's head, is treated as an *integrity constraint* axiom, like in *extended knowledge bases* [10]. Then such rules

Table 1 Categories of ontology rules used in this paper

(1)	$A(x) \to B(x)$	Subtype (subsumption)
(2)	$R(x,y) \to A(x)$	Domain
(3)	$R(x,y) \to B(y)$	Range
(4)	$R(x,a) \to A(x)$	Specialization (controlled by constant)
(5)	$R(x,y) \wedge B(y) \to A(x)$	Specialization (controlled by type)
(6)	$R(x,y) \wedge S(y,z) \to T(x,z)$	Chain
(7)	$R(x,y_1) \wedge R(x,y_2) \to y_1 \approx y_2$	Functionality
(8)	$R(x_1,y) \wedge R(x_2,y) \to x_1 \approx x_2$	Key (functionality of inversion)
(9)	$A(x) \to \exists y\, R(x,y)$	Difineability

do not affect complexity of query evaluation. Since the rest of rules are in OWL 2 RL, our ontology-enhanced system with faceted queries guarantees query answering in polynomial time.

Satisfaction of integrity constrains is considered globally in our system. In particular, the constraint $Author(x) \to \exists y\ authorOf(x,y)$ is not satisfied in G_1, since *authorOf* is not defined for *Eva*. However, it is satisfied in $G_1 \cup G_2 \cup G_3$.

5.2 Query Rewriting

Rewriting of queries follows from the following two reasons:

1. All predicates occurring in local databases are *extensional*. We assume that extensional predicates create a subset Σ_E of the signature Σ of the ontology. Non-extensional predicates are referred to as *intentional*, and are defined by means of extensional and other intensional ones.
2. Individuals of some types can be named by labeled nulls. It means that arguments of unary predicates can be from LabNull, e.g., $Paper(p_1)$ and $Paper(a_1)$, where $p_1, a_1 \in$ LabNull. Then different names can denote the same individual.

Replacing intentional predicates Any intentional predicate occurring in the query must be replaced by its extensional resolution. For example, *ACMAuthor*(x) in (1) is replaced with

$$\exists y_1(authorOf(x,y_1) \wedge \exists y_2(atConf(y_1,y_2) \wedge \exists y_3(org(y_2,y_3) \wedge y_3 = ACM))).$$

Dealing with labeled nulls Let us consider a boolean query asking whether a given conference is an ACM conference, i.e., we expect the logical value of the boolean query:

$$\exists y_1(Conf(y_1) \wedge \exists y_2(org(y_1,y_2) \wedge y_2 = ACM)). \quad (4)$$

Using the *active domain semantics* [1], existentially quantified variables are mapped to elements of Const. In the case of (4) these constants must by of type *Conf*. Assuming that (4) is evaluated in G_2, we could be interested in logical values of $\exists y_2(org(y_1, y_2) \land y_2 = ACM)$ for y_1 ranging over $\{c_1, c_2, c_3\}$. The value for c_2 is TRUE, for c_3 is FALSE, but for c_1 is unknown, since *org* is not defined on c_1. Then, the agent A_2 can decide to consult other agents. However, when A_3 receives the boolean query $\exists y_2(org(c_1, y_2) \land y_2 = ACM)$, its answer would be FALSE, since c_1 does not even exist in G_3. This trouble follows from the fact that c_1 is a labeled null and its identification property holds only in a local database.

We assume that for any type C, with elements identified by labeled nulls, there must be a *key* in the ontology, i.e., a binary predicate P with the domain C, for which the key-defining rule (rule (8) in Table 1) has to be specified. The key term is added to the boolean query. A key for *Conf* is *name*, so the considered boolean query is rewritten into

$$\exists v(name(v, ACM14) \land \exists y_2(org(v, y_2) \land y_2 = ACM)), \quad (5)$$

since $name(c_1, ACM14)$ holds in G_2, and *name* is the key for *Conf*. Now, this rewritten query can be propagated by A_2 to its partner agents.

Note that rewriting intentional predicates can be performed on the stage of query translation, while rewritings needed for resolving labeled nulls must be performed during the query execution.

5.3 Evaluating Queries

Let $C(x) \land \varphi(x)$ be a FO FQ, where all predicates are extensional. Manager agent sends asynchronously the request GETANSWER(x,C,φ) to all server agents. Each agent A_i uses ISSATISFIED(φ) to check satisfaction of $\varphi(a)$ for all a satisfying $C(x)$. The satisfaction is verified in ontology $\mathcal{O}_i$ (i.e., A_i's local database). If a satisfies the formula, then a is inserted into the answer.

ISSATISFIED(*expr*) acts recursively. If *expr* is not satisfied in $\mathcal{O}_i$ and propagation conditions hold, then A_i propagates *expr* to all partner agents by means of FROMPROPAGATION(*expr*). These two procedures are defined below, where the meaning of conditions in **if** instructions are (see the syntactic tree in Fig. 5): (a) *expr.isUP*—*expr* begins with UP formula, $C(a)$; (b) *expr.isBPConst*—*expr* begins with existentially quantified BP formula having a constant as its first argument; (c) *expr.isBPVar*—*expr* begins with existentially quantified BP formula having a variable as its first argument; (d) *expr.isEQ*—*expr* is equality; (e) *expr.isAND*—*expr* begins with $\land$ operator; (f) *expr.isOR*—*expr* begins with $\lor$ operator. The formula $\psi(x)[x/a]$ arises from $\psi(x)$ by replacing all occurrences of variable x with the constant a.

```
function IsSatisfied(Expr expr)
    if expr.isUP then                                        ▷ C(a) ∧ ψ(x)
        if O_i ⊨ C(a) then
            return IsSatisfied(ψ[x/a])
        else
            return FromPropagation(expr)
    if expr.isBPConst then                                   ▷ ∃x(R(a,x) ∧ ψ(x))
        if O_i ⊨ ∃xR(a,x) then
            return IsSatisfied(ψ[x/val(x)])                  ▷ where O_i ⊨ R(a, val(x))
        else
            return FromPropagation(expr)
    if expr.isBPVar then                       ▷ ∃v(R(v,a) ∧ ψ(v,x)) (only from propagation)
        if O_i ⊨ ∃vR(v,a) then
            return IsSatisfied(ψ(v,x)[v/val(v)])
        else
            return FALSE
    if expr.isEQ then                                        ▷ x = a
        return O_i ⊨ (x = a)
    if expr.isAND then                                       ▷ ∧(ψ1, ψ2)
        return IsSatisfied(ψ1) and IsSatisfied(ψ2)
    if expr.isOR then                                        ▷ ∨(ψ1, ψ2)
        return IsSatisfied(ψ1) or IsSatisfied(ψ2)
function FromPropagation(Expr expr)
    if !PropagationConditions(expr) then
        return FALSE
    if expr.isUP then                                        ▷ C(a) ∧ ψ(x)
        if LabNulls(C) then                  ▷ Elements of C are identified by labeled nulls
            keyVal ← such that O_i ⊨ R_Key(a, keyVal)        ▷ R_Key – the key for C
            expr ← ∃v(R_Key(v, keyVal) ∧ ψ(v,x))
    if expr.isBPConst then                                   ▷ ∃(R(a,x) ∧ ψ(x))
        if LabNulls(C = dom(R)) then
            keyVal ← such that O_i ⊨ R_Key(a, keyVal)        ▷ R_Key – the key for C
            expr ← ∃v(R_Key(v, keyVal) ∧ ∃x(R(v,x) ∧ ψ(v,x)))
    for each S in agentList do
        boolVals[S] ← S.IsSatisfied(expr)                    ▷ propagation to S
    for each S in agentList do
        boolVal ← boolVal or boolVals[S]
    return boolValue
```

6 Summary and Conclusions

We discussed execution of queries in a multiagent system, where a set of agents cooperates while evaluating queries against a system of graph (RDF) databases. There is a global schema consisting of a terminological part (i.e., signature and a set of rules) of an OWL 2 RL ontology. The schema is used by a manager agent to support users in interactive query formulation. This interaction results in a faceted query (FQ) which is next translated into FO FQ. The manager agent rewrites the query accordingly and sends asynchronously to all local server agents for evaluation. The local agent work in parallel and can consult with each another in answering the query. We proposed algorithms concerning the strategy of agents during query rewriting, execution, and propagation. We indicated what kind of ontological knowledge is necessary to perform those operations. The discussed model has been preliminary implemented and is under integration with our data integration system SixP2P [5, 12, 13], originally developed for integrating XML data. The future research will focus on issues concerning user-friendly interactive query formulation. We are also planning to incorporate another database models, including NoSQL models. This research has been supported by Polish Ministry of Science and Higher Education under grant 04/45/DSPB/0149.

References

1. Abiteboul, S., Hull, R., Vianu, V.: Foundations of Databases. Addison-Wesley, Reading, Massachusetts (1995)
2. Arenas, M., Grau, B.C., Kharlamov, E., Marciuska, S., Zheleznyakov, D.: Faceted search over ontology-enhanced RDF data. In: ACM CIKM 2014, pp. 939–948. ACM (2014)
3. Barceló, P., Fontaine, G.: On the data complexity of consistent query answering over graph databases. In: ICDT 2015. LIPIcs, vol. 31, pp. 380–397. Schloss Dagstuhl - Leibniz-Zentrum fuer Informatik (2015)
4. Beneventano, D., Bergamaschi, S.: The MOMIS methodology for integrating heterogeneous data sources. In: IFIP Congress Topical Sessions, pp. 19–24 (2004)
5. Brzykcy, G., Bartoszek, J., Pankowski, T.: Schema mappings and agents' actions in P2P data integration system. J. Univ. Comput. Sci. **14**(7), 1048–1060 (2008)
6. Calvanese, D., Horrocks, I., Jiménez-Ruiz, E., Kharlamov, E., Meier, M., Rodriguez-Muro, M., Zheleznyakov, D.: On rewriting, answering queries in OBDA systems for big data. In: OWLED. CEUR Workshop Proceedings, vol. 1080 (2013)
7. Dimartino, M.M., Calì, A., Poulovassilis, A., Wood, P.T.: Peer-to-peer semantic integration of linked data. In: Proceedings of the Workshops of the EDBT/ICDT 2015. CEUR Workshop Proceedings, vol. 1330, pp. 213–220. CEUR-WS.org (2015)
8. Gottlob, G., Orsi, G., Pieris, A.: Ontological queries: rewriting and optimization (extended version), pp. 1–25 (2011). arXiv:1112.0343
9. Lenzerini, M.: Data integration: a theoretical perspective. In: Popa, L. (ed.) PODS, pp. 233–246. ACM (2002)
10. Motik, B., Horrocks, I., Sattler, U.: Bridging the gap between OWL and relational databases. J. Web Semant. **7**(2), 74–89 (2009)
11. OWL 2 Web Ontology Language Profiles. http://www.w3.org/TR/owl2-profiles (2009)
12. Pankowski, T.: Query propagation in a P2P data integration system in the presence of schema constraints. In: Data Management in Grid and P2P Systems DEXA/Globe'08, LNCS 5187, pp. 46–57 (2008)
13. Pankowski, T.: Keyword search in P2P relational databases. In: Agent and Multi-Agent Systems: Technologies and Applications (KES-AMSTA 2015). Smart Innovation Systems and Technologies, vol. 38, pp. 325–335. Springer (2015)
14. Resource Description Framework (RDF) Model and Syntax Specification. http://www.w3.org/TR/PR-rdf-syntax/ (1999)
15. SPARQL Query Language for RDF. http://www.w3.org/TR/rdf-sparql-query (2008)
16. Ullman, J.D.: Information integration using logical views. In: Database Theory—ICDT 1997. Lecture Notes in Computer Science 1186, pp. 19–40 (1997)
17. Wache, H., Vgele, T., Visser, U., Stuckenschmidt, H., Schuster, G., Neumann, H., Hbner, S.: Ontology-based integration of information—a survey of existing approaches. IJCAI **2001**, 108–117 (2001)
18. Wang, H., Aggarwal, C.C.: A survey of algorithms for keyword search on graph data. In: Managing and Mining Graph Data. Advances in Database Systems, pp. 249–273. Springer (2010)

SWARM: A Multi-agent System for Layout Automation in Analog Integrated Circuit Design

Daniel Marolt, Jürgen Scheible, Göran Jerke and Vinko Marolt

Abstract Despite 30 years of Electronic Design Automation, analog IC layouts are still handcrafted in a laborious fashion today due to the complex challenge of considering all relevant design constraints. This paper presents *Self-organized Wiring and Arrangement of Responsive Modules* (SWARM), a novel approach addressing the problem with a multi-agent system: autonomous layout modules interact with each other to evoke the emergence of overall compact arrangements that fit within a given layout zone. SWARM's unique advantage over conventional optimization-based and procedural approaches is its ability to consider crucial design constraints both explicitly and implicitly. Several given examples show that by inducing a synergistic flow of self-organization, remarkable layout results can emerge from SWARM's decentralized decision-making model.

Keywords Integrated circuits · Electronic design automation · Analog layout · Constraint-driven design · Parameterized cells · Self-organization

1 Introduction

Microelectronic products are increasingly controlling, connecting, and changing our world, and today, market demands for low-cost, multifunctional, and densely integrated circuits (ICs) drive the trend to systems-on-chip with both analog and digital

D. Marolt (✉) · J. Scheible
Robert Bosch Center for Power Electronics,
Reutlingen University, Reutlingen, Germany
e-mail: daniel.marolt@reutlingen-university.de

J. Scheible
e-mail: juergen.scheible@reutlingen-university.de

G. Jerke · V. Marolt
Robert Bosch GmbH, Automotive Electronics, Reutlingen, Germany
e-mail: goeran.jerke@de.bosch.com

V. Marolt
e-mail: vinko.marolt@de.bosch.com

G. Jezic et al. (eds.), *Agent and Multi-Agent Systems: Technology and Applications*, Smart Innovation, Systems and Technologies 58,
DOI 10.1007/978-3-319-39883-9_2

content. But while the task of *digital* IC design follows highly automated synthesis flows based on optimization algorithms, Electronic Design Automation (EDA) is still struggling to apply such approaches in the *analog* domain. For over three decades, such attempts have repeatedly failed to find industrial acceptance and thus, most design steps are manually accomplished by expert designers with very little support by automation. In particular, the step of *layout design*, where a circuit schematic has to be turned into a graphical description of the detailed circuit geometries—needed for the photolithographical manufacturing of the IC—remains a laborious and severely time-consuming bottleneck in the design flow.

This paper presents a novel layout automation approach for analog IC design: *Self-organized Wiring and Arrangement of Responsive Modules* (SWARM) [1]. To our knowledge, SWARM is the first approach to address the design problem by implementing a multi-agent system, in which the individual agents are layout modules that interact with each other. Steered by a supervising control organ, the self-interested modules autonomously move, rotate and deform themselves inside an increasingly tightened layout zone, vying for the available space. By inducing a synergistic flow of self-organization, the decentralized decision-making of the modules is supposed to provoke the *emergence* of compact layout arrangements.

The presented approach is especially interesting in two aspects. First, it joins ideas from several different disciplines such as cybernetics, game theory, biology, geometry, and electrical engineering. Second, SWARM addresses a serious practical problem which could not yet be satisfactorily solved with layout algorithms equivalent to those that are successfully used for digital circuit designs—despite a substantial amount of EDA work. In contrast to these achievements, SWARM's decisive asset is that it facilitates an explicit *and* implicit consideration of crucial *design constraints*, as both are essential for achieving the degree of layout quality that is uncompromisingly demanded in the analog domain. Since SWARM is an interdisciplinary and quite extensive approach, this paper focuses on providing (1) an elementary introduction to the problem of IC layout design, (2) a presentation of the SWARM approach cast in the light of multi-agent systems, and (3) a couple of examples demonstrating what results can be achieved with SWARM.

The paper is organized as follows. Section 2 gives a basic introduction to the problem of IC layout design. Section 3 discusses existing works on analog layout automation, discerning algorithmic and generator approaches. Then, Sect. 4 covers the multi-agent system of SWARM from a high-level point of view, not going into all details. Section 5 exemplifies the application of SWARM and shows some achieved results. Section 6 finishes the paper with a summary and an outlook.

2 The Problem of Integrated Circuit Layout Design

As shown in Fig. 1, any IC layout design problem expects three inputs: (1) a set of *design rules*, inherently given by the semiconductor technology, (2) a structural description of the circuit, i.e., a *schematic* diagram or a *netlist*, and (3) a set of

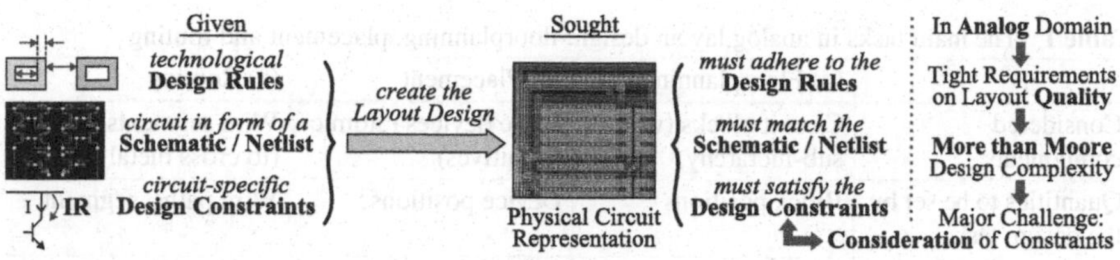

Fig. 1 The problem of IC layout design: turn a circuit into a physical representation

circuit-specific *design constraints*. Sought is a physical representation describing the detailed chip geometries of the circuit's devices and their interconnections on all *process layers*, used as photolithography masks in the manufacturing process. To ensure manufacturability and functionality, such a layout must (1) adhere to the design rules, (2) match the given circuit, and (3) satisfy all design constraints.

In terms of problem complexity, one has to discern *digital* from *analog* design. In digital design, with its primary objective to cram more and more components (nowadays millions and billions of logic gates) onto a chip—a desire referred to as *More Moore* [2]—the design problem is mainly a matter of *quantity*. In contrast, the major difficulty in the continuous-valued analog domain is to maintain signal integrity in the face of nonlinearities, parasitic effects, thermal gradients, high voltages, external physical influences and other *More than Moore* [2] challenges. This makes analog design complexity rather an issue of *quality*, and thus, obtaining a functional layout involves many, diverse, and correlated design constraints.

Analog layout design includes several tasks. As detailed in Table 1, the main tasks are: *floorplanning* (to specify positions, aspects ratios, and pin positions for the top-level layout blocks of a chip), *placement* (to set the position, orientation, and layout variant of all electronic devices, e.g., transistors, resistors, capacitors, inside a layout block), and *routing* (creating electrical wires, using so-called *vias* to connect wires across different metal layers). Overall, layout design is an *optimization problem* defined by restrictions (e.g., minimal wire widths and spaces, given by the design rules) and objectives (usually: reducing area and wirelength). For each specific layout problem, function-relevant restrictions and objectives are captured by the respective design constraints. This set of constraints (and thus, a definition of the optimization problem itself) in turn depends on a multitude of attributes, including the chosen semiconductor technology as well as the type, application, mission profile, and particular reliability requirements of the circuit.

While some design constraints relate to high-level aspects, e.g., the available layout space, most constraints deal with low-level details, in particular to achieve electrical symmetry for layout modules that perform critical electrical functions. This so-called *matching* usually requires equal sizes and consistent orientations of the module devices in a compact, interdigitated, common-centroid placement [3]. Such restrictions reduce the degrees of freedom, but—as shown by the examples in Fig. 2—they typically still leave much layout *variability*. This is a characteristic trait of analog design which often has to be exploited to satisfy all requirements.

Table 1 The main tasks in analog layout design: floorplanning, placement and routing

	(a) Floorplanning	(b) Placement	(c) Routing
Considered components	Circuit blocks (with sub-hierarchy)	Basic devices (atomic primitives)	Wire segments + Vias (to cross metal layers)
Quantities to be set by the design task	Block positions	Device positions	Wire paths, segment
	Aspect ratios	Device orientations	Layers and widths +
	Pin positions	Layout variants	Via positions and sizes
Typical restrictions	Layout boundary	Layout boundary	Available metal layers
	Maximum distances	Space for routing	No wires above devices
Primary objectives	Minimize total area	Device matching	Minimize number of vias
	Minimize wirelength	Overall symmetry	Homogenize wire density

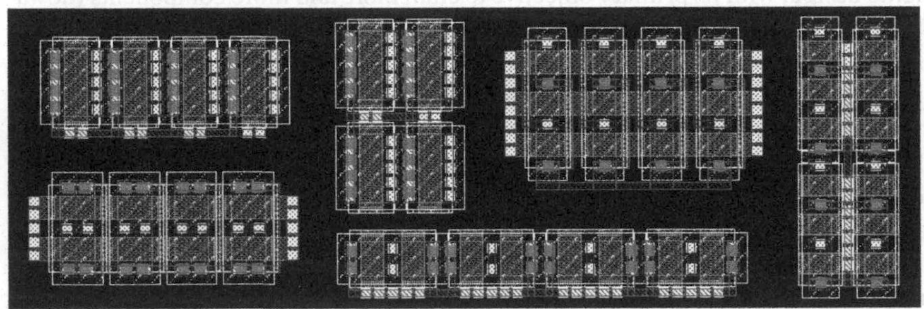

Fig. 2 Different variants of an analog layout module with equal electrical function

In today's flows of manual layout design, two basic forms of constraint consideration are found: high-level constraints usually need to be *explicitly* formulated to be really taken into account, while low-level matching requirements are often *implicitly* taken care of by experienced layout engineers. But as will be explained in Sect. 3, so far no automatism supports *both* forms of constraint consideration.

3 State of the Art in Analog Layout Automation

Although EDA research has put forth a rich variety of analog layout automation approaches over the past decades, these can be divided into two fundamental categories [4]: optimization algorithms (Fig. 3a, Sect. 3.1) and procedural generators (Fig. 3b, Sect. 3.2). They follow entirely different automation strategies, while SWARM (Fig. 3c, Sect. 4) is an attempt to conflate the two paradigms.

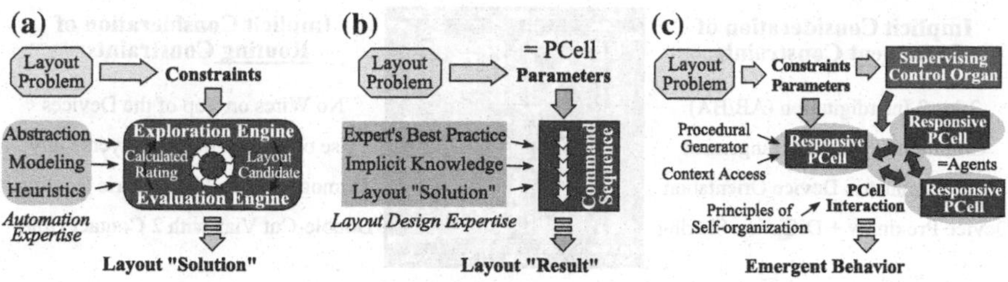

Fig. 3 Layout automation strategies: **a** algorithmic, **b** procedural, **c** SWARM

3.1 Optimization-Based Layout Automation Algorithms

Referring to [5], the use of optimization algorithms in EDA largely concentrates on a canonical form depicted in Fig. 3a, where a single candidate layout is repeatedly refined in a loop of solution space *exploration* and solution *evaluation*. For example, the most widely used algorithm for analog *placement* is Simulated Annealing [6] where exploration is done via a random modification of the current candidate placement, while every evaluation rates the new placement according to a formal cost function and validates if it satisfies all design restrictions. As an example for *routing*, the exploration-evaluation style of optimization is particularly apparent in techniques that *rip-up and reroute* existing wires as in [7].

For layout automation, a particular strength of such approaches is their ability to take into account formal expressions of high-level design constraints. However, optimization algorithms work by translating the optimization problem into an abstract mathematical model, in some cases based on a physical analogy (in the case of SA: the annealing of solids), and specifically optimize just the modeled aspects. This implies, that *all* relevant solution requirements and optimization goals *must* be *explicitly* expressed in a formal, comprehensive, unambiguous, and consistent representation of constraints that can be processed by the algorithm.

In analog layout however, it is tremendously intricate and still not possible today to *efficiently* and *sufficiently* describe the entire diversity, various impacts and correlated dependencies of *all* crucial design constraints in an *explicit* fashion due to the *More than Moore* complexity (see Sect. 2). The inherent quality loss is the major reason for the rejection of optimization-based automation in practice. Another shortcoming is that most of these approaches are not deterministic, thus yielding nonreproducible solutions. Furthermore, algorithmic layout automation typically concentrates on just one particular design step (e.g., placement *or* routing), although—in practice—these steps are heavily interrelated with each other.

In spite—or because—of these issues, analog EDA research continues to work on a vast range of topics in the context of optimization algorithms. These include (a) the consideration of different constraint types such as boundary, fixed-outline, or symmetry-island [8] constraints, (b) topological floorplan representations like SKB-trees [9], (c) performance improvements as in [10], (d) the simultaneous execution

Implicit Consideration of Placement Constraints:
2-by-2 Interdigitation (AB/BA)
Common-Centroid Arrangement
Centrosymmetric Device Orientation
Device Proximity + Diffusion Sharing
A B
B A
Implicit Consideration of Routing Constraints:
No Wires on Top of the Devices
Use of Metal1/Metal2 Layers Only
Homogenous and Symmetric Routing
Double-Cut Vias with 2 Contact Holes

Fig. 4 PCell instance and implicitly considered placement and routing constraints

of different design tasks [11], and (e) constraint propagation [12]. Please note that the given references are not comprehensive, but merely representative. Approaches beyond the canonical form of Fig. 3a will be covered in Sect. 3.3.

3.2 Procedural Generators (Parameterized Cells, PCells)

The predominantly manual design style in practice employs an entirely different kind of automation at the lowest design levels: procedural generators, also called *Parameterized Cells* (PCells). A PCell is an instantiable library component: instantiated in a design, it takes a set of parameter values and follows a predefined, successive sequence of commands to produce a customized layout result inside. Primarily, PCells automate the layout creation of *primitive devices* (e.g., transistors), but recently, industrial flows can be observed to pursue an advancement of PCells towards more powerful *hierarchical modules* which generate entire analog basic circuits (e.g., current mirrors). For such circuits, feasible layout variants are often already known from practical experience, and PCells simply represent instruments that embody this invaluable *expert knowledge* as layout generators. For example, Fig. 4 shows a "Quad" PCell instance that creates a *differential pair* layout for two transistors A and B, split into a cross-coupled AB/BA array.

In academia, PCells have been of little interest as they merely replicate a human expert's best practice: they generate a layout *result*, but the actual *solution* was preconceived by the design expert who implemented the PCell (Fig. 3b). However, this trait gives module PCells the advantageous ability to consider design constraints *implicitly*, i.e., without the need to formalize them. For example, as pointed out in Fig. 4, the Quad module innately takes care of all requirements crucial to achieve a high matching, without having been *explicitly* told to do so.

In contrast to the weaknesses of optimization algorithms mentioned before, PCells elude the need for abstract models and formal floorplan representations, work fast and deterministic, and can naturally do both placement and routing (as in Fig. 4). These assets are based on the fact that every PCell targets one specific type of device (or module), but the immanent downside is the effort required to implement a PCell for each desired type, and the inability to consider constraints explicitly. Due to this tradeoff, the use of PCells in industrial environments has not yet shown to be profitable enough above the level of analog basic circuits.

Amongst other topics, PCell-related work deals with (a) implementing higher-level layout modules as in [13], (b) developing advanced tools for PCell programming [14], (c) circuit PCells [15], and (d) auxiliary concepts such as hierarchical parameter value editing [16]. Again, the given references are only representative.

3.3 *Other Approaches*

After 30 years of EDA, research on layout automation still keeps a strong focus on the single-candidate style of optimization in Fig. 3a. However, population-based approaches using a pool of candidate solutions have also been applied, including genetic algorithms (e.g., for floorplan area optimization in [10]) and methods of swarm intelligence (such as the hybrid ant colony and particle swarm optimization algorithm in [17]). Still, existing approaches are not as sophisticated as optimization techniques found for example in computer science, e.g., the combination of multi-objective evolutionary learning and fuzzy modeling in [18] and the random forest based model of [19]. Compared to layout automation, a much richer spectrum of optimization techniques has been used for *circuit sizing* [20]. Fuzzy logic, despite gaining popularity in various types of applications (e.g., for the tuning of servo systems in [21]), has been of almost no interest for IC layout design so far, with just very few exceptions such as the placement approach [22].

One might argue, that many layout automation tools incorporate both algorithmic and procedural aspects (in the sense of approaches described in Sect. 3.1 and Sect. 3.2, respectively). That is right, but on closer examination, such tools always reveal a clear emphasis on one of the two automation strategies and the respective form of constraint consideration, which is either explicit or implicit. The same is true for template-based approaches such as [23] that extract implicit design knowledge from existing layout designs. Although this looks like a hybrid approach, the design knowledge is in fact turned into a formal representation and processed by a conventional optimization algorithm in a purely explicit fashion.

It can be said, that there is no layout automation approach which facilitates a balanced consideration of both explicit and implicit constraints, although today's manual design style suggests that both forms are indispensable to cope with the *More than Moore* design complexity in the analog domain. This is exactly where the multi-agent approach of SWARM is supposed to make a difference. Here, it should be noticed, that SWARM is not a method of swarm intelligence (in the algorithmic sense), for it does not employ a pool of candidate solutions. Instead, every agent in SWARM is a layout module and—as such—a *part* of the solution. This should become clear from the discussion of SWARM that follows in Sect. 4.

4 The Multi-agent System of SWARM

Although layout *algorithms* (again, referring to the prevalent, canonical form of optimization from Sect. 3.1) and procedural *generators* (as covered in Sect. 3.2) follow entirely different EDA strategies, both share a common drawback: to work in a totally *centralized* fashion. Either the algorithm must solve the entire design problem, or the solution must be completely preconceived by the design expert who develops the generator. SWARM addresses this problem by implementing a multi-agent system in which *module PCells* for analog basic circuits interact with each other to constitute an overall layout block in a flow of self-organization. This *decentralization* not only breaks down the complexity of the design problem, but allows low-level design constraints to be *implicitly* considered by each PCell, while high-level constraints can be *explicitly* considered during the interaction.

After referring some relevant agent-related works in the next subsection, the remaining subsections first give an overview of the SWARM system and its flow of PCell interaction, and then delve into some detailed features. Although the approach cannot be covered in its entirety here, the subsequent discussion should illustrate that SWARM is a relatively elaborate multi-agent system—mainly due to the following aspects: the interaction territory continually changes throughout the flow (Sect. 4.2), the agents don't obey a simple payoff function (Sect. 4.3), and the agents' potential actions always depend on their current situation (Sect. 4.4). Some notes on the design and application of SWARM are also given (Sect. 4.5).

4.1 Related Work

The SWARM approach is encouraged by the observation that multi-agent systems can outperform classical approaches in handling complexity [24]. For the design of such systems, principles of self-organization (like the *edge of chaos* [25]) can be accounted for to obtain operating points that are both stable and dynamic enough to effectively stir up *emergent behavior*. This phenomenon of *emergence* [26], which can be observed in many natural processes (e.g., crystallization), but also in artificial systems such as Conway's cellular automaton Game of Life [27], has become a topic of interest in many fields of science, philosophy, and art.

SWARM's idea of PCells that perform areal movements across a geometrical layout plane, is closely related to Reynold's Boids [28]. Both approaches feature a vivid biological analogon: while Boids simulates the motion of a flock of birds, SWARM imitates the roundup of an animal herd. As will be shown, the interaction in SWARM is not one of direct communication, but of *stigmergy* [29]—the kind of indirect coordination found among animal societies such as ant colonies. Existing agent simulators are discarded by SWARM due to its specific purpose so instead, the system has been conceived and implemented from the ground up.

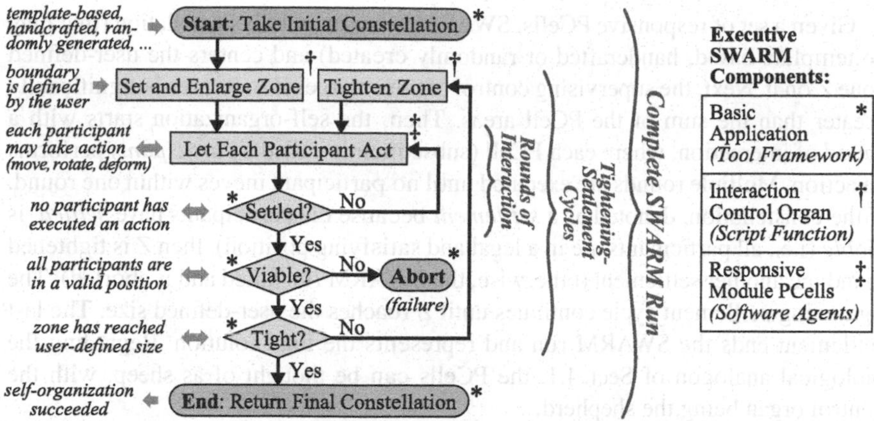

Fig. 5 Control flow in the multi-agent system SWARM for analog layout automation

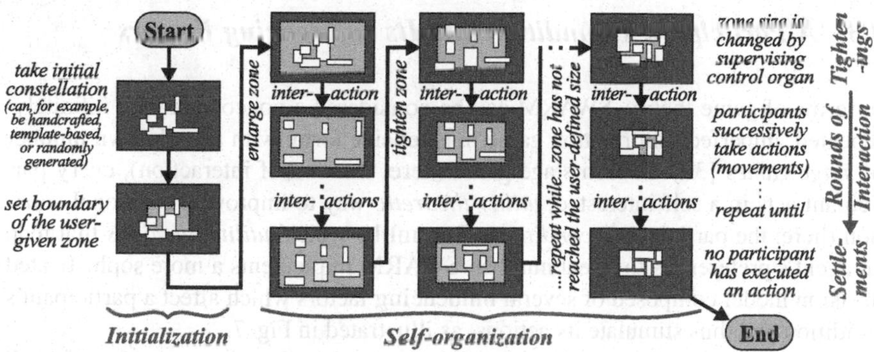

Fig. 6 Exemplary depiction of SWARM's PCell interaction and self-organization flow

4.2 Overview of SWARM and Its Interaction Flow

Due to conceptual limitations, PCells cannot access their design context. So, to use PCells as agents (as in Fig. 3c), SWARM enhances their native, "introversive" purpose (i.e., creating an internal layout) with an "extroversive" behavior by which a PCell can take an action (move, rotate, or deform itself) in response to changes of its environment. On this basis, SWARM allows a set of such *responsive PCells* to interact with each other and arrange themselves inside a given rectilinear zone. SWARM also implements a *supervising control organ* that recursively tightens the zone to steer the interaction towards a compact layout arrangement without depriving the PCells of the leeway needed to organize themselves. The overall flow of control in such a so-called SWARM *run* is depicted in Fig. 5 and exemplarily illustrated in Fig. 6. A more algorithmic presentation is given in [1].

Given a set of responsive PCells, SWARM takes an initial constellation (that may be template-based, handcrafted or randomly created) and centers the user-defined zone Z on it. Next, the supervising control organ enlarges Z so its area is significantly greater than the sum of the PCell areas. Then, the self-organization starts with a *round* of interaction, where each PCell (subsequently called a *participant*) performs an action. Multiple rounds are executed until no participant moves within one round. If the constellation, denoted as a *settlement* because the participants have *settled*, is *viable* (i.e., all participants are in a legal and satisfying position), then Z is tightened to induce another settlement (otherwise, the SWARM run failed and is aborted). The tightening-settlement cycle continues until Z reaches the user-defined size. The last settlement ends the SWARM run and represents the final solution. Regarding the biological analogon of Sect. 4.1, the PCells can be thought of as sheep, with the control organ being the shepherd.

4.3 A Participant's Condition and Its Influencing Factors

In terms of game theory, SWARM can be considered a noncooperative, infinitely-repeated, imperfect-information game in extensive form, with an unknown number of *stage games* [30]. In each stage game (here: a *round* of interaction), every participant acts in a self-interested, *utility-theoretic* way to improve its personal situation (here: the participant's *condition*). But unlike typical *utility functions* that map an agent's preferences to a real number, SWARM implements a more sophisticated decision model composed of several influencing factors which affect a participant's condition and thus stimulate its actions, as illustrated in Fig. 7.

The currently implemented influencing factors largely rely on the fact that all participating PCells are geometric objects with a rectangular bounding box and thus have an area, in contrast to dot-like particles found in some other systems. A participant P is forced to take an action if it suffers at least one of the following factors:

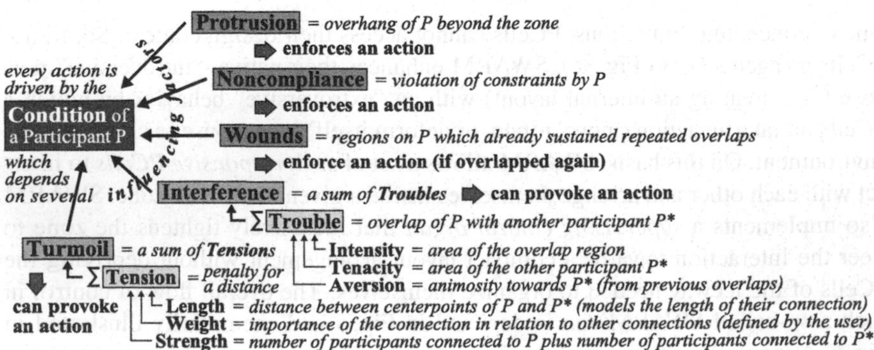

Fig. 7 A participant (PCell agent) is influenced by five factors with several subfactors

protrusion (overhang beyond the zone boundary), *noncompliance* (violation of constraints), or *wounds* (regions on *P* which repeatedly overlapped with other participants in previous rounds of interaction). Further factors, which *can* (depending on their magnitude) provoke an action, are *interference* and *turmoil.*

Interference is a sum of *troubles*, each of which is caused by an overlap of *P* with another participant *P**. As detailed in Fig. 7, each trouble again involves several subfactors. One of these is the *aversion* of *P* towards *P**. Like wounds, aversions are inflicted (or aggravated) by overlaps and can be seen as concepts of memory. Aversion has a long-term effect that hinders a perpetual interference of two participants, while a wound has an immediate impact and is more effective for preventing marginal interferences. Both are essential aspects of the decision model and a careful balance of these is crucial for a fluent progress of interaction.

Turmoil is a concept relevant if participants should move close to each other because they need to be electrically connected afterwards, thus minimizing the total wirelength. Turmoil is a sum of *tensions*, with each tension penalizing the distance between the acting participant *P* and another one to which it is directly connected. Like trouble, tension involves multiple subfactors as Fig. 7 points out. For some tasks, turmoil is essential (e.g., for the floorplanning task in Fig. 10), but for others it may be ignored (as in the place-and-route problem of Fig. 11).

4.4 How a Participant Determines and Performs an Action

Many classes of problems, to which multi-agent systems have been applied, involve agents whose possible actions are simply given as a predefined, discrete, limited set of alternative options, from which the agent has to choose one. In SWARM, the situation is more complicated because the participants can freely move across a continuous layout plane. So, to take an action, a participant must first determine its set of possible actions depending on the current constellation. A participant's decision-making always follows a common *action scheme* which consists of the following four measures: (1) assessing the participant's condition, (2) perceiving its so-called *free peripheral space*, (3) exploring and evaluating all possible actions, and (4) executing the preferred action (or staying idle). Each of these measures, exemplarily illustrated in Fig. 8, will be subsequently described.

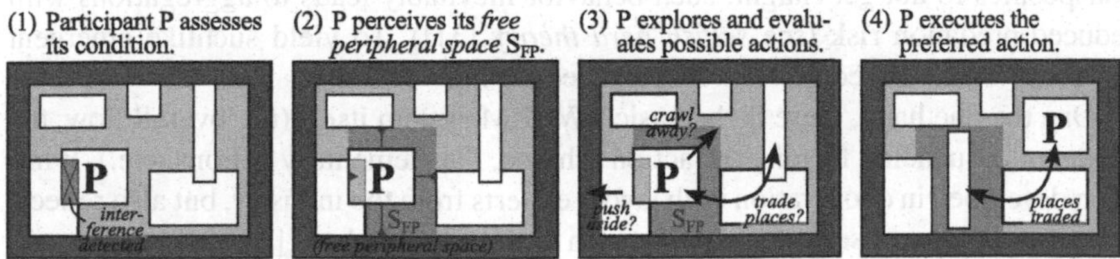
(1) Participant P assesses its condition.
(2) P perceives its free peripheral space S_FP.
(3) P explores and evaluates possible actions.
(4) P executes the preferred action.
P
inter-ference detected
P
S_FP
(free peripheral space)
crawl away?
push aside?
P
S_FP
trade places?
P
places traded

Fig. 8 A participant's actions follow an action scheme consisting of four measures

Considering a constellation of six participants as in Fig. 8 (1), assume that it is participant *P*'s turn to take an action. Following the action scheme, *P* begins by assessing its condition. While *P* is not in a state of *protrusion* nor *noncompliance* here, *P* detects *interference* with another participant. For that reason, *P* is said to be *discontented* and strives for an action that improves its condition.

To do so, *P* perceives the vacant area around it, since most of a participant's possible actions are based on this so-called *free peripheral space* S_{FP}. As shown in Fig. 8 (2), SWARM determines S_{FP} as a rectangle obtained by extending each of *P*'s four edges in its respective direction until another participant—or the zone boundary—is encountered. Surely, other definitions of S_{FP} are also conceivable.

Next, as indicated in Fig. 8 (3), *P* explores all actions available in the current situation. Although a participant could implement its own actions, SWARM provides nine different elementary actions that are automatically explored (see [1]). Six of these actions (called *centering*, *budging*, *evasion*, *yielding*, *re-entering*, and *lingering*) only affect the acting participant itself (or none at all), while three actions (called *pairing*, *swapping*, *hustling*) also involve other participants. For the latter class of actions, only translational moves are considered. Otherwise, rotations and deformations are also explored. In this context, deformation means to assume another layout variant with equal electrical behavior (such as in Fig. 2).

The actions are evaluated so the one which improves *P*'s condition the most can be chosen and executed. In Fig. 8 (4), the executed action is even synergistic: *P* trades places with another participant and *both* get rid of their interference.

4.5 Final Remarks About the Design and Application of SWARM

The participants' desire to improve their condition and the repeated layout zone tightenings are the driving forces behind SWARM's self-organization. It should again be emphasized that the participants act in a *utilitaristic* way and that each tightening represents just a *stigmergic* change of their environment. So, the inherent optimization of the overall layout solution is not incited directly (like with a global cost function, as is typical for EDA algorithms), and is thus not actively pursued via *collaborative* agents but emerges from their *competitive* interaction. This effect can also be observed in nature, where animals within a population seek cover behind other conspecifics to not get caught: such behavior inevitably leads to aggregations with reduced predation risk (see *selfish herd theory* [31]). To yield suchlike emergent behavior, SWARM requires a well-balanced taring.

One the one hand, there is the basic SWARM system itself (the overall flow, the model of influencing factors, the action scheme, the elementary actions, etc.). It has been developed in cooperation with layout experts from the industry, but also reflects several principles of self-organization such as the edge of chaos [25] (addressed with wounds and aversion), Ashby's law of requisite variety [32] (i.e., exploiting a PCell's layout variants during the interaction), and the reduction of friction [24] (by inducing synergistic moves as in Fig. 8 (4)). That basic system is supposed to be generally feasible, independent of the particular layout problem.

On the other hand, the system involves numerous actuating variables which allow a layout designer to fine-tune SWARM when applying it to a specific problem. These variables, described in detail in [1], mostly pertain to the influencing factors (Fig. 7) and the zone tightening. They can thus be used to adjust the participants' decision-making (between tentative and vivid) and to set an adequate tightening policy (between relaxed and aggressive). A more sensitive approach would be to make the tightening—and also the participants' behavior—*adaptive* so it changes during the interaction. Going one step further, this might even be achieved using an overlay network of synchronized agents as presented in [33].

Such advanced techniques, as well as the dependence of the self-organization on the actuating variables, on the initial module constellation, and on the order in which the participants act, are subject to further investigation. Another idea not examined so far is to equip the participants with different *attitudes* (between dominant and subordinate). In these regards, SWARM provides many opportunities for future enhancements. Still, remarkable results can already be achieved with the current implementation of SWARM, as will be demonstrated in Sect. 5.

5 Implementation and Examples

With the exclusive aim of layout automation, targeting the IC design framework *Cadence Virtuoso®*, SWARM has been implemented from scratch in Virtuoso's own script language SKILL, using PCell Designer [16] for developing the PCells.

Figure 9 has a theoretic example with eight participants (PCell agents) inside a rectangular zone (a) disregarding PCell connections. The PCells are just symbolic (i.e., they contain no real layout) and perform only translational movements (b), (c). The final constellation (d) demonstrates, that—without having knowledge of the overall placement problem—the agents are, in principle, able to find even the *optimal* solution (from which this example was initially constructed from).

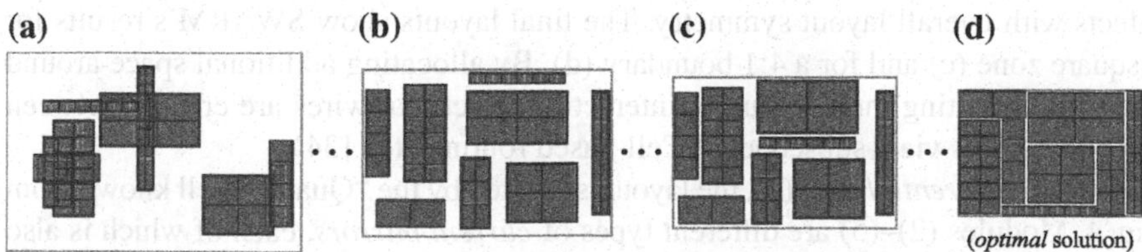

Fig. 9 Example of applying SWARM to a hypothetical analog layout design problem. **a** Initial SWARM constellation. **b** Intermediate constellation. **c** Intermediate constellation. **d** Final constellation

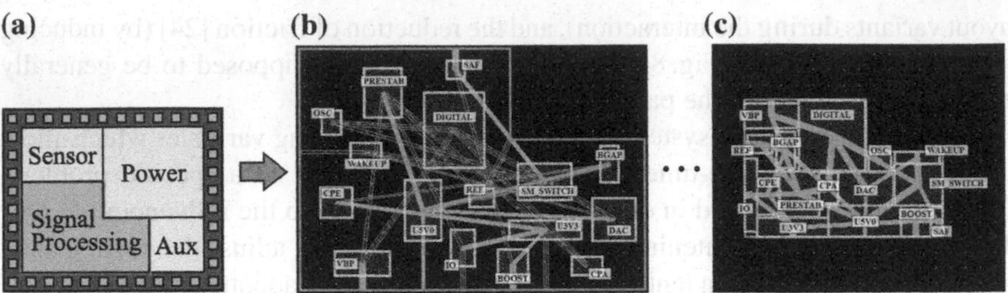

Fig. 10 Example of applying SWARM to a practical IC layout floorplanning problem. **a** Section of an IC to be floorplanned. **b** Initial SWARM Constellation. **c** Final constellation

Figure 10 uses SWARM for *floorplanning* a mixed-signal chip section (a) with digital and analog *blocks*. Each block is realized as a simple "hull" PCell, leaving the internal layout to subsequent layout steps. As indicated in Table 1 (a), the aspect ratios of such blocks may cover *full* variability (not discrete variants as in Fig. 2) and furthermore, the layout zone can be *rectilinear* (not just rectangular). SWARM supports both, and also takes block distances into account here. In the initial constellation (b), 19 out of 30 connections violate a maximum-length constraint (see patterned lines). In the compact arrangement of SWARM's final constellation (c), these constraints are all satisfied (solid lines) and all blocks lie inside the given layout zone. In floorplanning, the final routing of the blocks is usually done in layers above them, so no space has to be reserved between them.

The 'dead' space and runtime (*quantitative* criteria that stem from digital design automation) are comparable to other works, but more importantly, SWARM surpasses existing approaches in a *qualitative* way, reflecting the notion of quantitative and qualitative complexity in the digital and analog IC domain (Sect. 2): to our knowledge, SWARM is the first floorplanner that (1) minimizes both area and wirelength, (2) supports *nonslicing* floorplan structures, (3) allows for *fully* variable block dimensions, (4) pays respect to a user-definable *rectilinear* outline, *and* (5) works completely *deterministic* because no randomization is involved.

The primary purpose of SWARM is *placement* and *routing* (Table 1 (b), (c)). As an example, Fig. 11a shows the schematic of an amplifier circuit found in high-precision automotive IC designs. It consists of several analog basic circuit modules for which a desired arrangement (b) is defined via constraints to even out parasitic effects with overall layout symmetry. The final layouts show SWARM's results for a square zone (c) and for a 4:1 boundary (d). By allocating additional space around every participating module during interaction, electrical wires are created between the participants via a subsequent PCell-based routing step [34].

For the *differential pair* (1), the layout is created by the "Quad" PCell known from Fig. 4. Modules (2)–(5) are different types of *current mirrors*, each of which is also covered by a respective (and responsive) PCell agent. In fact, the flow is closer to that in manual design: first, the circuit's *primitive devices* are instantiated in the layout, and then the PCell agents are imposed on them as so-called *governing modules* which manage the devices by placing and routing them [1].

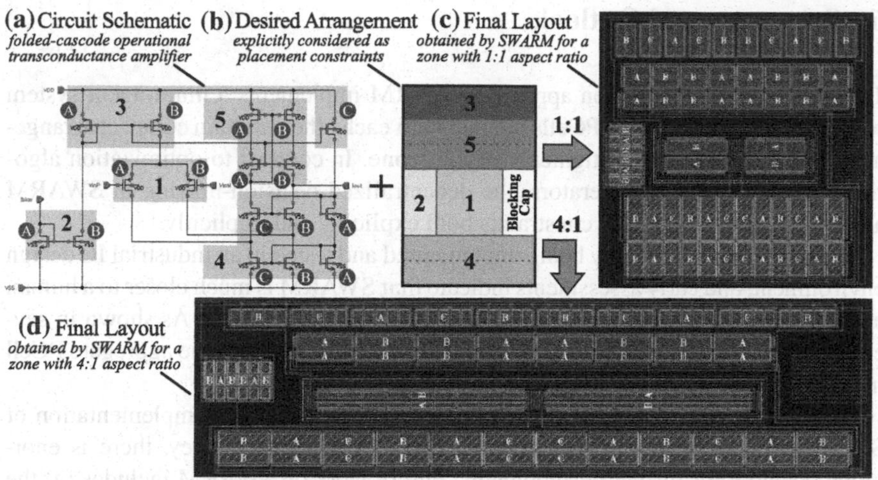

Fig. 11 Example of applying SWARM to a practical place-and-route layout problem

Table 2 Comparison of SWARM with algorithmic and procedural layout automation

	Algorithmic (Optimizer)	Procedural (Generator)	The SWARM approach (Multi-agent system)
Layout solution	Found by the engine	Preconceived by human expert	Emerges from Interaction of PCell Modules (= Agents)
Constraint considera-tion	Explicit (formalized)	Implicit (nonformalized)	Explicit and implicit (formalized and nonformalized)
Paradigm	"Top-down"	"Bottom-up"	"Bottom-up Meets Top-down"

This example displays SWARM's *qualitative* asset: while existing approaches are bound to either an explicit *or* implicit consideration of constraints, SWARM supports *both*. So, high-level requirements, such as the zone boundary and desired arrangement are *explicitly* enforced during interaction. Also, in this example the *Quad* module (1) is explicitly prevented from changing its AB/BA layout into a single-row variant. Furthermore, the *Blocking Cap* obeys a *following* move to mimic the actions of *current mirror* (2) and thus obtain overall layout symmetry. Simultaneously, each PCell module *implicitly* takes care of all detailed low-level *matching* restrictions and objectives, such as those listed in Fig. 4 for the *Quad*.

Besides all of these technical merits, Table 2 evaluates SWARM from a more strategic point of view. While algorithmic and procedural layout automatisms (in the sense of Sects. 3.1 and 3.2) follow different paradigms referred to as *top-down* and *bottom-up* automation [4], SWARM can be regarded as a combination of the two. And in the long run, such *bottom-up meets top-down* approaches may be one essential key to finally close the automation gap in analog layout design.

6 Summary and Outlook

The novel layout automation approach SWARM implements a multi-agent system where autonomous module PCells interact with each other to attain compact arrangements inside a recursively tightened layout zone. In contrast to optimization algorithms and procedural generators, the decentralized decision-making in SWARM allows it to consider design constraints both explicitly and implicitly.

Our approach is currently being implemented and tested in an industrial IC design environment, and early assessments indicate that SWARM is much closer to a human expert's manual design style than existing automation strategies. As shown in several given examples, remarkable layout results can emerge from the aggregate PCell interaction by inducing a synergistic flow of self-organization.

The presented results have been achieved with the very first implementation of SWARM. So, considering that the approach is still in its infancy, there is enormous potential for further developments. Future work on SWARM includes (a) the realization of multiple concurrent control organs, (b) adaptive control policies, (c) hierarchically nested interaction flows, (d) modules with learning aptitude, (e) improvements of convergence and robustness, (f) parallelization of module activity via multi-threading, and even (g) real-time human intervention.

References

1. Marolt, D., Scheible, J., Jerke, G., Marolt, V.: SWARM: a self-organization approach for layout automation in analog IC design. Int. J. Electron. Electr. Eng. **4**(5), 374–385 (2016)
2. Arden, W., Brillouët, M., Cogez, P., Graef, M., Huizing, B., Mahnkopf, R.: 'More-than-Moore' white paper. International Technical Roadmap for Semiconductors (2010)
3. Hastings, A.: The Art of Analog Layout, 2nd edn. Prentice Hall (2005)
4. Scheible, J., Lienig, J.: Automation of analog IC layout—challenges and solutions. In: Proceedings of the ACM International Symposium on Physical Design, pp. 33–40 (2015)
5. Rutenbar, R.: Analog CAD: not done yet. Pres. NSF Workshop: Electronic Design Automation—Past, Present, and Future, Slide 6 (2009)
6. Kirkpatrick, S., Gelatt Jr., C., Vecchi, M.: Optimization by simulated annealing. Science **220**(4598), 671–680 (1983)
7. Bowen, O.: Rip-up, reroute strategy accelerates routing process. IEEE Potentials **25**(2), 18–23 (2006)
8. Lin, P., Chang, Y., Lin, S.: Analog placement based on symmetry-island formulation. IEEE Trans. Comp.-Aid. Des. Integr. Circ. Syst. **28**(6), 791–804 (2009)
9. Lin, J., Hung, Z.: SKB-tree: a fixed-outline driven representation for modern floorplanning problems. IEEE Trans. VLSI Syst. **20**(3), 473–484 (2012)
10. Tang, M., Lau, R.: A parallel genetic algorithm for floorplan area optimization. In: 7th International Conference on Intelligent Systems Design and Applications, pp. 801–806 (2007)
11. Ou, H., Chien, H., Chang, Y.: Simultaneous analog placement and routing with current flow and current density considerations. In: Proceedings of the ACM/IEEE 50th Design Automation Conference, pp. 1–6 (2013)
12. Mittag, M., Krinke, A., Jerke, G., Rosenstiel, W.: Hierarchical propagation of geometric constraints for full-custom physical design of ICs. In: Proceedings of the Design, Automation and Test in Europe Conference, pp. 1471–1474 (2012)

13. Reich, T., Eichler, U., Rooch, K., Buhl, R.: Design of a 12-bit cyclic RSD ADC sensor interface IC using the intelligent analog IP library. In: ANALOG, pp. 30–35 (2013)
14. Graupner, A., Jancke, R., Wittmann R.: Generator based approach for analog circuit and layout design and optimization. In: Design, Automation and Test in Europe Conference Exhibition, pp. 1–6 (2011)
15. Marolt, D., Greif, M., Scheible, J., Jerke, G.: PCDS: a new approach for the development of circuit generators in analog IC design. In: Proceedings of the 22nd Austrian Workshop on Microelectronics (Austrochip), pp. 1–6 (2014)
16. Jerke, G., Burdick, T., Herth, P., Marolt, V., Bürzele, C., et al.: Hierarchical module design with Cadence PCell Designer. In: Pres. CDNLive! EMEA 2015, Munich, CUS02
17. Kaur, P.: An enhanced algorithm for floorplan design using hybrid ant colony and particle swarm optimization. Int. J. Res. Appl. Sci. Eng. Technol. **2**(IX), 473–477 (2014)
18. Gacto, M., Galende, M., Alcalá, R., Herrera, F.: METSK-HD^e: a multiobjective evolutionary algorithm to learn accurate TSK-fuzzy systems in high-dimensional and large-scale regression problems. Inf. Sci. **276**, 63–79 (2014)
19. Tomin, N., Zhukov, A., Sidorov, D., Kurbatsky, V., Panasetsky, D., Spiryaev, V.: Random forest based model for preventing large-scale emergencies in power systems. Int. J. Artif. Intell. **13**(1), 211–228 (2015)
20. Rocha, F., Martins, R., Lourenço, N., Horta, N.: Electronic Design Automation of Analog ICs combining Gradient Models with Multi-Objective Evolutionary Algorithms. Springer International Publishing (2014)
21. Precup, R., David, R., Petriu, E., Preitl, S., Rădac, M.: Fuzzy logic-based adaptive gravitational search algorithm for optimal tuning of fuzzy-controlled servo systems. IET Control Theory Appl. **7**(1), 99–107 (2013)
22. Lin, R., Shragowitz, E.: Fuzzy logic approach to placement problem. In: Proceedings of the ACM/IEEE 29th Design Automation Conference, pp. 153–158 (1992)
23. Chin, C., Pan, P., Chen, H., Chen, T., Lin, J.: Efficient analog layout prototyping by layout reuse with routing preservation. In: International Conference on Computer-Aided Design, pp. 40–47 (2013)
24. Gershenson, C.: Design and control of self-organizing systems. Ph.D. Dissertation, Vrije Universiteit Brussel (2007)
25. Langton, C.: Computation at the edge of chaos: phase transitions and emergent computation. Physica D: Nonlinear Phenomena **42**(1–3), 12–37 (1990)
26. Johnson, S.: Emergence: The Connected Lives of Ants, Brains, Cities, and Software. Scribner, New York, NY, USA (2001)
27. Gardner, M.: Mathematical games—the fantastic combinations of John Conway's new solitaire game 'Life'. Sci. Am. **223**, 120–123 (1970)
28. Reynolds, C.: Flocks, herds, and schools: a distributed behavioral model. In: Proceedings of 14th Annual Conference on Computer Graphics and Interactive Techniques, pp. 25–34 (1987)
29. Marsh, L., Onof, C.: Stigmergic epistemology, stigmergic cognition. Cogn. Syst. Res. Elsevier B.V. 1–15 (2007)
30. Shoham, Y., Leyton-Brown, K.: Multiagent Systems: Algorithmic, Game-Theoretic, and Logical Foundations. Cambridge University Press (2009)
31. Hamilton, W.: Geometry for the selfish herd. J. Theor. Biol. **31**(2), 295–311 (1971)
32. Ashby, W.: An Introduction to Cybernetics. Wiley, New York, NY, USA (1956)
33. Bojic, I., Podobnik, V., Ljubi, I., Jezic, G., Kusek, M.: A self-optimizing mobile network: auto-tuning the network with firefly-synchronized agents. Inf. Sci. **182**, 77–92 (2012)
34. Marolt, D., Scheible, J., Jerke, G., Marolt, V.: Analog layout automation via self-organization: enhancing the novel SWARM approach. In: Proceedings of the 7th IEEE Latin American Symposium on Circuits and Systems, pp. 55–58 (2016)

Assignment Problem with Preference and an Efficient Solution Method Without Dissatisfaction

Kengo Saito and Toshiharu Sugawara

Abstract We formulate an assignment problem-solving framework called single-object resource allocation with preferential order (SORA/PO) to incorporate values of resources and individual preferences into assignment problems. We then devise methods to find semi-optimal solutions for SORA/PO problems. The assignment, or resource allocation, problem is a fundamental problem-solving framework used in a variety of recent network and distributed applications. However, it is a combinatorial problem and has a high computational cost to find the optimal solution. Furthermore, SORA/PO problems require solutions in which participating agents express no or few dissatisfactions on the basis of the relationship between relative values and the agents' preference orders. The algorithms described herein can efficiently find a semi-optimal solution that is satisfactory to almost all agents even though its sum of values is close to that of the optimal solution. We experimentally evaluate our methods and the derived solutions by comparing them with tho optimal solutions calculated by CPLEX. We also compare the running times for the solution obtained by these methods.

Keywords Resource allocation problem · Assignment problem · Preference · Cardinal and ordinal values

1 Introduction

The assignment, or resource allocation, problem is a general problem-solving framework [10] that is used in a variety of recent network and distributed applications such as e-commerce and computerized services in which resources/tasks are assigned to agents, i.e., models of persons, computer systems, or software programs. Thus,

---

K. Saito (✉) · T. Sugawara
Department of Computer Science, Waseda University, Tokyo 1698555, Japan
e-mail: k.saito@isl.cs.waseda.ac.jp; kt@isl.cs.waseda.ac.jp

T. Sugawara
e-mail: sugawara@waseda.jp

G. Jezic et al. (eds.), *Agent and Multi-Agent Systems: Technology and Applications*, Smart Innovation, Systems and Technologies 58,
DOI 10.1007/978-3-319-39883-9_3

many assignments for these applications have been developed. However, because they usually impose high computational costs for frequent use, many studies have tried instead to design assignment algorithms to find semi-optimal solutions. Notable methods in the multi-agent systems context are based on game-theoretic and market-like approaches [1, 2, 5, 9]. However, their studies usually aim at finding solutions to maximize social welfare; only a few focus on the preferences of individual agents. Users obviously often have their own preferences that are not correlated with socially-defined values that are usually decided in markets or statistics based on past data, thus cannot be decided individually [3]. Furthermore, they are affected by the internal and external conditions of the problem setting, such as the budget and time restrictions [8].

For such problems, we have proposed a type of assignment problem called *single-object resource allocation with preferential order* (SORA/PO) to reflect agents' preferences [8]. In the SORA/PO framework, agents can be assigned only one resource from the set of resources, but can specify any number of possible resources with *values* and a preference order. The values here represent different aspects, such as monetary prices that agents will accept to pay, benefits had by receiving the resources, capabilities described by numerical values (e.g., GPA scores, ranking and other comparable assessment values from certain viewpoints [12]), and utilities in mechanism design. As such, the total values should be maximized from a social viewpoint. On the other hand, the preference order belongs to an individual and may not be compared with those of other agents. Thus, higher values (prices or benefits) do not always express stronger preferences of individual agents. Accordingly, the solution is an assignment that maximizes the total value of participating agents while taking into account as many individual preferences as possible. Because SORA/PO is a combinatorial problem, we have proposed a number of efficient algorithms to find near-optimal solutions in terms of the total value derived by the assignments [8]. However, they often produce assignments with which a few agents are dissatisfied, meaning that the preferred resources are assigned to other agents with lower values, and this is a kind of "envy" in game theory and economics [6].

Here, we propose new algorithms that produce near-optimal assignments without dissatisfying agents by extending the previous algorithms [8]. In these algorithms, when any dissatisfaction is found during generation of the assignment, it is remedied by canceling the assignment in the causal part and performing a reassignment later. We demonstrate that the proposed algorithms do not result in dissatisfaction. We also experimentally evaluate the solutions and efficiency of the algorithms by comparing them by the previous methods [8] and optimal solutions obtained by CPLEX.

This paper is organized as follows. First, we describe related work in Sect. 2 and formulate SORA/PO within the integer programming framework in Sect. 3. Section 4 describes the algorithms to find near-optimal solutions and discusses their features. Finally, in Sect. 5, we experimentally evaluate the quality of the near-optimal solutions by comparing them with the optimal solutions obtained by CLPEX. We also compare the CPU times taken to find these solutions in order to evaluate the efficiency of the proposed algorithms.

2 Related Work

There have been a number of studies related to assignment and resource allocations like the SORA/PO problem. One approach that is within a multi-agent context is the use of game theoretic and auction-like protocols [10], and many studies have been conducted on semi-optimal allocations that are incentive compatible and strategy-proof. For example, Edelman et al. [5] proposed generalized second-price auctions that could be used by search engines to allocate online advertisements. Devanur and Hayes [4] formalized the adwords problem, which is an extension of keyword auctions, by adding the feature of random keyword arrivals, and they devised an efficient algorithm motivated by *probably approximately correct* (PAC) learning. Cavallo [3] took into account two types of resources, ones which are plentiful or scarce in situations where no money flows since money is often an inappropriate medium to describe user incentives. These approaches are effective in multi-agent settings, but since they rely on game theory, they usually take social welfare into consideration by assuming that the participants' preferences are consistent with the prices. Market-based approaches have been used to apply resource/task allocation problems to self-interested agents [1, 2]. An et al. [1], for example, proposed a market-based resource allocation in environments where only incomplete information is available.In our research, agents are self-interested and have their own preferences that may be inconsistent with their values. However, we believe that our algorithms can be used in these methods when resources are allocated after auctioneers have gathered bids and requirements.

3 Single-Object Resource Allocation

For a set of agents $A = \{1, \dots, n\}$ and a set of resources $G = \{g_1, \dots, g_m\}$, an assignment between G and A is a subset L of the direct product $G \times A$ s.t. $\forall g \in G$ and $\forall i \in A$ that appears at most once in L. Let $\mathcal{L}$ be the set of all assignments between G and A. For any assignment $L \in \mathcal{L}$, two functions can be defined.

$$g_L : A \longrightarrow G \cup \{\emptyset\} \text{ and } a_L : G \longrightarrow A \cup \{\emptyset\},$$

where $g_L(i) = \emptyset$ and $a_L(g) = \emptyset$ when i and g do not appear in L.

Agent i has a sequence of disjoint subsets of resources, $G^i_1, \dots, G^i_{N_i}$ ($\subset G$), and the associated value, V^i_m, for each G^i_m. Note that the values defined here express a number of aspects, such as the monetary price that i accepts to pay and utility by being assigned the resource, which may not mean "utility" in utility value theory. Subscript m is i's preferred number of G^i_m, so G^i_1 and G^i_2, for example, are the sets of the first and second choices for agent i. We also define $d_L(i) = m_i$ if $g_L(i) \in G^i_{m_i}$ for $L \in \mathcal{L}$.

Let us consider an assignment with which i is dissatisfied. Suppose that resource $g \in G^i_{m_0}$ is assigned to i, and that $g' \in G^i_{m_1}$ is done to $j \in A$ and $m_0 > m_1$. When $g' \in G^j_{m_2}$, then j pays $V^j_{m_2}$. However, if $V^j_{m_2} < V^i_{m_1}$, i complains about the assignment. Now, we can define the SORA/PO by using integer programming.

Definition 1 Let G be the set of resources and $B = \{B^i\}_{i \in A}$ be the collection of ordered sets of declared resources preferred by $i \in A$, where $B^i = \{(G^i_1, V^i_1), \ldots, (G^i_{N_i}, V^i_{N_i})\}$ is the ordered set. A SORA/PO is defined as a problem to find an assignment $L^* \in \mathcal{L}$ that maximizes

$$TV(G, A, B, L^*) = \sum_{(g,i) \in L^*} V^i_{d_{L^*}(i)}, \tag{1}$$

subject to

$$\left.\begin{array}{l} \text{For } \forall (g,i) \in L^* \land 0 < \forall k < d_{L^*}(i), \\ \text{if } \exists g' \in G^i_k \text{ s.t. } a_{L^*}(g') = j \land i \neq j, \text{ then } V^j_{d_{L^*}(j)} \geq V^i_k, \end{array}\right\} \tag{2}$$

This problem is denoted by SORA/PO(G,A,B).

The value of expression (1) is referred to as the *total value* of assignment L^* and the total value for $\forall L \in \mathcal{L}$ is denoted by $TV(G, A, B, L)$. Expression (2) ensures that no agents are dissatisfied with L^*.

Definition 2 Agent i is *dissatisfied with assignment* L if the satisfiable conditions (2) are not met, i.e., $\exists (g, i)$ and $\exists (g', j) \in L, \exists k < d_L(i)$ s.t. $g' \in G^i_k$ and $V^j_{d_L(j)} < V^i_k$. The number of agents dissatisfied with assignment L is denoted by $Dis(G, A, B, L)$.

For example, for $|A| = 3$ and $G = \{g_1, g_2\}$, if $B^1 = \{(\{g_2\}, 8300), (\{g_1\}, 10000)\}$, $B^2 = \{(\{g_2\}, 7500), (\{g_1\}, 9500)\}$, $B^3 = \{(\{g_1\}, 9600), (\{g_2\}, 8200)\}$, then assignment $L^* = \{(g_1, 3), (g_2, 1)\} \subset G \times A$ is the solution to SORA/PO and $TV(G, A, B, L^*) = 17900$. Although $L' = \{(g_1, 1), (g_2, 3)\}$ generates more total value, i.e., $TV(G, A, B, L') = 18200$, agent 1 is dissatisfied with L'. Naive algorithms for finding optimal solutions to SORA/PO require exponential time with respect to the size of A and G to find a solution in the worst case. We want to define a more intuitive subclass of SORA/PO problems.

Definition 3 Problem SORA/PO(G, A, B) is consistent when the values in B are always amenable to the associated preferences, i.e., when the following conditions are satisfied: $\forall i \in A, k, k' \in \mathbb{N}$ ($\mathbb{N}$ is the set of natural numbers), if $k < k'$, then $V^i_k \geq V^i_{k'}$.

Note that the consistency between the values and preferences of this definition is only that of a single agent: the definition does not include non-local consistencies.

We will now explain an example that can be formalized using SORA/PO. Suppose that the tickets to a piano-solo concert are to be sold at auction [7] from an online shop. Agents (customers) usually buy center seats, so their price is higher than other seats. However, a few agents try to buy seats on the left side where concert goers can see the pianist and his/her fingers on the keyboard, even seats at the far left and to the rear. Thus, there are two types of agents who have different preferences, and the pricing system determined by standard agents is not consistent with that of the second type of agents. For a standard agent $i \in A$, i's preference and associated prices are, e.g., $B^i = \{(G_1^i, V_1^i), (G_2^i, V_2^i), (G_3^i, V_3^i)\}$, where G_1^i is the set of front center seats, G_2^i is the set of the front non-center seats, and G_3^i is the set of seats near the rear and at the sides. Of course, standard costumers may have different ideas about which seats are in the center or at the front, so $G_k^i \neq G_k^{i'}$ even if $i' \in A$ is also a standard agent. However, their declared values (the value is the acceptable price in this example) are consistent with their preference, so $V_1^i \geq V_2^i \geq V_3^i$. The preferences and associated prices of the second type of agent may be more complicated; for $j \in A$, they could be $B^j = \{(G_1^j, V_1^j), (G_2^j, V_2^j), (G_3^j, V_3^j)\}$, where G_1^j, for example, is the set of front seats on the left-hand side, G_2^j is the set of rear seats on the left, and G_3^j is the set of center seats. Obviously, center seats are more expensive than rear seats, so $V_2^j < V_3^j$ and this is the case where values are not correlated with its preference. After all the prices and preferences are declared, SORA/PO generates an assignment L^* that maximizes sales earnings with no or at most a few agent complaints. Note that because $V_3^j > V_2^j$ for the second type of customer j, so that this problem is not a consistent SORA/PO.

Other examples, including consistent SORA/PO, are described in Saito and Sugawara [8].

4 Proposed Methods for Finding Near-Optimal Solutions

4.1 *Efficient Exploration*

First, let us introduce a number of functions: for $g \in G$ and $S \subseteq A$,

$$p_S(g) = \min\{j \mid \forall i \in S, g \in G_j^i\}, \tag{3}$$

$$v_{1st}(g) = \max\{V_j^i \mid \forall i \in A, g \in G_j^i\}, \text{and} \tag{4}$$

$$v_{2nd}(g) = \max\{V_j^i \mid \forall i \in A \setminus \{a_{first}(g)\}, g \in G_j^i\}, \tag{5}$$

where $a_{first}(g) \in A$ is one of the agents who declared the highest value for g. Functions $p(g)$, $v_{1st}(g)$, and $v_{2nd}(g)$ correspond to the smallest preferential number, the first (largest) value and the second (second largest) value declared by agents for g. The function $N_{bid}(g)$ outputs the number of agents who bid for g. If no agents bid for g, $p_A(g) = v_{1st}(g) = v_{2nd}(g) = \text{undef}$.

Since the original SORA/PO(G, A, B) is a combinatorial problem, we propose a number of methods of obtaining semi-optimal solutions by relaxing Conditions (1) into soft constraints. First, we propose a simple and efficient algorithm that assigns resources one by one according to the values and preferences declared by the agents. It iterates the following steps until $G = \emptyset$ or no resources can be assigned to agents.

Step 1 (Resource Selection): We first assign one of resources whose highest value is declared by agents with higher preferences. Let $H(g)$ be the set of agents who declare the highest values for $g \in G$. Then $G_j = \{g \in G \mid p_{H(g)}(g) = j\} \setminus G_{j-1}$, where we set $G_0 = \emptyset$. Here, $\tilde{G}'$ is one of $G_1, G_2, \ldots$ that is not empty and whose associated subscript is the smallest. Then, we define

$$\tilde{G} = \{g \in \tilde{G}' \mid v_{1st}(g) = m(\tilde{G}')\},$$

where $m(\tilde{G}') = \max_{g\in\tilde{G}'}\{v_{1st}(g)\}$. If $\tilde{G}$ is a singleton, let g_0 be the element in $\tilde{G}$. Otherwise, $g_0(\in \tilde{G})$ is selected using the strategy called *smallest request number first* (SRNF):

$$g_0 \in \tilde{G}_2^{SRNF} = \arg\min_{g\in\tilde{G}}(N_{bid}(g)).$$

If $\tilde{G}^{SRNF}$ is not a singleton, g_0 is randomly selected from $\tilde{G}^{SRNF}$. SRNF assigns a resource g required by a smaller number of agents first, to avoid situations with no assignments. Other strategies such as *lowest second value first*, *largest second value first* and *largest request number first* [8] can be considered, but as the qualities of their solutions were not as high as SRNF in our experiments described below, so will not deal with them here.

Step 2 (Resource assignment): Next, g_0 is assigned to agent $i_0 \in A$ who declared the largest value with the highest preference (i.e., the smallest number of preferences) for g_0. If multiple agents declared it with the highest preference, one of them is randomly selected.

Step 3 (Resolution of Dissatisfaction, RD): Here, suppose that $\exists j_0 \in A$ has been assigned g_1, and it expresses dissatisfaction with the algorithm allocating g_0 to i_0 with the value of v_0. This indicates that for the assignment so far L', $\exists k \in \mathbb{N}$ s.t. $d_{L'}(j_0) > k \geq 1$, $g_0 = g_{L'}(i_0) \in G_k^{j_0}$, and $V_k^{j_0} > V_{d_{L'}(i_0)}^i$. If multiple agents express dissatisfaction, the one whose value for g_0 is the highest is selected as j_0. Then, g_0 is reassigned to j_0, and the assignment to i_0 is canceled. Furthermore, $g_{L'}(j_0)$ becomes free (the assignment of $g_{L'}(j_0)$ to j_0 is also canceled). The requested data for $g_{L'}(j_0)$ that were eliminated in the previous Step 4 are also restored.

Step 4 (Data Elimination): The requested data, B^{i_0}, declared by i_0 are eliminated and $A \leftarrow A \setminus \{i_0\}$. Resource g_0 is marked "assigned." Then, if all elements in G_k^i are marked for $i(\neq i_0) \in A$, (G_k^i, V_k^i) is temporarily eliminated from B^i, and the preference of i's requests are moved over; so $N_i \leftarrow N_i - 1$, $G_k^i \leftarrow G_{k+1}^i$, and $V_k^i \leftarrow V_{k+1}^i$ for $k = j, \ldots, N_i$. Note that because some of the requested data may be restored in a subsequent Step 3, the eliminated data have to be stored.

The proposed method is referred to as *SRNF exploration while resolving dissatisfaction* (SRNF/RD). The previous method [8] is a simpler one in which Step 3 is omitted from SRNF/RD; it is called *SRNF exploration* (or simply SRNF).

4.2 Features of SRNF and SRNF/RD

Although SRNF and SRNF/RD cannot generate optimal solutions, they are applicable to situations in which agent's declarations B^i arrive intermittently, because they assign resources one by one according to the current declarations and available resources. This feature is very useful in real applications where complete information, such as user requests on the Internet, cannot be obtained, and thus, optimal solutions at a certain time may be not optimal at some other time.

Furthermore, if a SORA/PO problem is consistent, SRNF and SRNF/RD can efficiently generate solutions without dissatisfying agents.

Theorem 1 *For any consistent SORA/PO(G,A,B) problem, SRNF generates an assignment, L, s.t.* $Dis(G,B,A,L) = 0$.

Proof [8] We will only sketch the proof. Suppose that $Dis(G,B,A,L) > 0$. Then an agent $i \in A$ exists that is dissatisfied with L. Therefore, $\exists j \in A, \exists k \in \mathbb{N}, d_L(i) > k \geq 1, g_L(j) \in G^i_k$, and $V^i_k > V^j_{d_L(j)}$. Suppose that $g_L(j)$ was assigned to j before $g_L(i)$ was assigned to i. Since $V^i_k > V^j_{d_L(j)}$, $g_L(j) \in G^i_k$, and $g_L(j) \in V^j_{d_L(j)}$, $g_L(j)$ was never assigned to j. Thus, $g_L(j)$ was assigned to j after $g_L(i)$ was assigned to i. Let $\tilde{A}$ be the set of agents to which no resources were assigned at the time $g_L(i)$ was going to be assigned to i. At this time, $g_L(j)$ was not assigned to i, so $\exists j_1 \in \tilde{A} \setminus \{i,j\}$ s.t. $g_L(j) \in G^{j_1}_{k_1}$ and j_1's value for $g_L(j)$ is the largest.

$$V^{j_1}_{k_1} > V^i_k > V^i_{d_L(j)}.$$

Note that $V^{j_1}_{k_1} \neq V^i_k$, since g_1 was not assigned to j_1. However, $g_L(j)$ was not assigned to j_1 either, so $k_1 > d_L(i) > k \geq 1$. Thus, $\exists g_1 \in G^{j_1}_1$ and $V^{j_1}_1 \geq V^{j_1}_{k_1}$. However, as g_1 was not assigned to j_1, $\exists j_2 \in \tilde{A} \setminus \{i,j,j_1\}$ $g_1 \in G^{j-2}_{k_2}$ s.t. $k_2 > d_L(i)$ and $V^{j_2}_{k_2} > V^{j_1}_1$. We can repeat this forever, but A and G are finite. □

This theorem also indicates that SNRF and SNRF/RD are identical for any consistent SORA/PO. Furthermore, we can show that SRNF/RD always stops and finds an assignment without dissatisfaction. Note that it is obvious that SRNF stops in a finite amount of time.

Theorem 2 *For any SORA/PO(G,A,B) problem, SRNF/RD halts with assignment, L, s.t.* $Dis(G,B,A,L) = 0$.

Proof Suppose that $Dis(G,A,B,L') > 0$ right after allocating g to agent i, where L' is the assignment generated by SRNF/RD so far. Let $j \in A$ be an agent that expressed dissatisfaction and whose value for g was the highest. Then, $\exists k \in \mathbb{N}$ s.t. $d_{L'}(j) > k \geq 1$, $g = g_{L'}(i) \in G^j_k$ and $V^j_k > V^i_{d_{L'}(i)}$. SRNF/RD reassigned g to j, and canceled the assignment of $g_{L'}(j)$ in L'. After this reassignment, we assume $Dis(G,A,B,L') > 0$, which means another agent $j' \in A$ was dissatisfied with it as well. Then, $\exists k' \in \mathbb{N}$ s.t. $d_{L'}(j') > k' \geq 1$, $g \in G^{j'}_{k'}$ and $V^{j'}_{k'} > V^j_k > V^i_{d_{L'}(i)}$. Thus, j' also expressed dissatisfaction with SRNF/RD allocating g to i. However, because j's value for g is the highest, this results in a contradiction. Next, we assume that SRNF/RD does not halt for a certain SORA/PO(G, A, B); then $\exists i \in A$ s.t. i expresses dissatisfaction infinitely often. Any reassignment of a resource to i occurs only a finite number of times because reassignment always strictly lowers the preferential number (or raises the preference order) whose minimum value is 1. This means that infinite reassignments occur for other agents in A and always strictly raise their preference order of the assigned resources. However, A and G are finite, and this also produces a contradiction. □

5 Experimental Evaluation

5.1 Experimental Setting

We investigated the qualities of the solutions found by SRNF/RD and SRNF by comparing them with the optimal solutions derived with IBM's ILOG CPLEX using integer programming. Note that the SRNF(/RD) was implemented in Java.

For the sets of resources, $G = \{g_1, \ldots, g_m\}$, and agents, $A = \{1, \ldots, n\}$, the agents' preferences for these resources were generated as follows: For agent $\forall i \in A$, eleven integers, $1 \leq k_i \leq m$, $0 \leq l^i_1 \leq 2$, and $0 \leq l^i_j \leq 3$ (for $2 \leq j \leq 10$) were randomly selected. Then, i's required resources with the preferences were defined as

$$\begin{aligned} G^i_1 &= \{g_{k_i - l^i_1}, \ldots, g_{k_i}, \ldots, g_{k_i + l^i_1}\}, \\ G^i_2 &= \{g_{k_i - l^i_1 - l^i_2}, \ldots, g_{k_i + l^i_1 + l^i_2}\} \setminus G^i_1, \\ &\ldots \\ G^i_{10} &= \{g_{k_i - \sum_{j=1}^{10} l^i_j}, \ldots, g_{k_i + \sum_{j=1}^{10} l^i_j}\} \setminus \cup_{j=1}^{9} G^i_j \end{aligned}$$

We set $G^i_j = \emptyset$ for $j > m$. If $G^i_{j_0} = \emptyset$, we set $G^i_j \leftarrow G^i_{j+1}$ for $j_0, \leq \forall j \leq 9$.

We considered two cases concerning the relationship between values and preferences. In the first case (Case 1), all agents had values consistent with their preferences; thus, the resulting SORA/PO is consistent. i's values were determined as follows. V^i_1 was determined according to a normal distribution, $N(3000, 500)$, whose average and variance were 3000 and 500. Then, integers, $10 \leq b_j \leq 20$ (for $1 \leq j \leq 9$) were randomly selected, and V^i_{j+1} was set to $(1 - b_j/100)V^i_j$. In the second case (Case 2) agents' values were not co-related with preferences, and their values V^i_j for $1 \leq j \leq 10$ were randomly selected to be between 1000 and 5000. The experimental data shown below are the mean values of 50 runs.

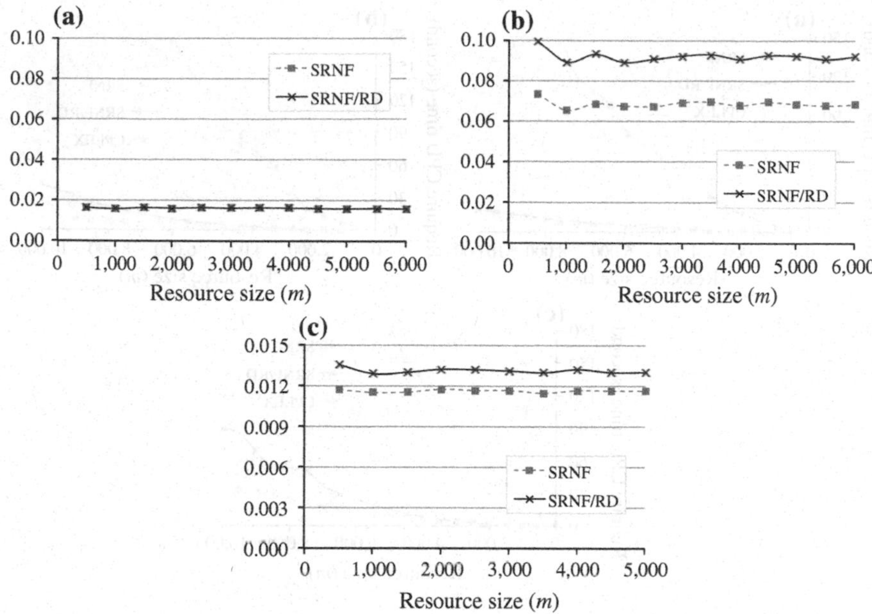

Fig. 1 Ratios of difference in rewards. **a** Exp. 1 (Case 1, $n = m$). **b** Exp. 1 (Case 2, $n = m$). **c** Exp. 2 (Case 2, $n = 2m$)

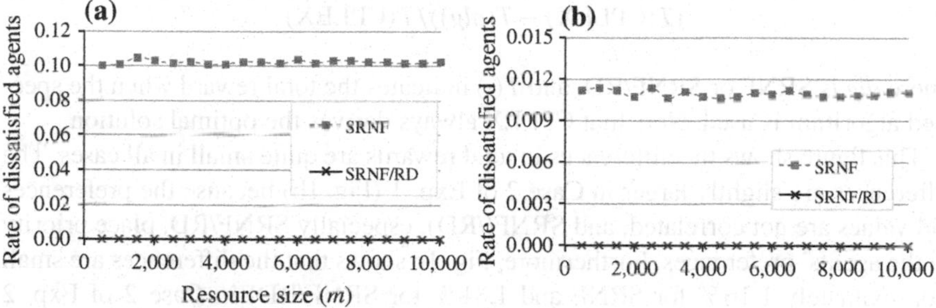

Fig. 2 Ratios of dissatisfied agents. **a** Exp. 1 (Case 2, $n = m$). **b** Exp. 2 (Case 2, $n = 2m$)

5.2 Experimental Results

We conducted two experiments; we set $n = m$ in the first experiment (Exp. 1) and $n = 2m$ in the second experiment (Exp. 2). Thus, in Exp. 2 the number of agents was twice of that of the resources.

Figure 1 plots the difference ratios of social walfare (total rewards) between the optimal solutions obtained by CPLEX and those found by SRNF(/RD), wherein each difference ratio is calculated as

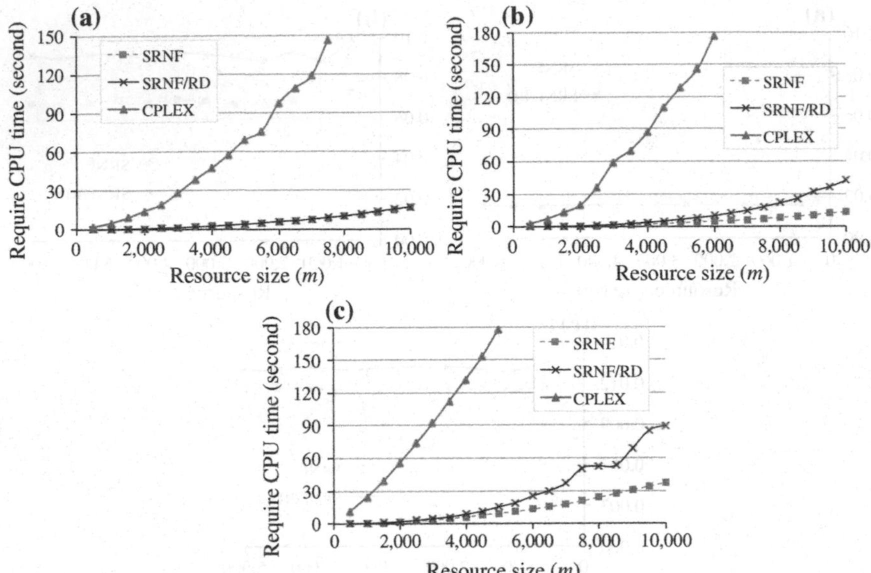

Fig. 3 Ratios of unassigned resources. **a** Exp. 1 (Case 1, $n = m$). **b** Exp. 1 (Case 2, $n = m$). **c** Exp. 2 (Case 2, $n = 2m$)

$$(T(\text{CPLEX}) - T(alg))/T(\text{CPLEX}),$$

where *alg* is SRNF or SRNF/RD, and $T(\cdot)$ indicates the total reward when the specified algorithm is used. Note that CPLEX always derives the optimal solution.

This figure shows the differences in total rewards are quite small in all cases. The differences are slightly larger in Case 2 of Exp. 1 (Fig. 1b) because the preferences and values are not correlated, and SRNF(/RD), especially SRNF/RD, place priority on the agents' preferences. Furthermore, Fig. 1c shows that the differences are small (approximately 1.16 % for SRNF and 1.31 % for SRNF/RD) in Case 2 of Exp. 2 because $n = 2m$ and the requirement are relaxed.

Figure 2 plots $Dis(G, A, B, L)/|A|$, the ratio of agents dissatisfied with the resulting solutions. We have already proved that all algorithms in this paper do not generated dissatisfied agents in Case 1 (a consistent SORA/PO), so the associated figure is omitted. We can see that SRNF left approximately 10 % of the agents dissatisfied in Case 2 of Exp. 1 but left only 1.2 % of the agents dissatisfied in Case 2 in Exp. 2, regardless of the number of resources. The results also confirm that the SRNF/RD does not generate dissatisfied agents, as demonstrated in Theorem 2.

Finally, let us compare the CPU times for finding solutions. The results are shown shown in Fig. 3. Note that we only show the data for CPLEX for $m \leq 7500$ in Case 1 of Exp. 1, $m \leq 6000$ in Case 1 of Exp. 1, and $m \leq 5000$ in Case 1 of Exp. 1, because of its long computational time. The graphs in this figure clearly indicate that SRNF and SRNF/RD are much faster than CPLEX: SRNF/RD is approximately 100 times faster than CPLEX. Moreover, SRNF is faster than SRNF/RD, although it generates solutions with dissatisfied agents. We can see from Fig. 3a that the CPU times for SRNF and SRNF/RD are identical, because SRNF does not generate dissatisfied agents in Case 1. Comparing Fig. 3a, b, the CPU times for SRNS seem almost identical, but CPLEX and SRNF/RD require more CPU time in Case 2 (the values of the individual agents are random) than in Case 1 (consistent SORA/PO). Figure 3c also indicates that all algorithms require more CPU time in Case 2 because the number of agents is double.

5.3 *Remarks*

Overall, the experimental results show that SRNF/RD is quite efficient and it never generates dissatisfied agents, although the difference between the social walfare by the optimal allocation and those by SRNF and SRNF/RD are quite small. We think that one important feature of our algorithms is that they can be used in time-critical applications. This is because SRNF and SRNF/RD are not only efficient; they also determine an assignment one by one; even if the calculation ceases due to time limitations, they provide partial solutions. Furthermore, as new requirements from agents and new resource information arrive, SRNF and SRNF/RD can continue their calculation simply by including the new data. In contrast, in integer/linear-programming, new data are ignored until the current calculation stops or the calculation must restart from scratch [10].

6 Conclusion

We formulated an assignment problem, called SORA/PO, that takes into account agents preferences and devised the SRNF and SRNF/RD methods to find semi-optimal solutions for it. Each agent is assigned only one resource (one-to-one mapping), but agents can declare a number of resources together with their values and preferences. We introduced the conditions to ensure satisfaction; these are like the envy-freeness conditions of game theory for a particular problem setting and demonstrated that SRNF/RD never causes dissatisfaction, and that SRNF dissatisfies only a few agents and is more efficient. We experimentally compared the solutions obtained by these algorithms with the optimal solutions obtained with CPLEX. Our experiments revealed that the reductions in quality of the SRNF and SRNF/RD solutions were quite small, and yet their required CPU times were much shorter than those of CPLEX. We plan to apply our algorithms to realistic assignment applications in the future.

References

1. An, B., Miao, C., Shen, Z.: Market based resource allocation with incomplete information. IJCAI **7**, 1193–1198 (2007)
2. Bredin, J., Kotz, D., Rus, D., Maheswaran, R., Imer, C., Basar, T.: Computational markets to regulate mobile-agent systems. Auton. Agent. Multi-Agent Syst. **6**(3), 235–263 (2003)
3. Cavallo, R.: Incentive compatible two-tiered resource allocation without money. In: Proceedings of the 2014 International Conference on Autonomous Agents and Multi-agent Systems, pp. 1313–1320 (2014)
4. Devanur, N.R., Hayes, T.P.: The adwords problem: Online keyword matching with budgeted bidders under random permutations. In: ACM Conference on Electronic Commerce. ACM (2009)
5. Edelman, B., Ostrovsky, M., Schwarz, M.: Internet advertising and the generalized second-price auction: selling billions of dollars worth of keywords. Am. Econ. Rev. **97**(1), 242–259 (2007)
6. Foley, D.K.: Resource allocation an the public sector. Yale Econ. Essays **7**(1), 45–98 (1967)
7. Nishino, N., Fukuya, K., Ueda, K.: Service design in movie theaters using auction mechanism with seat reservations. In: Proceedings of the International Conference of Soft Computing and Pattern Recognition (SOCPAR'09), pp. 528–533. IEEE (2009)
8. Saito, K., Sugawara, T.: Single-object resource allocation in multiple bid declaration with preferential order. In: Proceedings of the 14th IEEE/ACIS International Conference on Computer and Information Science (ICIS 2015), pp. 341–348. IEEE, Las Vegas, USA (2015)
9. Sakurai, Y., Yokoo, M., Kamei, K.: An efficient approximate algorithm for winner determination in combinatorial auctions. In: Proceedings of the 2nd ACM Conference on Electronic Commerce, pp. 30–37 (2000)
10. Shoham, Y., Leyton-Brown, K.: Multiagent Systems: Algorithmic, Game-Theoretic, and Logical Foundations. Cambridge University Press, New York, NY, USA (2008)
11. Smith, R.G.: The contract net protocol: high-level communication and control in a distributed problem solver. IEEE Trans. Comput. **C-29**(12), 1104–1113 (1980)
12. Varian, H.R.: Equity, envy and efficiency. J. Econ. Theory **9**(1), 63–91 (1974)

Efficient Model Checking Timed and Weighted Interpreted Systems Using SMT and SAT Solvers

Agnieszka M. Zbrzezny, Andrzej Zbrzezny and Franco Raimondi

Abstract In this paper we define Satisfiability Modulo Theory based (SMT-based) and SAT-based bounded model checking (BMC) methods for the existential fragment of Weighted Epistemic Linear Time Logic (WLTLK) interpreted on Timed Weighted Interpreted Systems (TWISs). We provide an implementation based on Z3 and PicoSAT solvers and we present a comparison of the two methods on common instances that can be scaled up to for performance evaluation. A detailed analysis of results shows that in many instances the SMT-based approach can be significantly faster than the SAT-based approach.

1 Introduction

Model checking [1] is an automatic verification technique for concurrent systems. To be able to check automatically whether the system satisfies a given property, one must first create a model of the system, and then describe in a formal language both the created model and the property. Bounded model checking (BMC) for multi-agent systems (MAS) is a symbolic model checking method. It uses a reduction of the problem of truth of, among others, an epistemic formula [2] in a model of MAS to the problem of satisfiability of formulae. The reduction is typically to SAT, but more recently to SMT.

Weighted formalisms are well known in the model checking research area [3, 4] to reason about resource requirements. Another quantitative dimension is the temporal

---

A.M. Zbrzezny (✉) · A. Zbrzezny
IMCS, Jan Długosz University, Al. Armii Krajowej 13/15, 42-200 Częstochowa, Poland
e-mail: agnieszka.zbrzezny@ajd.czest.pl; aga.zbrzezny@gmail.com

A. Zbrzezny
e-mail: a.zbrzezny@ajd.czest.pl

F. Raimondi
School of Science and Technology, Middlesex University, London, England
e-mail: f.raimondi@mdx.ac.uk

G. Jezic et al. (eds.), *Agent and Multi-Agent Systems: Technology and Applications*, Smart Innovation, Systems and Technologies 58,
DOI 10.1007/978-3-319-39883-9_4

duration of actions. In this area were considered: the timed automata [5] the timed interpreted systems [6], and the weighted timed automata [3].

To the best of our knowledge, there is no work that considers model checking of resources in systems where time is modelled explicitly. In this paper we provide a BMC method to reason about resources in timed systems. More in detail, we check the existential fragment of weighted LTLK (WELTLK) in timed weighted interpreted systems (TWIS).

We do not compare our results with other model checkers for MASs, e.g. MCMAS [7] or MCK [8], simply because they do not support the WELTLK language and the timed weighted interpreted systems. PRISM [9] is the only available probabilistic (epistemic) model checker that allows for quantitative reasoning in MAS, however it is based on BDD structures, and it does not provide any bounded model checking methods.

In this paper, we make the following contributions. Firstly, we define and implement an SMT-based BMC method for WELTLK and for TWISs. Secondly, we define and implement a SAT-based BMC method for WELTLK and for TWISs. Next, we report on the initial experimental evaluation of our SMT and SAT-based BMC methods.

2 Preliminaries

Timed Weighted Interpreted Systems were proposed in [10] to extend interpreted systems (ISs) in order to make possible reasoning about real-time aspects of MASs and to extend ISs to make the reasoning possible about not only temporal and epistemic properties, but also agents's quantitative properties.

Clocks. Let $\mathbb{N}$ be a set of natural numbers, $\mathbb{N}_+ = \mathbb{N} \setminus \{0\}$, and $\mathcal{X}$ be a finite set of *clocks*. Each clock is a variable ranging over a set of non-negative natural numbers. A clock valuation v of $\mathcal{X}$ is a total function from $\mathcal{X}$ into the set of natural numbers. The set of all the clock valuations is denoted by $\mathbb{N}^{|\mathcal{X}|}$. For $\mathcal{X}' \subseteq \mathcal{X}$, the valuation which assigns the value 0 to all clocks is defined as: $\forall_{x \in \mathcal{X}'} v'(x) = 0$ and $\forall_{x \in \mathcal{X} \setminus \mathcal{X}'} v'(x) = v(x)$. For $v \in \mathbb{N}^{|\mathcal{X}|}$ and $\delta \in \mathbb{N}$, $v + \delta$ is the clock valuation of $\mathcal{X}$ that assigns the value $v(x) + \delta$ to each clock x.

Clock constraints over $\mathcal{X}$ are conjunctions of comparisons of a clock with a time constant from the set of natural numbers $\mathbb{N}$. The grammar $\varphi := \mathbf{true} \mid x < c \mid x \leq c \mid x = c \mid x \geq c \mid x > c \mid \varphi \wedge \varphi$ generates the set $C(\mathcal{X})$ of clock constraints over $\mathcal{X}$, where $x \in \mathcal{X}$ and $c \in \mathbb{N}$. A clock valuation v satisfies a clock constraint φ, written as $v \models \varphi$, iff φ evaluates to true using the clock values given by v.

TWISs. Let $\mathcal{A} = \{1, \ldots, n\}$ denotes a non-empty and finite set of agents, and $\mathcal{E}$ be a special agent that is used to model the environment in which the agents operate and $\mathcal{PV} = \bigcup_{\mathbf{c} \in \mathcal{A} \cup \{\mathcal{E}\}} \mathcal{PV}_{\mathbf{c}}$ be a set of propositional variables, such that $\mathcal{PV}_{\mathbf{c}_1} \bigcap \mathcal{PV}_{\mathbf{c}_2} = \emptyset$ for all $\mathbf{c}_1, \mathbf{c}_2 \in \mathcal{A} \cup \{\mathcal{E}\}$. A *timed weighted interpreted system* is a tuple $(\{L_{\mathbf{c}}, Act_{\mathbf{c}}, \mathcal{X}_{\mathbf{c}}, P_{\mathbf{c}}, \mathcal{V}_{\mathbf{c}}, \mathcal{I}_{\mathbf{c}}, d_{\mathbf{c}}\}_{\mathbf{c} \in \mathcal{A} \cup \{\mathcal{E}\}}, \{t_{\mathbf{c}}\}_{\mathbf{c} \in \mathcal{A}}, \{t_{\mathcal{E}}\}, \iota)$, where $L_{\mathbf{c}}$ is a non-empty set of *local states* of the agent $\mathbf{c}$, $S = L_1 \times \cdots \times L_n \times L_{\mathcal{E}}$ is the set of all global states, $\iota \subseteq S$ is a non-empty set of initial states, $Act_{\mathbf{c}}$ is a non-empty set of *possible actions* of the agent $\mathbf{c}$, $Act = Act_1 \times \cdots \times Act_n \times Act_{\mathcal{E}}$ is the set of *joint actions*, $\mathcal{X}_{\mathbf{c}}$ is a non-empty set of *clocks*, $P_{\mathbf{c}} : L_{\mathbf{c}} \to 2^{Act_{\mathbf{c}}}$ is a *protocol function*, $t_{\mathbf{c}} : L_{\mathbf{c}} \times L_{\mathcal{E}} \times C(\mathcal{X}_{\mathbf{c}}) \times 2^{\mathcal{X}_{\mathbf{c}}} \times Act \to L_{\mathbf{c}}$ is a (partial) *evolution function* for agents, $t_{\mathcal{E}} : L_{\mathcal{E}} \times C(\mathcal{X}_{\mathcal{E}}) \times 2^{\mathcal{X}_{\mathcal{E}}} \times Act \to L_{\mathcal{E}}$ is a (partial) *evolution function* for environment, $\mathcal{V}_{\mathbf{c}} : L_{\mathbf{c}} \to 2^{\mathcal{PV}_{\mathbf{c}}}$ is a *valuation function* assigning to each local state a set of propositional variables that are assumed to be true at that state, $\mathcal{I}_{\mathbf{c}}: L_{\mathbf{c}} \to C(\mathcal{X}_{\mathbf{c}})$ is an *invariant function*, that specifies the amount of time the agent $\mathbf{c}$ may spend in a given local state, and $d_{\mathbf{c}} : Act_{\mathbf{c}} \to \mathbb{N}$ is a *weight function.*

We assume that if $\epsilon_{\mathbf{c}} \in P_{\mathbf{c}}(l_{\mathbf{c}})$, then $t_{\mathbf{c}}(l_{\mathbf{c}}, l_{\mathcal{E}}, \varphi_{\mathbf{c}}, \mathcal{X}, (a_1, \ldots, a_n, a_{\mathcal{E}})) = l_{\mathbf{c}}$ for $a_{\mathbf{c}} = \epsilon_{\mathbf{c}}$, any $\varphi_{\mathbf{c}} \in C(\mathcal{X})$, and any $\mathcal{X} \in 2^{\mathcal{X}_{\mathbf{c}}}$. An element $< l_{\mathbf{c}}, l_{\mathcal{E}}, \varphi_{\mathbf{c}}, \mathcal{X}, a, l'_{\mathbf{c}} >$ represents a transition from the local state $l_{\mathbf{c}}$ to the local state $l'_{\mathbf{c}}$ of agent $\mathbf{c}$ labelled with action a. The invariant condition allows the TWIS to stay at the local state l as long only as the constraint $\mathcal{I}_{\mathbf{c}}(l_{\mathbf{c}})$ is satisfied. The guard φ has to be satisfied to enable the transition.

Let c_{max} be a constant such that some clock x is compared with c in some constraint appearing in an invariant or a guard of TWIS, and $v, v' \in \mathbb{N}^{|\mathcal{X}|}$ two clock valuation. We say that $v \simeq v'$ iff the following condition holds for each $x \in \mathcal{X}$: $v(x) > c_{max}$ and $v'(x) > c_{max}$ or $v(x) \leq c_{max}$ and $v'(x) \leq c_{max}$ and $v(x) = v'(x)$. The clock valuation v' such that for each clock $x \in \mathcal{X}$, $v'(x) = v(x) + 1$ if $v(x) \leq c_{max}$, and $v'(x) = c_{max} + 1$ otherwise, is called a time successor of v (written $succ(v)$).

For a given TWIS we define a *timed weighted model* (or a *model*) as a tuple $\mathcal{M} = (Act, S, \iota, T, \mathcal{V}, d)$, where: $Act = Act_1 \times \cdots \times Act_n \times Act_{\mathcal{E}}$ is the set of all the joint actions, $S = (L_1 \times \mathbb{N}^{|\mathcal{X}_1|}) \times \cdots \times (L_n \times \mathbb{N}^{|\mathcal{X}_n|}) \times (L_{\mathcal{E}} \times \mathbb{N}^{|\mathcal{X}_{\mathcal{E}}|})$ is the set of all the *global states*, $\iota = (\iota_1 \times \{0\}^{|\mathcal{X}_1|}) \times \cdots \times (\iota_n \times \{0\}^{|\mathcal{X}_n|}) \times (\iota_{\mathcal{E}} \times \{0\}^{|\mathcal{X}_{\mathcal{E}}|})$ is the set of all the *initial* global states, $\mathcal{V} : S \to 2^{\mathcal{PV}}$ is the valuation function defined as $\mathcal{V}(s) = \bigcup_{\mathbf{c} \in \mathcal{A} \cup \{\mathcal{E}\}} \mathcal{V}_{\mathbf{c}}(l_{\mathbf{c}}(s))$, $T \subseteq S \times (Act \cup \mathbb{N}) \times S$ is a transition relation defined by action and time transitions. For $a \in Act$ and $\delta \in \mathbb{N}$: action transition is defined as $(s, a, s') \in T$ (or $s \xrightarrow{a} s'$) iff for all $\mathbf{c} \in \mathcal{A} \cup \mathcal{E}$, there exists a local transition $t_{\mathbf{c}}(l_{\mathbf{c}}(s), \varphi_{\mathbf{c}}, \mathcal{X}', a) = l_{\mathbf{c}}(s')$ such that $v_{\mathbf{c}}(s) \models \varphi_{\mathbf{c}} \wedge \mathcal{I}(l_{\mathbf{c}}(s))$ and $v'_{\mathbf{c}}(s') = v_{\mathbf{c}}(s)[\mathcal{X}' := 0]$ and $v'_{\mathbf{c}}(s') \models \mathcal{I}(l_{\mathbf{c}}(s'))$; time transition is defined as $(s, \delta, s') \in T$ iff for all $\mathbf{c} \in \mathcal{A} \cup \mathcal{E}$, $l_{\mathbf{c}}(s) = l_{\mathbf{c}}(s')$ and $v'_{\mathbf{c}}(s') = v_{\mathbf{c}}(s) + \delta$ and $v'_{\mathbf{c}}(s') \models \mathcal{I}(l_{\mathbf{c}}(s'))$. $d : Act \to \mathbb{N}$ is the "joint" weight function defined as follows: $d((a_1, \ldots, a_n, a_{\mathcal{E}})) = d_1(a_1) + \cdots + d_n(a_n) + d_{\mathcal{E}}(a_{\mathcal{E}})$.

Given a TWIS, one can define for any agent $\mathbf{c}$ the indistinguishability relation $\sim_{\mathbf{c}} \subseteq S \times S$ as follows: $s \sim_{\mathbf{c}} s'$ iff $l_{\mathbf{c}}(s') = l_{\mathbf{c}}(s)$ and $v_{\mathbf{c}}(s') \simeq v_{\mathbf{c}}(s)$. We assume the following definitions of epistemic relations: $\sim^E_\Gamma \stackrel{def}{=} \bigcup_{\mathbf{c} \in \Gamma} \sim_{\mathbf{c}}$, $\sim^C_\Gamma \stackrel{def}{=} (\sim^E_\Gamma)^+$ (the transitive closure of $\sim^E_\Gamma$), $\sim^D_\Gamma \stackrel{def}{=} \bigcap_{\mathbf{c} \in \Gamma} \sim_{\mathbf{c}}$, where $\Gamma \subseteq \mathcal{A}$.

A run in $\mathcal{M}$ is an infinite sequence $\rho = s_0 \xrightarrow{\delta_0, a_0} s_1 \xrightarrow{\delta_1, a_1} s_2 \xrightarrow{\delta_2, a_2} \ldots$ of global states such that the following conditions hold for all $i \in \mathbb{N} : s_i \in S, a_i \in Act, \delta_i \in \mathbb{N}_+$, and there exists $s'_i \in S$ such that $(s_i, \delta_i, s'_i) \in T$ and $(s_i, a_i, s_{i+1}) \in T$. The definition of a run does not permit two consecutive joint actions to be performed one after the other, i.e., between each two joint actions some time must pass; such a run is called *strongly monotonic*.

Abstract model. The set of all the clock valuations is infinite which means that a model has an infinite set of states. We need to abstract the proposed model before we can apply the BMC technique. Let $\mathbb{D}_{\mathbf{c}} = \{0, \ldots, c_{\mathbf{c}} + 1\}$ with $c_{\mathbf{c}}$ be the largest constant appearing in any enabling condition or state invariants of agent $\mathbf{c}$ and $\mathbb{D} = \bigcup_{\mathbf{c} \in \mathcal{A} \cup \mathcal{E}} \mathbb{D}_{\mathbf{c}}^{|\mathcal{X}_{\mathbf{c}}|}$. A tuple $\widehat{\mathcal{M}} = (Act, \widehat{S}, \widehat{\iota}, \widehat{T}, \widehat{\mathcal{V}}, d)$, is an *abstract model*, where $\widehat{\iota} = \prod_{\mathbf{c} \in \mathcal{A} \cup \mathcal{E}} (\iota_{\mathbf{c}} \times \{0\}^{|\mathcal{X}_{\mathbf{c}}|})$ is the set of all the initial global states, $\widehat{S} = \prod_{\mathbf{c} \in \mathcal{A} \cup \mathcal{E}} (L_{\mathbf{c}} \times \mathbb{D}_{\mathbf{c}}^{|\mathcal{X}_{\mathbf{c}}|})$ is the set of all the abstract global states. $\widehat{\mathcal{V}} : \widehat{S} \to 2^{\mathcal{PV}}$ is the valuation function such that: $p \in \widehat{\mathcal{V}}(\widehat{s})$ iff $p \in \bigcup_{\mathbf{c} \in \mathcal{A} \cup \mathcal{E}} \widehat{\mathcal{V}}_{\mathbf{c}}(l_{\mathbf{c}}(\widehat{s}))$ for all $p \in \mathcal{PV}$; and $\widehat{T} \subseteq \widehat{S} \times (Act \cup \tau) \times \widehat{S}$. Let $a \in Act$. Then, action transition is defined as $(\widehat{s}, a, \widehat{s}') \in \widehat{T}$ iff $\forall_{\mathbf{c} \in \mathcal{A}} \exists_{\phi_{\mathbf{c}} \in \mathcal{C}(\mathcal{X}_{\mathbf{c}})} \exists_{\mathcal{X}'_{\mathbf{c}} \subseteq \mathcal{X}_{\mathbf{c}}} (t_{\mathbf{c}}(l_{\mathbf{c}}(\widehat{s}), \phi_{\mathbf{c}}, \mathcal{X}'_{\mathbf{c}}, a) = l_{\mathbf{c}}(\widehat{s}')$ and $v_{\mathbf{c}} \models \phi_{\mathbf{c}} \wedge \mathcal{I}(l_{\mathbf{c}}(\widehat{s}))$ and $v'_{\mathbf{c}}(\widehat{s}') = v_{\mathbf{c}}(\widehat{s})[\mathcal{X}'_{\mathbf{c}} := 0]$ and $v'_{\mathbf{c}}(\widehat{s}') \models \mathcal{I}(l_{\mathbf{c}}(\widehat{s}')))$; time transition is defined as $(\widehat{s}, \tau, \widehat{s}') \in \widehat{T}$ iff $\forall_{\mathbf{c} \in \mathcal{A} \cup \mathcal{E}} (l_{\mathbf{c}}(\widehat{s}) = l_{\mathbf{c}}(\widehat{s}'))$ and $v_{\mathbf{c}}(\widehat{s}) \models \mathcal{I}(l_{\mathbf{c}}(\widehat{s}))$ and $succ(v_{\mathbf{c}}(\widehat{s})) \models \mathcal{I}(l_{\mathbf{c}}(\widehat{s})))$ and $\forall_{\mathbf{c} \in \mathcal{A}} (v'_{\mathbf{c}}(\widehat{s}') = succ(v_{\mathbf{c}}(\widehat{s}')))$ and $(v'_{\mathcal{E}}(\widehat{s}') = succ(v_{\mathcal{E}}(\widehat{s})))$. Given an abstract model one can define for any agent $\mathbf{c}$ the indistinguishability relation $\sim_{\mathbf{c}} \subseteq \widehat{S} \times \widehat{S}$ as follows: $\widehat{s} \sim_{\mathbf{c}} \widehat{s}'$ iff $l_{\mathbf{c}}(\widehat{s}') = l_{\mathbf{c}}(\widehat{s})$ and $v_{\mathbf{c}}(\widehat{s}') = v_{\mathbf{c}}(\widehat{s})$. A path π in an abstract model is a sequence $\widehat{s}_0 \xrightarrow{b_1} \widehat{s}_1 \xrightarrow{b_2} \widehat{s}_2 \xrightarrow{b_3} \ldots$ of transitions such that for each $i > 1$, $b_i \in Act \cup \{\tau\}$ and $b_1 = \tau$ and for each two consecutive transitions at least one of them is a time transition. Next, $\pi[j..m]$ denotes the finite sequence $\widehat{s}_j \xrightarrow{\delta_{j+1}, a_{j+1}} \widehat{s}_{j+1} \xrightarrow{\delta_{j+2}, a_{j+2}} \ldots \widehat{s}_m$ with $m - j$ transitions and $m - j + 1$ states, and $D\pi[j..m]$ denotes the (cumulative) weight of $\pi[j..m]$ that is defined as $d(a_{j+1}) + \cdots + d(a_m)$ (hence 0 when $j = m$). The set of all the paths starting at $\widehat{s} \in \widehat{S}$ is denoted by $\Pi(\widehat{s})$, and the set of all the paths starting at an initial state is denoted by $\Pi = \bigcup_{\widehat{s}^0 \in \widehat{\iota}} \Pi(\widehat{s}^0)$.

WELTLK. Let I be an interval in $\mathbb{N}$ of the form: $[a, b)$ or $[a, \infty)$, for $a, b \in \mathbb{N}$ and $a \neq b$. WELTLK is the existential fragment of WLTLK [11], defined by the grammar: $\varphi ::= \mathbf{true} \mid \mathbf{false} \mid p \mid \neg p \mid \varphi \wedge \varphi \mid \varphi \vee \varphi \mid \mathbf{X}_I \varphi \mid \varphi \mathbf{U}_I \varphi \mid \varphi \mathbf{R}_I \varphi \mid \overline{\mathrm{K}}_{\mathbf{c}} \varphi \mid \overline{\mathrm{E}}_{\Gamma} \varphi \mid \overline{\mathrm{D}}_{\Gamma} \varphi \mid \overline{\mathrm{C}}_{\Gamma} \varphi$.

The semantics of WELTLK is the following. A WELTLK formula φ is true along the path π in the abstract model $\widehat{\mathcal{M}}$ (in symbols $\widehat{\mathcal{M}}, \pi \models \varphi$) iff $\widehat{\mathcal{M}}, \pi^0 \models \varphi$, where:

- $\widehat{\mathcal{M}}, \pi^m \models \mathbf{X}_I \alpha$ iff $D\pi[m..m+1] \in I$ and $\widehat{\mathcal{M}}, \pi^{m+1} \models \alpha$,
- $\widehat{\mathcal{M}}, \pi^m \models \alpha \mathbf{U}_I \beta$ iff $(\exists i \geq m)(D\pi[m..i] \in I$ and $\widehat{\mathcal{M}}, \pi^i \models \beta$ and $(\forall m \leq j < i) \widehat{\mathcal{M}}, \pi^j \models \alpha)$,
- $\widehat{\mathcal{M}}, \pi^m \models \alpha \mathbf{R}_I \beta$ iff $(\forall i \geq m)(D\pi[m..i] \in I$ implies $\widehat{\mathcal{M}}, \pi^i \models \beta)$ or $(\exists i \geq m)$ $(D\pi[m..i] \in I$ and $\widehat{\mathcal{M}}, \pi^i \models \alpha$ and $(\forall m \leq j \leq i) \widehat{\mathcal{M}}, \pi^j \models \beta)$,
- $\widehat{\mathcal{M}}, \pi^m \models \overline{\mathrm{K}}_{\mathbf{c}} \alpha$ iff $(\exists \pi' \in \Pi)(\exists i \geq 0)(\pi'(i) \sim_{\mathbf{c}} \pi(m)$ and $\widehat{\mathcal{M}}, \pi'^i \models \alpha)$.
- $\widehat{\mathcal{M}}, \pi^m \models \overline{Y}_{\Gamma} \alpha$ iff $(\exists \pi' \in \Pi)(\exists i \geq 0)(\pi'(i) \sim_{\Gamma}^{Y} \pi(m)$ and $\widehat{\mathcal{M}}, \pi'^i \models \alpha)$, where $Y \in \{\mathrm{D}, \mathrm{E}, \mathrm{C}\}$.

Theorem 1 *Let TWIS be a timed weighted interpreted system, $\mathcal{M}$ be a concrete model for TWIS, φ WELTLK formula, and $\widehat{\mathcal{M}}$ the abstract model for $TWIS_\varphi$. Then, $\mathcal{M} \models \mathbf{E}\varphi$ iff $\widehat{\mathcal{M}} \models \mathbf{E}\varphi$.*

3 Bounded Model Checking

Let $\widehat{\mathcal{M}}$ be an abstract model, and $k \in \mathbb{N}$ a bound. A *k-path* π_l is a pair (π, l), where π is a finite sequence $\widehat{s}_0 \xrightarrow{a_1} \widehat{s}_1 \xrightarrow{a_2} \dots \xrightarrow{a_k} \widehat{s}_k$ of transitions such that for each $1 < i < k$, $a_i \in Act \cup \{\tau\}$ and $a_1 = \tau$ and for each two consecutive transitions at least one is a time transition. A k-path π_l is a *loop* if $l < k$ and $\pi(k) = \pi(l)$. If a k-path π_l is a loop, then it represents the infinite path of the form uv^ω, where $u = (\widehat{s}_0 \xrightarrow{a_1} \widehat{s}_1 \xrightarrow{a_2} \dots \xrightarrow{a_l} \widehat{s}_l)$ and $v = (\widehat{s}_{l+1} \xrightarrow{a_{l+2}} \dots \xrightarrow{a_k} \widehat{s}_k)$. $\Pi_k(\widehat{s})$ denotes the set of all the k-paths of $\widehat{\mathcal{M}}$ that start at $\widehat{s}$, and $\Pi_k = \bigcup_{\widehat{s}^0 \in \iota} \Pi_k(\widehat{s}^0)$. The bounded satisfiability relation $\models_k$ which indicates k-truth of a WELTLK formula in the model $\mathcal{M}$ at some state s of $\mathcal{M}$ is defined in [11]. A WELTLK formula φ is *k-true* in the abstract model $\widehat{\mathcal{M}}$ (in symbols $\widehat{\mathcal{M}} \models_k \varphi$) iff φ is k-true at some initial state of the abstract model $\widehat{\mathcal{M}}$.

Theorem 2 *Let $\widehat{\mathcal{M}}$ be an abstract model and φ a WELTLK formula. Then, the following equivalence holds: $\widehat{\mathcal{M}} \models \varphi$ iff there exists $k \leq |\widehat{\mathcal{M}}| \cdot |\varphi| \cdot 2^{|\varphi|}$ such that $\widehat{\mathcal{M}} \models_k \varphi$.*

Note however that from the BMC point of view the bound k that makes the bounded and unbounded semantics equivalent is insignificant. This is because the BMC method for large k is unfeasible.

Translation to SMT. Let $\widehat{\mathcal{M}}$ be an abstract model, φ a WELTLK formula, and $k \geq 0$ a bound. The presented SMT encoding of the BMC problem for WELTLK and for TWIS is based on the SAT encoding presented in [11, 12] and it relies on defining a quantifier-free first-order formula.

It is well known that the main idea of the SMT-based BMC method consists in translating the bounded model checking problem, i.e., $\widehat{\mathcal{M}} \models_k \varphi$, to the problem of checking the satisfiability of the following quantifier-free first-order formula: $[\widehat{\mathcal{M}}, \varphi]_k^{\mathrm{SMT}} := [\widehat{\mathcal{M}}^{\varphi,\widehat{\iota}}]_k^{\mathrm{SMT}} \wedge [\varphi]_{\widehat{\mathcal{M}},k}^{\mathrm{SMT}}$. The definition of the formula $[\widehat{\mathcal{M}}, \varphi]_k^{\mathrm{SMT}}$ assumes that each abstract global state $\widehat{s} \in \widehat{S}$ of $\widehat{\mathcal{M}}$ can be represented by a valuation of a symbolic state $\overline{\mathbf{w}} = ((w_1, v_1), \dots, (w_n, v_n), (w_\mathcal{E}, v_\mathcal{E}))$ that consists of symbolic local states. Each symbolic local state is a pair $(w_\mathbf{c}, v_\mathbf{c})$ of individual variables ranging over the natural numbers that consists of a local state of the agent $\mathbf{c}$ and a clock valuation. Similarly, each action can be represented by a valuation of a symbolic joint action $\overline{a}$ that is a vector of the individual variables ranging over the natural numbers, each sequence of weights associated with the joint action can be represented by a valuation of a symbolic weights and each symbolic local weight $d_\mathbf{c}$ is an individual variable ranging over the natural numbers.

The formula $[\widehat{\mathcal{M}}^{\varphi,\hat{\iota}}]_k^{\mathrm{SMT}}$ constrains the $f_k(\varphi)$ symbolic k-paths to be valid k-paths of $\widehat{\mathcal{M}}$, while the formula $[\varphi]_{\widehat{\mathcal{M}},k}$ encodes a number of constraints that must be satisfied on these sets of k-paths for φ to be satisfied. Note that the exact number of necessary symbolic k-paths depends on the checked formula φ, and it can be calculated by means of the function f_k : WELTLK $\rightarrow \mathbb{N}$ which is an auxiliary function defined in [11]. Now, since in the BMC method we deal with existential validity, the number of k-paths sufficient to validate φ is given by the function $\widehat{f}_k$: WELTLK $\rightarrow \mathbb{N}$ that is defined as $\widehat{f}_k(\varphi) = f_k(\varphi) + 1$.

Let $\overline{\mathbf{w}}$ and $\overline{\mathbf{w}}'$ be two different symbolic states, $\overline{d}$ a sequence of symbolic weighs, $\overline{a}$ a symbolic action, δ a symbolic time passage, and u be a symbolic number. We assume definitions of the following auxiliary quantifier-free first-order formulae: $I_s(\overline{\mathbf{w}})$—it encodes the state s of the abstract model $\widehat{\mathcal{M}}$, $\mathcal{T}_{\mathbf{c}}(w_{\mathbf{c}}, ((a_{\mathbf{c}}, d_{\mathbf{c}}), \delta), w'_{\mathbf{c}})$ encodes the local evolution function of agent $\mathbf{c}$; We assume that the first transition is the time one, and between each two action transitions at least one time transition appears. $\mathcal{A}(\overline{a})$ encodes that each symbolic local action $a_{\mathbf{c}}$ of $\overline{a}$ has to be executed by each agent in which it appears, and $\mathcal{T}(\overline{\mathbf{w}}, ((\overline{a}, \overline{d}), \delta), \overline{\mathbf{w}}') := \mathcal{A}(\overline{a}) \wedge \bigwedge_{\mathbf{c} \in \mathcal{A} \cup \{\mathcal{E}\}} \mathcal{T}_{\mathbf{c}}(w_{\mathbf{c}}, ((a_{\mathbf{c}}, d_{\mathbf{c}}), \delta, w'_{\mathbf{c}})$.

Let $\boldsymbol{\pi}_j$ denote the jth symbolic k-path, i.e. the sequence of symbolic transitions: $\overline{\mathbf{w}}_{0,j} \xrightarrow{(\overline{a}_{1,j}, \overline{d}_{1,j}), \delta_{1,j}} \overline{\mathbf{w}}_{1,j} \xrightarrow{(\overline{a}_{2,j}, \overline{d}_{2,j}), \delta_{2,j}} \dots \xrightarrow{(\overline{a}_{k,j}, \overline{d}_{k,j}), \delta_{k,j}} \overline{\mathbf{w}}_{k,j}$. Thus, given the above, one can define the formula $[\widehat{\mathcal{M}}^{\varphi,\hat{\iota}}]_k^{\mathrm{SMT}}$ as follows:

$$[\widehat{\mathcal{M}}^{\varphi,\hat{\iota}}]_k^{\mathrm{SMT}} := \bigvee_{s \in \hat{\iota}} I_s(\overline{\mathbf{w}}_{0,0}) \wedge \bigvee_{j=1}^{\widehat{f}_k(\varphi)} \overline{\mathbf{w}}_{0,0} = \overline{\mathbf{w}}_{0,j} \wedge \bigwedge_{j=1}^{\widehat{f}_k(\varphi)} \bigvee_{l=0}^{k} l = u_j \wedge \bigwedge_{j=1}^{\widehat{f}_k(\varphi)} \bigwedge_{i=0}^{k-1} \mathcal{T}(\overline{\mathbf{w}}_{i,j}, ((\overline{a}_{i,j}, \overline{d}_{i,j}), \delta_{i,j}), \overline{\mathbf{w}}_{i+1,j})$$

where $\overline{\mathbf{w}}_{i,j}$, $\overline{a}_{i,j}$, $\overline{d}_{i,j}$, and $\delta_{i,j}$ are, respectively, symbolic states, symbolic actions, symbolic weights, and symbolic time passage for $0 \leq i \leq k$ and $1 \leq j \leq \widehat{f}_k(\varphi)$.

The formula $[\varphi]_{\widehat{\mathcal{M}},k}^{\mathrm{SMT}}$ encodes the bounded semantics of a WELTLK formula φ, and it is defined on the same sets of individual variables as the formula $[\widehat{\mathcal{M}}^{\varphi,\hat{\iota}}]_k^{\mathrm{SMT}}$. Moreover, it uses the auxiliary quantifier-free first-order formulae defined in [11].

Let $F_k(\varphi) = \{j \in \mathbb{N} \mid 1 \leq j \leq \widehat{f}_k(\varphi)\}$, and $[\varphi]_k^{[m,n,A]}$ denote the translation of φ along the nth symbolic path $\boldsymbol{\pi}_n^m$ with the starting point m by using the set $A \subseteq F_k(\varphi)$. Then, the next step is a translation of a WELTLK formula φ to a quantifier-free first-order formula $[\varphi]_{\widehat{\mathcal{M}},k}^{\mathrm{SMT}} := [\varphi]_k^{[0,1,F_k(\varphi)]}$ that was presented in [13].

Theorem 3 *Let $\widehat{\mathcal{M}}$ be an abstract model, and φ a WELTLK formula. Then for every $k \in \mathbb{N}$, $\widehat{\mathcal{M}} \models_k \varphi$ if, and only if, the quantifier-free first-order formula formula $[\widehat{\mathcal{M}}, \varphi]_k^{SMT}$ is satisfiable.*

Translation to SAT. Let $\widehat{\mathcal{M}}$ be an abstract model, φ a WELTLK formula, and $k \geq 0$ a bound. The propositional encoding of the bounded model checking problem for WELTLK was presented in [11]. The formula $[\varphi]^{\mathrm{SAT}}_{\widehat{\mathcal{M}},k}$ encodes the the bounded semantics of the WELTLK formula φ, and it is defined on the same sets of individual variables as the formula $[\widehat{\mathcal{M}}^{\varphi,\hat{\iota}}]^{\mathrm{SAT}}_k$.

The definition of $[\widehat{\mathcal{M}}^{\varphi,\iota}]^{\mathrm{SAT}}_k$ assumes that the states and the join actions of $\widehat{\mathcal{M}}$, and the sequence of weights associate to the join actions, the time passage are encoded symbolically, which is possible, since both the set of states and the set of joint actions are finite. Formally, each state $\hat{s} \in \widehat{S}$ is represented by a vector $\overline{\mathbf{w}} = ((w_1, v_1), \ldots, (w_r, v_r))$ (called a symbolic state) of propositional variables whose length r depends on the number of local states of agents and the possible maximal value of weights appearing in the given MASs. Then, each joint action $a \in Act$ is represented by a vector $\overline{a} = (a_1, \ldots, a_t)$ (called a symbolic action) of propositional variables whose length t depends on the number of local actions of agents. Next, each sequence of weights associate to a join action is represented by a sequence $\overline{d} = (d_1, \ldots, d_{n+1})$ of symbolic weights. The symbolic weight $d_{\mathbf{c}}$ is a vector $(d_1, \ldots, d_x)$ of propositional variables (called *weight variables*), whose length x depends on the weight functions $d_{\mathbf{c}}$ for each $\mathbf{c} \in \mathcal{A} \cup \{\mathcal{E}\}$. Let $\boldsymbol{\pi}_j$ denote the jth symbolic k-path, i.e. the sequence of symbolic transitions: $(\overline{\mathbf{w}}_{0,j} \xrightarrow{(\overline{a}_{1,j},\overline{d}_{1,j}),\delta_{1,j}} ws_{1,j} \xrightarrow{(\overline{a}_{2,j},\overline{d}_{2,j}),\delta_{2,j}} \ldots \xrightarrow{(\overline{a}_{k,j},\overline{d}_{k,j}),\delta_{k,j}} \overline{\mathbf{w}}_{k,j}, u)$, where $\overline{\mathbf{w}}_{i,j}$ are symbolic states, $\overline{a}_{i,j}$ are symbolic actions, $\overline{d}_{i,j}$ are sequences of symbolic weights, for $0 \leq i \leq k$ and $1 \leq j \leq \widehat{f}_k(\varphi)$, and u be a symbolic number that is a vector $u = (u_1, \ldots, u_y)$ of propositional variables with $y = max(1, \lceil \log_2(k+1) \rceil)$. The definition of $[\widehat{\mathcal{M}}^{\varphi,\iota}]^{\mathrm{SAT}}_k$ assumes that the states and the join actions of $\widehat{\mathcal{M}}$, and the sequence of weights associate to the join actions are encoded symbolically, which is possible, since both the set of states and the set of joint actions are finite.

Let w and w' be two different symbolic states, $\overline{d}$ a sequence of symbolic weights, $\overline{a}$ a symbolic action, δ a symbolic time passage, and u be a symbolic number. Moreover, it uses the following auxiliary propositional formulae: $H(\overline{\mathbf{w}}, \overline{\mathbf{w}}')$—encodes equality of two global states and $\mathcal{N}^{=}_j(u)$—encodes that the value j is equal to the value represented by the symbolic number u.

The formula $[\widehat{\mathcal{M}}^{\varphi,\hat{\iota}}]^{\mathrm{SAT}}_k$, which encodes the unfolding of the transition relation of the abstract model $\widehat{\mathcal{M}}\, \widehat{f}_k(\varphi)$-times to the depth k, is defined as follows:

$$[\widehat{\mathcal{M}}^{\varphi,\hat{\iota}}]^{\mathrm{SAT}}_k := \bigvee_{s\in\hat{\iota}} I_s(\overline{\mathbf{w}}_{0,0}) \wedge \bigvee_{j=1}^{\widehat{f}_k(\varphi)} H(\overline{\mathbf{w}}_{0,0}, \overline{\mathbf{w}}_{0,j}) \wedge \bigwedge_{j=1}^{\widehat{f}_k(\varphi)} \bigvee_{l=0}^{k} \mathcal{N}^{=}_l(u_j) \wedge$$
$$\bigwedge_{j=1}^{\widehat{f}_k(\varphi)} \bigwedge_{i=0}^{k-1} \mathcal{T}(\overline{\mathbf{w}}_{i,j}, ((\overline{a}_{i,j}, \overline{d}_{i,j}), \overline{\delta}_{i,j}), \overline{\mathbf{w}}_{i+1,j})$$

where $w_{i,j}$, $\overline{a}_{i,j}$, $\overline{d}_{i,j}$, $\delta_{i,j}$, and u_j are, respectively, symbolic states, symbolic actions, sequences of symbolic weights, symbolic time passages, and symbolic number, for $0 \leq i \leq k$ and $1 \leq j \leq \widehat{f}_k(\varphi)$. Then, the next step is a translation of a WELTLK formula φ to a propositional formula $[\varphi]^{\mathrm{SAT}}_{\widehat{\mathcal{M}},k} := [\varphi]^{[0,1,F_k(\varphi)]}_k$ that was presented in [11].

Theorem 4 *Let $\widehat{\mathcal{M}}$ be an abstract model, and φ a WELTLK formula. Then for every $k \in \mathbb{N}$, $\widehat{\mathcal{M}} \models_k \varphi$ if, and only if, the propositional formula $[\widehat{\mathcal{M}}, \varphi]_k^{SAT}$ is satisfiable.*

4 Experimental Results

In this section we experimentally evaluate the performance of our SMT-based BMC and SAT-based BMC encoding for WELTLK over the TWIS semantics. We compare our experimental results with each other. We have conducted the experiments using two benchmarks: the timed weighted generic pipeline paradigm (TWGPP) TWIS model [10] and we modified the timed train controller system (TTCS) TIS model [6] by adding weights. We called it the timed weighted train controller system (TWTCS). We would like to point out that both benchmarks are very useful and scalable examples.

TWGPP. The specifications we consider are as follows:

- $\varphi_1 = \mathrm{K}_P\mathbf{G}(\mathit{ProdSend} \rightarrow \mathrm{K}_C\mathrm{K}_P\mathbf{F}_{[0,Min-d_P(Produce))}\mathit{ConsFree})$, *which states that Producer knows that always if she/he produces a commodity, then Consumer knows that Producer knows that Consumer has received the commodity and the cost is less than Min − d_P(Produce).*
- $\varphi_2 = \mathrm{K}_C\mathbf{G}(\mathit{ProdReady} \rightarrow \mathbf{X}_{[d_P(Produce),d_P(Produce)+1)}\mathit{ProdSend})$, *which expresses that Consumer knows that the cost of producing of a commodity by Producer is d_P(Produce).*

TWTCS. The TWTCS consists of n (for $n \geq 2$) trains $T_1, \ldots, T_n$, each one using its own circular track for travelling in one direction and containing its own clock x_i, together with controller C used to coordinate the access of trains to the tunnel through all trains have to pass at certain point, and the environment $\mathcal{E}$. There is only one track in the tunnel, so trains arriving from each direction cannot use it in this same time. There are signals on both sides of the tunnel, which can be either red or green. All trains notify the controller when they request entry to the tunnel or when they leave the tunnel. The controller controls the colour of the displayed signal, and the behaviour of the scenario depends on the values δ and Δ ($\Delta > \delta + 3$ makes it incorrect—the mutual exclusion does not hold).

Controller C has $n + 1$ states, denoting that all trains are away (state 0), and the numbers of trains, i.e., $1, \ldots, n$. Controller C is initially at state 0. The action Start_i of train T_i denotes the passage from state away to the state where the train wishes to obtain access to the tunnel. This is allowed only if controller C is in state 0. Similarly, train T_i synchronises with controller C on action $\mathit{approach}_i$, which denotes setting C to state i, as well as out_i, which denotes setting C to state 0. Finally, action in_i denotes the entering of train T_i into the tunnel. For environment, we shall consider just one local state: $L_{\mathcal{E}} = \{\cdot\}$. The set of actions for $\mathcal{E}$ is $Act_{\mathcal{E}} = \{\epsilon_{\mathcal{E}}\}$. The local protocols of $\mathcal{E}$ is the following: $P_{\mathcal{E}}(\cdot) = Act_{\mathcal{E}}$. The set of clocks of $\mathcal{E}$ is empty, and the invariant function is $\mathcal{I}_{\mathcal{E}}(\cdot) = \{\emptyset\}$.

The set of all the global states $\widehat{S}$ for the scenario is defined as the product $\prod_{i=1}^{n}(L_{T_i} \times \mathbb{D}_{T_i}^{|\mathcal{X}_{T_i}|}) \times (L_C \times \mathbb{D}_{L_C}^{|\mathcal{X}_{L_C}|}) \times L_{\mathcal{E}}$. The set of the initial states is defined as $\hat{\iota} = \{s^0\}$, where $s^0 = (away_1, 0), \ldots, (away_n, 0), (0, 0), (\cdot))$.

Moreover, we assume the following set of propositional variables: $\mathcal{PV} = \{tunnel_1, \ldots, tunnel_n\}$ with the following definition of local valuation functions for $i \in \{1, \ldots, n\}$: $\widehat{\mathcal{V}}_{T_i}(tunnel_i) = \{tunnel_i\}$.

Let $Act = \prod_{i=1}^{n} Act_{T_i} \times Act_C \times Act_{\mathcal{E}}$, with $Act_C = \{start_1, \ldots, start_n, approach_1, \ldots, approach_n, in_1, \ldots, in_n, out_1, \ldots, out_n\}$, $Act_{T_i} = \{start_1, \ldots, start_n, approach_1, \ldots, approach_n, in_1, \ldots, in_n, out_1, \ldots, out_n\}$, and $Act_{\mathcal{E}} = \{\epsilon_{\mathcal{E}}\}$ defines the set of joint actions for the scenario. For $\widetilde{a} \in Act$ let $act_{T_i}(\widetilde{a})$ denotes an action of $Train_i$, $act_C(\widetilde{a})$ denotes an action of Controller, and $act_{\mathcal{E}}(\widetilde{a})$ denotes an action of environment $\mathcal{E}$.

We assume the following local evolution functions for $i \in \{1, \ldots, n\}$: $t_{T_i}(away_i, \cdot, true, \{x_i\}, \widetilde{a}) = try_i$, if $act_{T_i}(\widetilde{a}) = start_i$ and $act_C(\widetilde{a}) = start_i$; $t_{T_i}(try_i, \cdot, \Delta > x_i, \{x_i\}, \widetilde{a}) = wait_i$, if $act_{T_i}(\widetilde{a}) = approach_i$ and $act_C(\widetilde{a}) = approach_i$; $t_{T_i}(wait_i, \cdot, x_i > 6, \{\emptyset\}, \widetilde{a}) = tunnel_i$, if $act_{T_i}(\widetilde{a}) = in_i$ and $act_C(\widetilde{a}) = in_i$; $t_{T_i}(tunnel_i, \cdot, true, \{\emptyset\}, \widetilde{a}) = away_i$, if $act_{T_i}(\widetilde{a}) = out_i$ and $act_C(a) = out_i$.

Finally, we assume the following two local weight functions for each agent: $d_{T_{n1}}(start_n) = 1$, $d_{T_{n1}}(approach_n) = 2$, $d_{T_{n1}}(in_n) = 4$, $d_{T_{n1}}(out_n) = 1$; $d_{T_{n10^6}}(start_n) = 1000000$, $d_{T_{n10^6}}(approach_n) = 2000000$, $d_{T_{n10^6}}(in_n) = 4000000$, $d_{T_{n10^6}}(out_n) = 1000000$.

The specifications we consider are as follows for $w \in \{1, 10^6\}$:

- $\phi_1 = \mathbf{G}_{[28\cdot w, \infty)}\left(\bigwedge_{j=1}^{n-1}\bigwedge_{j=i+1}^{n}(tunnel_i \vee tunnel_j)\right)$, *which expresses that the system satisfies mutual exclusion property.*
- $\phi_2 = \mathbf{G}_{[0,15\cdot w)}\left(tunnel_1 \rightarrow \mathbf{K}_{T_1}\left(\mathbf{G}(\bigwedge_{j=2}^{n} \neg tunnel_j)\right)\right)$, *which expresses that always if the* $Train_1$ *enters its critical section, then it knows that always in the future no other train will enter its critical section.*

Performance evaluation

We have performed our experimental results on a computer equipped with I7-3770 processor, 32 GB of RAM, and the operating system Arch Linux with the kernel 4.2.5. We set the CPU time limit to 3600 s. Our SMT-based and SAT-based BMC algorithms are implemented as standalone programs written in the programming language C++. For SMT-BMC we used the state of the art SMT-solver Z3 [14] (https://github.com/Z3Prover) and for the SAT-BMC we used the state of the art SAT-solver PicoSAT [15]. All the benchmarks together with an instruction how to reproduce our experimental results can be found at the web page http://tinyurl.com/bmc4twis.

TWGPP. The experimental results show that the SMT-BMC and the SAT-BMC are sensitive to scaling up the size of the benchmarks, but they are not sensitive to scaling up the weights, while the SAT-based BMC is more sensitive to scaling up the weights. The SAT-BMC is able to verify the formula φ_1 for TWGPP with 200 nodes for the `bw` and for the `bw` multiple by 10^6, and the SMT-BMC is able to verify the formula for TWGPP with 1000 nodes for the `bw` and for the `bw` multiple by 10^6. The memory usage for the SMT-BMC is lower than for SAT-BMC. The

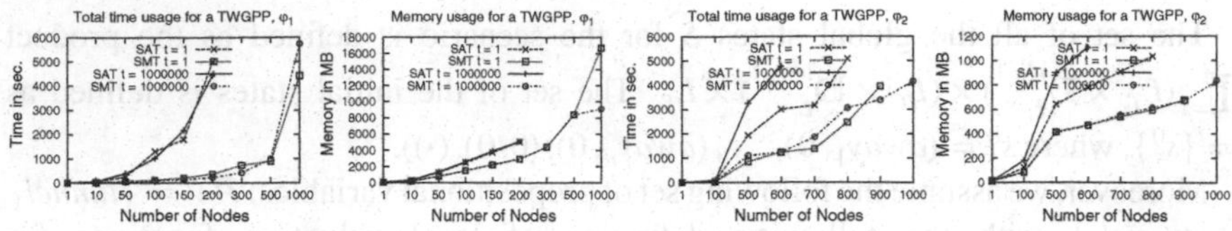

Fig. 1 φ_1: SAT- and SMT-based BMC: TWGPP with n nodes

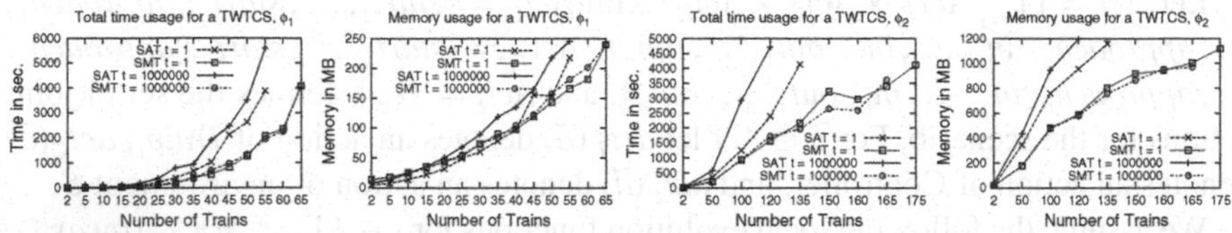

Fig. 2 SAT- and SMT-based BMC: TWTCS with n trains

experimental results for the formula φ_2 for TWGPP are very interesting. SMT-BMC is able to verify only 900 nodes for bw and 1000 nodes for the bw multiple by 10^6. SAT-BMC is able to verify 800 nodes for bw and 600 nodes for the bw multiple by 10^6. The memory usage for SMT-BMC is lower than for SAT-BMC (Fig. 1).

TWTCS. As one can see from the line charts (Fig. 2) for the TWTCS system, in the case of this benchmark the SMT-based BMC performs much better in terms of the total time and the memory consumption for both the tested formulae. Moreover, the SMT-based method is able to verify more nodes for both tested formulae. In particular, in the time limit set for the benchmarks, the SMT-based BMC is able to verify the formula ψ_1 for 65 nodes while the SAT-based BMC can handle 55 nodes. For ψ_2 the SMT-based BMC is still more efficient—it is able to verify 175 nodes for the bw and 165 for the bw multiple by 10^6, whereas the SAT-based BMC verifies only 135 and 120 nodes respectively.

5 Conclusions

We have proposed, implemented, and experimentally evaluated the SMT-based BMC approach and the SAT-based BMC for WELTLK interpreted over the timed weighted interpreted systems. We have compared both methods. The experimental results show that the SMT-based BMC in more cases is better than SAT-based BMC. In general the SMT-based BMC approach appears to be superior for the both systems, while the SAT-based approach appears to be superior only for two formulae for the TWGPP system. This is a novel and interesting result, which shows that the choice of the BMC method should depend on the considered system and complexity of the considered formula.

We would like to use other SAT- and SMT-solvers in our implementations and compare experimental results. The BMC for WELTLK and for TWISs may also be performed by means of Ordered Binary Diagrams (OBDD). This will be explored in the future.

Acknowledgments Partly supported by National Science Centre under the grant No. 2014/15/N/ST6/05079.

References

1. Clarke, E., Grumberg, O., Peled, D.: Model Checking. MIT Press (1999)
2. Fagin, R., Halpern, J.Y., Moses, Y., Vardi, M.Y.: Reasoning About Knowledge. MIT Press, Cambridge (1995)
3. Bouyer, P., Markey, N., Sankur, O.: Robust weighted timed automata and games. In: Proceedings of FORMATS 2013, pp. 31–46 (2013)
4. Larsen, K.G., Mardare, Radu: Complete proof systems for weighted modal logic. Theor. Comput. Sci. **546**, 164–175 (2014)
5. Alur, R., Dill, D.L.: The theory of timed automata. In: Proceedings of REX Workshop, pp. 45–73 (1991)
6. Woźna-Szcześniak, B., Zbrzezny, A.: Checking EMTLK properties of timed interpreted systems via bounded model checking. Stud. Logica 1–38 (2015)
7. Lomuscio, A., Qu, H., Raimondi, F.: MCMAS: a model checker for the verification of multi-agent systems. In: Proceedings of CAV'2009. LNCS, vol. 5643, pp. 682–688. Springer (2009)
8. Gammie, P., van der Meyden, R.: MCK: model checking the logic of knowledge. In: Proceedings of CAV'2004, LNCS, vol. 3114, pp. 479–483. Springer (2004)
9. Kwiatkowska, M.Z., Norman, G., Parker, D.: PRISM: probabilistic symbolic model checker. In: Proceedings of TOOLS 2002, pp. 200–204 (2002)
10. Zbrzezny, A.M., Zbrzezny, A.: Checking WECTLK properties of timed real-weighted interpreted systems via SMT-based bounded model checking. In: Proceedings of EPIA 2015. LNCS, vol. 9273, pp. 638–650. Springer (2015)
11. Woźna-Szcześniak, B., Zbrzezny, A.M., Zbrzezny, A.: SAT–based bounded model checking for weighted interpreted systems and weighted linear temporal logic. In: Proceedings of PRIMA'2013. LNAI, vol. 8291, pp. 355–371. Springer (2013)
12. Zbrzezny, A.: A new translation from ECTL* to SAT. Fundamenta Informaticae **120**(3–4), 377–397 (2012)
13. Zbrzezny, A.M., Zbrzezny, A.: Checking WELTLK properties of weighted interpreted systems via SMT–based bounded model checking. In: Proceedings of PRIMA 2015. LNCS, vol. 9387, pp. 660–669. Springer (2015)
14. De Moura, L., Bjørner, N.: Z3: an efficient SMT solver. In: Proceedings of TACAS'2008. LNCS, vol. 4963, pp. 337–340. Springer (2008)
15. Biere, A.: PicoSAT essentials. J. Satisfiability, Boolean Model. Comput. (JSAT) **4**, 75–97 (2008)

Building a Realistic Data Environment for Multiagent Mobility Simulation

Feirouz Ksontini, Mahdi Zargayouna, Gérard Scemama and Bertrand Leroy

Abstract Transport systems are increasingly complex and are made of more and more connected entities. It becomes critical to develop micro-simulation tools to understand the new transport systems dynamics. However, the data for building mobility simulation are quite hard to get, and simulations on new areas are not easy to set up. In this paper, we propose methods for building a realistic data environment for multimodal mobility simulators. We also propose a method to integrate travel patterns (patterns of travelers' origins and destinations). The methods presented in this paper can be used when dealing with new areas for which we have few and incomplete data.

1 Introduction

The development of new information and communications technology contributes to the emergence of a new generation of multimodal real-time services that assist travelers throughout their trip (mobile devices, localized vehicles, trackable goods, etc.). These information flows increasingly impact travelers behaviors and the traffic generated by their local decision making. In this context, dynamic simulation of travelers mobility is a crucial step for understanding, analyzing and predicting this evolving dynamics of transport networks.

F. Ksontini · M. Zargayouna (✉) · G. Scemama
Université Paris-Est, IFSTTAR, GRETTIA, Boulevard Newton,
77447 Champs sur Marne, Marne la Vallée Cedex 2, France
e-mail: hamza-mahdi.zargayouna@ifsttar.fr; zargayouna@ifsttar.fr

F. Ksontini
e-mail: feirouz.ksontini@vedecom.fr

G. Scemama
e-mail: gerard.scemama@ifsttar.fr

F. Ksontini · B. Leroy
Institut VeDeCom, Versailles, France
e-mail: bertrand.leroy@vedecom.fr

G. Jezic et al. (eds.), *Agent and Multi-Agent Systems: Technology and Applications*, Smart Innovation, Systems and Technologies 58,
DOI 10.1007/978-3-319-39883-9_5

The multiagent paradigm is relevant for the simulation of urban transport systems [1]. It indeed facilitates an approach by analogy in the transport domain which one of the objectives is the coordination of distributed entities [2]. This is why the multiagent approach is often chosen to model, solve and simulate transport problems [3]. The authors in [4] list several reasons for the privileged use of multiagent systems in these applications, such as the natural and intuitive problem solving, the ability of autonomous agents for the modeling of heterogeneous systems, the ability to capture complex constraints connecting all problem-solving phases, etc. Indeed, the concept of an agent is well suited for the representation of travelers in transit or road traffic scenarios [5]. They are autonomous entities which are situated in an environment, adapt their behaviors to the dynamics they perceive and interact with others agents in order to achieve specific goals. For Parunak [6], “Agent-based modelling is most appropriate for domains characterized by a high degree of localization and distribution”, which is the case for complex and dynamic transport applications.

We have designed and implemented the multimodal travel simulator SM4T (Simulator for Multiagent MultiModal Mobility of Travelers) in the context of the EC-funded project Instant Mobility [7]. The simulator allows for the understanding and the prediction of future status of the networks and it can be also used for testing new applications that track individual travelers. For instance, it has been used to evaluate the impact of individualized real-time data on the behaviors of traveler agents [8]. SM4T is a fully agent-based tool for multimodal travelers mobility. It enables for the rapid prototyping and execution of simulations for several kinds of online applications. The application simulates the movements of travelers on the different transport modes and networks while taking into account the changes in travel times and the status of the networks. Since it assumes the continuous localization of travelers, SM4T can notably simulate and evaluate the impact of a wide range of community transportation apps, such as user-submitted travel times and route details, community-based driver assistance, community parking, etc.

However, multimodal mobility simulators such as SM4T need a lot of data about the considered geographic region and the different existing transport modes. For instance, SM4T has been deployed on the city of Toulouse (France) with the geographic data of the city, the description of the road network, the description of all the available public transport network, the timetables of the vehicles, etc. These data were made available to us from the operating support system of public transport operators and from road transport support systems of road transport operators. This great amount of data of different nature limits the applicability of multimodal mobility simulators to new areas. To this end, this paper focuses on building a realistic data environment for multimodal mobility simulators in the absence of access to these proprietary data. We also propose a method to integrate travel patterns (patterns of travelers' origins and destinations) in these simulators.

The remainder of this paper is structured as follows. In Sect. 2, we discuss the choice of the simulation platform and previous proposals for travelers mobility simulation. In Sect. 3, we briefly present SM4T. Section 4 presents the methods to use when dealing with a new area to simulate. In Sect. 5, we describe some experiments before to conclude and describe some further work we are conducting.

2 Related Work

There exists several multiagent simulators for travelers mobility. For instance, MATSim [9] is a widely known platform for mobility micro-simulation. However, the mobile entities in MATSim are passive and their state is modified by central modules, which limits its flexibility and its ability to integrate new types of (proactive) agents. Transims [10] simulates multimodal movements and evaluates impacts of policy changes in traffic or demographic characteristics. AgentPolis [11] is also a multiagent platform for multimodal transportation. The proposals of this paper might profit to these platforms since they also require the same kind of information than SM4T. SM4T has been developed on top of Repast Simphony [12]. In the context of transportation applications, one main choice criterion for the simulation platform is its ability to create geospatial agent-based models, i.e. its ability to integrate and process geographic data. Among all the available simulation platforms that would fit with our requirements (e.g. Gama [13]), Repast is the most mature one.

3 The SM4T Simulator

The purpose of SM4T is to represent travelers (drivers and passengers) and transport means (public transport vehicles and private cars) in a micro-level and to simulate their dynamic movements and their interaction (tracking, planning requests, plans update, etc.). In the following, we briefly describe the simulator, a more detailed description can be found in our previous paper [7].

The multiagent system is made of planner agents, car agents, public transport vehicle agents and traveler agents. The planner agents compute the best road itinerary for the car agents and the best multimodal itinerary for the traveler agents based on the latest status of the networks. The calculation method has been detailed in [14]. A planner agent is created when an agent request is submitted to the system, and leaves the system right after.

Each car agent has an origin and a destination when created, which are chosen randomly. The agent asks the planner for the best itinerary between his origin and his destination. At each simulation tick, the car agent checks if he has reached his destination. If so, he leaves the simulation.

The origins and the destinations of the public transport vehicle agents are provided by the predefined timetables. When created, each vehicle agent infers his itinerary from his timetable. If there are passengers onboard, they are moved to the same coordinates at the same time by the vehicle agent. That means that, when they are onboard a vehicle, traveler agents delegate the control of their movements to the vehicle agent. While the vehicle agent has not reached his destination, he travels at each tick the allowed distance, following his current speed. When the vehicle reaches a stop, he searches among his onboard travelers who has to leave at this stop. Then he searches among the waiting travelers at the stop who has to take him.

As for car agents, the origin and destination of the traveler agent are chosen randomly. When they are not walking, traveler agents do not travel on their own, but share rides with others, which are responsible of their movements. The traveler agent alternates between walking and waiting for a vehicle.

4 Network Data

As for any multimodal mobility simulator, SM4T needs a minimal set of data to function properly. The minimum input data (xml files) of the simulator are:

- the road network,
- the public transport network,
- the transfer mapping,
- the timetables of the public transport vehicles.

Optionally, the simulator might use travelers profiles and use them to infer certain agents characteristics (pedestrian speeds for instance). In the first version of SM4T, all these data were made available to us from the transport operators of the considered area (Toulouse, France). When considering new areas, these data have to be approximated otherwise.

4.1 Road Network

The road network is a description of the roads, crossroads and driving directions. Apart from the geographic description of the roads, mobility simulators need to have a mapping between traffic flows (vehicles/hour), the traffic density and the speeds. Indeed, to make vehicles move in a realistic way, they should not always move in free-flow speed, but should slow down when traffic becomes dense.

When dealing with a new territory, the geographic description of the network can be found in the form of free editable maps (such as OSM). However, the mapping between the number of vehicles and the speeds is generally missing in these maps. To make vehicles move in a realistic way, we approximate this mapping by analogy from data that we have about other areas. The objective is to have a realistic triangular fundamental diagram of traffic flow that gives a relation between the flow q (vehicles/hour) and the density k (vehicles/km) (cf. Fig. 1). The fundamental diagram suggests that if we exceed a critical density of vehicles k_c, the more vehicles there are on a road, the slower they will be. Here is the equation we use to model this phenomenon:

This equation is parametrized with α the free flow speed on this road, β the congestion wave speed and k_c the critical density. As $v = \frac{q}{k}$:

$$v = \begin{cases} \alpha & \text{if } k \leq k_c \\ \frac{-\beta(k-k_c)+\alpha k_c}{k} & \text{if } k > k_c \end{cases} \tag{1}$$

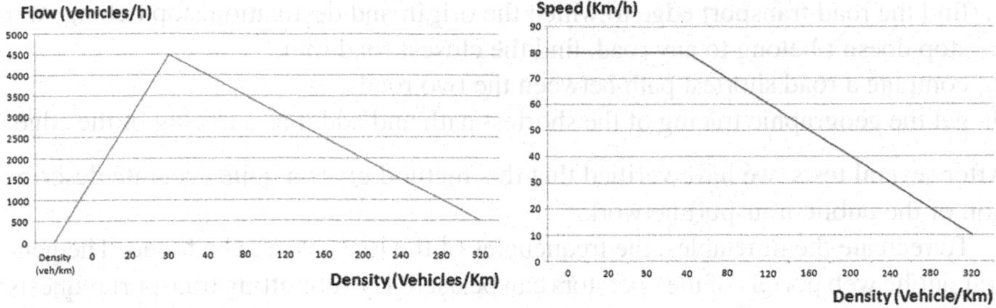

Fig. 1 Fundamental diagram with $\alpha = 112,5$, $\beta = 12,5$ and $k_c = 40$ (*left*). Speed in function of density (*right*)

Thus we can define a cost function that returns a travel time per distance units $(1/v)$ in function of the number of agents $|A_e|$ on this edge:

$$cost(|A_e|) = \begin{cases} \frac{1}{\alpha} & \text{if } |A_e| \leq k_c \\ \frac{|A_e|}{-\beta(|A_e|-k_c)+\alpha k_c} & \text{if } |A_e| > k_c \end{cases} \tag{2}$$

When a car agent joins an edge of the network, his speed is calculated following the equation above. Another method for defining a realistic behavior of cars flow is to specify drivers behaviors that depend of the surrounding cars behaviors, generally using a car-following model. Such model would need however a great number of parameters. Including car-following behaviors in SM4T is one of our ongoing works.

4.2 *Public Transport Network*

A public transport network is composed of two elements. The first element is the network, which is described by a set of transport lines, each of them composed of a set of itineraries, each itinerary is composed of a sequence of edges, each edge has a tracing in the form of a sequence of pairs $\langle longitude, latitude \rangle$, and is composed of an origin node and a destination node. Finally, every node is defined by its name and coordinates. The second element is the timetables of the vehicles, which describe the paths of the vehicles and the corresponding visit times. Each timetable is then a sequence of pairs $\langle stop, time \rangle$.

When not available from the public transport operator, these data are quite hard to recreate. However, increasingly, we can find some open data describing the network in terms of lines, but without the itinerary details, and without the geographic tracing between the stops. To approximate the geographic tracing between stops, we define the following procedure, using the road network defined in the previous subsection:

1. find the road transport edge to which the origin and destination stop belong. If a stop doesn't belong to any road, find the closest road to it,
2. compute a road shortest path between the two roads,
3. get the geographic tracing of the shortest path and add it as a tracing of the edge.

After several tests, we have verified that this method gives a quite accurate description of the public transport network.

To recreate the timetables, the frequencies of the lines have to be found. The general public web portals of the operators can be used. By submitting transport requests at several times of the day (peak and off-peak), frequencies can be approximated. We create vehicles departure times following the description of the lines accordingly, and with visit times at the stops that are coherent with the geographic tracing that we have defined in the previous step.

4.3 The Transfer Mapping

The transfer mapping is a table informing about the stops of the network for which a transfer by foot is possible and the road transport nodes that are reachable from the stops. This mapping is very important because passengers start and end their trip on the road network (while the main part of their trip is on the public transport network). They thus have to pass from one network to another. To recreate this file, we start at each stop of the public transport network and look for all the reachable stops and nodes (crossroads) that are at most 500 m away.

4.4 Travel Patterns

A travel pattern clusters the considered geographic region in zones and describes the number of persons asking to leave or to join each region. Travel patterns are very important because they allow to have a simulation that mimics more realistically the mobility behaviors of cars and passengers. In order for the simulator to integrate travel patterns, we propose the following procedure.

Let the travel pattern for the considered period of time in the form of a matrix $D = \{(d_{ij})\}$ of dimension $N \times N$ with N the number of regions; d_{ij} is the number of persons traveling from zone i to zone j. For instance, in Table 1, the number of travelers in the travel pattern is $M = 20$, 8 are going from zone 1 to zone 2, 4 are going from zone 2 to zone 1, 2 ar going from zone 3 to zone 1, etc. However, the number of actually simulated agents A is not necessarily equal to the number of persons M in the travel patterns. Our objective is to generate non-deterministic simulated origins and destinations that are proportional to the travel pattern. To this end, we first create a matrix $S = \{(s_{ij})\}$ of dimension $N \times N$; s_{ij} is the number of simulated agents that will be traveling from zone i to zone j, $s_{ij} = d_{ij} \times \frac{A}{M}$, where A is the number

Table 1 Example of D matrix ($N = 3$)

	Z_1	Z_2	Z_3
Z_1	0	8	4
Z_2	4	0	2
Z_3	2	0	0

Table 2 Example of S matrix with $|A| = 10$ and $|M| = 20$

	Z_1	Z_2	Z_3
Z_1	0	4	2
Z_2	2	0	1
Z_3	1	0	0

Table 3 Example of P table

Interval (%)	Origin-destination	Interval length calculation
$[0, 40.00[$	$Z_1 - Z_2$	$\frac{4}{10}$
$[40.00, 60.00[$	$Z_1 - Z_3$	$\frac{2}{10}$
$[60.00, 80.00[$	$Z_2 - Z_1$	$\frac{2}{10}$
$[80.00, 90.00[$	$Z_2 - Z_3$	$\frac{1}{10}$
$[90.00, 100[$	$Z_3 - Z_1$	$\frac{1}{10}$

Table 4 P after the choice of (Z_1, Z_2)

Interval (%)	Origin-destination	Interval length calculation
$[0, 33.33[$	$Z_1 - Z_2$	$\frac{3}{9}$
$[33.33, 55.55[$	$Z_1 - Z_3$	$\frac{2}{9}$
$[55.55, 77.77[$	$Z_2 - Z_1$	$\frac{2}{9}$
$[77.77, 88.88[$	$Z_2 - Z_3$	$\frac{1}{9}$
$[88.88, 100[$	$Z_3 - Z_1$	$\frac{1}{9}$

of simulated agents and M the number of actual travelers in the pattern, i.e. $M = \sum_{i=0}^{N} \sum_{j=0}^{N} d_{ij}$. Based on S, a dynamic mapping table P is created (cf. Table 3). The P table maps intervals with $(zone_{origin}, zone_{destination})$ pairs. The length of the interval is proportional to the current relative weight of the zones pairs in S (Table 2). When an agent is generated with an origin belonging to Z_1, and a destination belonging to Z_2, the (Z_1, Z_2) cell in S is decremented and P is updated accordingly (cf. Table 3).

At each tick of time, the simulator generates a number of new traveler agents for which we have to define an origin and a destination. For each new traveler agent, a random number $\rho \in [0, \ldots, 1]$ is chosen and P is used to choose the origin and destination zones. Let's say that $\rho = 0.15$, then following the P table in Table 3, the $Z_1 - Z_2$ pair is chosen for the agent. The origin is then chosen randomly in Z_1 while the destination is chosen randomly in Z_2. The new P table is given in Table 4 and the probability to chose that pair of zones again becomes lower ($\frac{3}{9}$). This way, even if the origin and destination are chosen nondeterministically, the chosen origin and destination zones remain proportional to the travel pattern all along the simulation.

However, travel patterns are usually defined after long surveys on big geographic regions, while the simulations often concern smaller areas. That means that we will have a huge D matrix with zones pairs that mostly do not concern the considered area. We could simply consider the submatrix $D' \subset D$ with zones pairs in the considered area, but in this case, we would have underestimated volumes in D'. Indeed, four cases for the values $d_{ij} \in D$ are possible:

1. both i and j are in the considered area
2. i is the considered area but not j
3. j is the considered area but not i
4. neither i nor j are in the considered area

For the first case, d_{ij} are simply copied in d'_{ij}, since these volumes completely concern the considered area. For the second case, we should add d_{ij} to a certain cell d'_{ik} where k is in the considered area. To do so, we execute a shortest path from the centroid of zone i to the centroid of zone j and report the sequence of zones $i, \dots, k, \dots, j$ that a traveler going from i to j would visit: k is the last zone in the considered area. The volumes in d_{ij} are then added to d'_{ik}. A similar procedure is followed for the third case, where we add d_{ij} to the cell d'_{kj} where k is the first zone in the zones shortest path sequence $i, \dots, k, \dots, j$. For the last case, we could think of ignoring them, since neither the origin nor the destination zone are in the considered area. However, some travelers, even if they are not departing from nor arriving in the considered area, could pass by the considered area, and should therefore be considered in the simulation because they impact traffic. Again, we execute a shortest path and report the sequence of zones $i, \dots, k, \dots, l, \dots j$ that a traveler going from i to j would visit: k is the first zone in the sequence that is in the considered area, while l is the last zone in the sequence that is in the considered area; d_{kl} is added to d'_{kl}.

Figure 2 summarizes the transformation process of the travel pattern. On the left, we have an example of the three last cases of travelers flows: an incoming flow to the

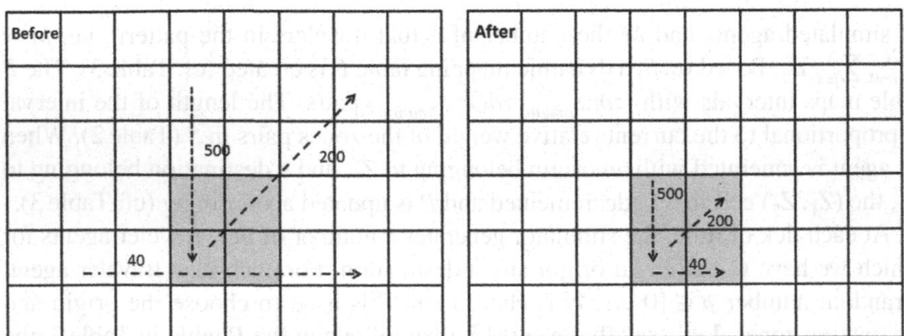

Fig. 2 Restriction of the travel pattern to the considered area (*grey zone*)

Fig. 3 Simulation Execution (cars: *pink circles*, travelers: *black crosses*, buses: *blue squares*) (Color figure online)

considered area, an outgoing flow and a traversing flow. On the right, we have the result of our procedure with a restriction of the travel pattern to origin and destination zones of the considered area.

5 Experiments and Results

We demonstrate the use of the SM4T simulator on a new area, for the Paris-Saclay region, France, for which we have built a realistic data environment. For public transport network, we have partially relied on open data[1] with GTFS[2] format. GTFS defines a common format for public transportation schedules and associated geographic information. These data represent all the Ile-de-France, we have extracted those that match with the simulated region. We have also travel patterns for all the Ile-de-France.[3] We have used all the procedures described in this paper to build the data environment of the simulator. The result is a road network with 23,594 roads, a public network with 7,431 stops and 7,431 edges. We have simulated 18,178 buses, 30,000 cars and 30,000 passengers. A screenshot of our current simulation is in Fig. 3 (blue squares are buses, pink circles are cars and black crosses are pedestrians).

[1]provided by the STIF: the organizing Authority of sustainable mobility in the Ile-de-France.

[2]General Transit Feed Specification.

[3]Provided by LVMT Lab from MODUS model which was developed in collaboration with DRIEA-IF.

The validation of the simulated dynamic status of the network is very important. For a multimodal traffic simulator, the simulated traffic status have to be confronted with the real traffic. This is however a very difficult process, because the data are difficult to get from the transport operators, and general public real-time information only concern main streets and highways and generally no information about public transport vehicles positions and occupancy is provided. We are currently investigating methods to validate the simulation results.

6 Conclusion and Perspectives

In this paper, we have focused on the reconstruction of input data needed for a multimodal mobility simulator and built a realistic data environment for it. This realistic data environment could be used by any other mobility simulator. We have applied these methods on the SM4T simulator with a new territory to simulate: the Paris-Saclay region. Now that we have a running simulation on this territory, the next step that we want to explore consists on testing new services on it. For instance, we are investigating the provision of an autonomous vehicles service and evaluate its impact on the surrounding traffic.

References

1. Badeig, F., Balbo, F., Scemama, G., Zargayouna, M.: Agent-based coordination model for designing transportation applications. In: 11th International IEEE Conference on Intelligent Transportation Systems, 2008. ITSC 2008, pp. 402–407. IEEE (2008)
2. Zargayouna, M., Balbo, F., Scémama, G.: A multi-agent approach for the dynamic vrptw. In: Proceedings of the International Workshop on Engineering Societies in the Agents World (ESAW 2008). Springer (2008)
3. Davidsson, P., Henesey, L., Ramstedt, L., Tornquist, J., Wernstedt, F.: An analysis of agent-based approaches to transport logistics. Transp. Res. Part C Emerg. Technol. **13**(4), 255–271 (2005)
4. Bazzan, A.L., Klügl, F.: A review on agent-based technology for traffic and transportation. Knowl. Eng. Rev. **29**(03), 375–403 (2014)
5. Bessghaier, N., Zargayouna, M., Balbo, F.: An agent-based community to manage urban parking. In: Advances on Practical Applications of Agents and Multi-Agent Systems, pp. 17–22 (2012)
6. Parunak, H.V.D., Savit, R., Riolo, R.L.: Agent-based modelling versus equation-based modelling: a case study and users' guide. In: Proceedings of Workshop on Modelling Agent Based Systems (MABS98), Paris (1998)
7. Zargayouna, M., Zeddini, B., Scemama, G., Othman, A.: Agent-based simulator for travellers multimodal mobility. Front. Artif. Intell. Appl. **252**, 81–90 (2013)
8. Zargayouna, M., Othman, A., Scemama, G., Zeddini, B.: Impact of travelers information level on disturbed transit networks: a multiagent simulation. In: IEEE 18th International Conference on Intelligent Transportation Systems (ITSC), pp. 2889–2894. IEEE (2015)

9. Maciejewski, M., Nagel, K.: Towards multi-agent simulation of the dynamic vehicle routing problem in matsim. In: Proceedings of the 9th International Conference on Parallel Processing and Applied Mathematics—Volume Part II. PPAM'11, pp. 551–560. Springer, Berlin (2012)
10. Nagel, K., Rickert, M.: Parallel implementation of the transims micro-simulation. Parallel Comput. **27**(12), 1611–1639 (2001)
11. Jakob, M., Moler, Z., Komenda, A., Yin, Z., Jiang, A.X., Johnson, M.P., Pechoucek, M., Tambe, M.: Agentpolis: towards a platform for fully agent-based modeling of multi-modal transportation (demonstration). In: International Conference on Autonomous Agents and Multiagent Systems, AAMAS 2012, Valencia, Spain, June 4–8, 2012, 3 vols. pp. 1501–1502 (2012)
12. Tatara, E., Ozik, J.: How to build an agent-based model iii—repast simphony. In: Applied Agent-based Modeling in Management Research, Academy of Management Annual Meeting, Chicago (2009)
13. Taillandier, P., Vo, D.A., Amouroux, E., Drogoul, A.: Gama: a simulation platform that integrates geographical information data, agent-based modeling and multi-scale control. In: PRIMA. Lecture Notes in Computer Science, vol. 7057, pp. 242–258. Springer (2012)
14. Zargayouna, M., Zeddini, B., Scemama, G., Othman, A.: Simulating the impact of future internet on multimodal mobility. In: 11th IEEE/ACS International Conference on Computer Systems and Applications, AICCSA 2014, Doha, Qatar, November 10–13, 2014, pp. 230–237. IEEE Computer Society (2014)

Agent-Based System for Reliable Machine-to-Machine Communication

Pavle Skocir, Mario Kusek and Gordan Jezic

Abstract Reliability is, along with energy efficiency and security, commonly referred to as one of the most important requirements which have to be met to successfully deploy Machine-to-Machine (M2M) communication systems. We propose an enhancement of agent-based system for M2M communication which ensures two reliability levels according to the intervals in which data obtained from sensors is sent to back-end system. Our assumption is that communication reliability is less important for reporting measurements sent with higher frequency than for reporting measurements sent with lower frequency. A mechanism is proposed which determines the appropriate reliability level. Introduced mechanism is implemented on two Libelium Waspmote devices which collect meteorological data in different intervals. Suitability for using one of the two proposed reliability levels is determined according to measured communication reliability within 24 h and energy consumption data for one operating cycle.

Keywords Energy efficiency · Machine-to-machine system · Reliability

1 Introduction

Machine-to-Machine (M2M) communication, which enables direct communication between devices with limited human intervention, boosts development of applications and services that use data provided by connected devices. By integrating M2M communication with the concept of Internet of Things (IoT), these devices and data

P. Skocir (✉) · M. Kusek · G. Jezic
Faculty of Electrical Engineering and Computing, Internet of Things Laboratory,
University of Zagreb, Unska 3, 10000 Zagreb, Croatia
e-mail: pavle.skocir@fer.hr

M. Kusek
e-mail: mario.kusek@fer.hr

G. Jezic
e-mail: gordan.jezic@fer.hr

G. Jezic et al. (eds.), *Agent and Multi-Agent Systems: Technology and Applications*, Smart Innovation, Systems and Technologies 58,
DOI 10.1007/978-3-319-39883-9_6

they provide can be accessed via Internet. When referring to the concept of IoT, the focus is largely on application design, on methods for data forwarding between devices and end users, and on data visualization [4]. Solutions which have M2M communication in focus usually consider challenges regarding network design and device grouping (e.g. creating clusters) [2].

In this paper we focus on core concept of M2M communication, i.e. communication between devices and a gateway. More specifically, M2M communication reliability is considered which is defined as a probability for successful transmission of the required amount of information from the source node to the destination [5]. We propose an approach which defines different reliability levels for reporting sensor measurement data. Reliability is more important for transmissions which occur in larger intervals than for transmissions which occur in smaller intervals. For instance, if an interval for reporting temperature values within an application that manages a heating system is one minute, and if one measurement has not reached its destination, there is no much harm because new value will be transmitted in another minute, and temperature usually cannot change drastically in that time. However, if the interval is a couple of hours in which temperature might change drastically, this one measurement should be sent with higher reliability so that fluctuation in temperature could be detected. We propose communication reliability levels based precisely on sensor reporting intervals defined within application or service parameters.

In our previous work, we focused on mechanisms for data exchange to synchronize M2M devices with M2M gateway [10], and on data filtering mechanisms based on current and historical measurement values which enabled reduction of energy consumption [13]. In this work the focus is on mechanisms for ensuring communication reliability of monitoring applications with constant measurement intervals. Along with reliability, we consider its influence on energy efficiency. Since higher reliability is usually ensured at the expense of energy efficiency, balance between those two requirements needs to be found.

Section 2 presents contemporary research and existing mechanisms for ensuring communication reliability in M2M systems. Section 3 describes a context-aware mobile agent network model, while Sect. 4 the introduces our mechanism within the aforementioned model for ensuring communication reliability. The benefits of using the mechanism are presented in Sect. 5. Concluding remarks are expressed in Sect. 6.

2 Related Work

Lu et al. [9] claim that reliability is, along with energy efficiency and security, one of the most important requirements which have to be met to successfully deploy M2M communication systems. Furthermore, they emphasize the significance of reliability because unreliable sensing, processing, and transmission can cause false monitoring data reports, long delays and data loss, which could result in reduced people's interest in M2M communication. The authors consider reliability issues in M2M communication systems for 3 types of tasks: sensing and processing, transmission, and at back-end servers. To improve the reliability of M2M communications, they exploit different redundancy technologies: information redundancy, spatial redundancy and temporal redundancy.

Prasad et al. [14] present quite a similar solution for dealing with reliability in M2M systems as in [9]. However, they neglect the security issues and introduce the term *Energy Efficiency Reliability* (EER), and declare it as a most important issue of concern when developing IoT systems.

Kim et al. [6] present an overview of existing M2M service platforms, along with their requirements and functionalities. They consider reliability as one of the Quality of Service (QoS) requirements, accompanied by delay, priority, and throughput. Additionally, the authors state that M2M core network should be able to guarantee reliability, along with other QoS requirements. The core network should be designed to satisfy QoS of a certain traffic pattern generated by applications. According to authors, applications can generate three types of traffic patterns: periodic, event-driven, and streaming.

According to Al-Fuqaha et al. [1], reliability is with availability, mobility, performance, scalability, interoperability, security, management, and trust one of the key challenges that needs to be addressed to enable service providers and application programmers the possibility to implement their services efficiently. The authors state that reliability aims to increase the success rate of IoT service delivery, which is specially important in emergency response applications. They point out that reliability must be implemented in software and hardware throughout all IoT layers. Reliability in underlying communication of IoT services is particularly important because unreliable perception, data gathering, processing and transmission could lead to long delays, loss of data and ultimately wrong decisions. Additionally, the authors present an overview of standardized protocols used for IoT. On application layer Constrained Application Protocol (CoAP), which enables a simpler way to exchange data than Hypertext Transfer Protocol (HTTP), includes operational modes which guarantee different levels of communication reliability. Another application protocol, Message Queue Telemetry Transport (MQTT) also specifies three levels of QoS, which are closely linked with reliability. Infrastructure protocols at lower level, such as IEEE 802.15.4 also provide reliable communication, along with a high level of security, encryption and authentication services.

In this paper we consider reliability in transmission for periodic traffic pattern, in which data from M2M devices is being sent periodically to M2M gateway. Reliability is ensured on link level, by using IEEE 802.15.4 (XBee). XBee provides multiple reliability levels, and in the next sections we present a solution for deciding when to use which level of reliability.

3 Context-Aware Mobile Agent Network Model

This section presents a context-aware mobile agent network model introduced in [10] which is extended in this paper with a mechanism for choosing the appropriate level of communication reliability. The model of M2M system is amended to be in

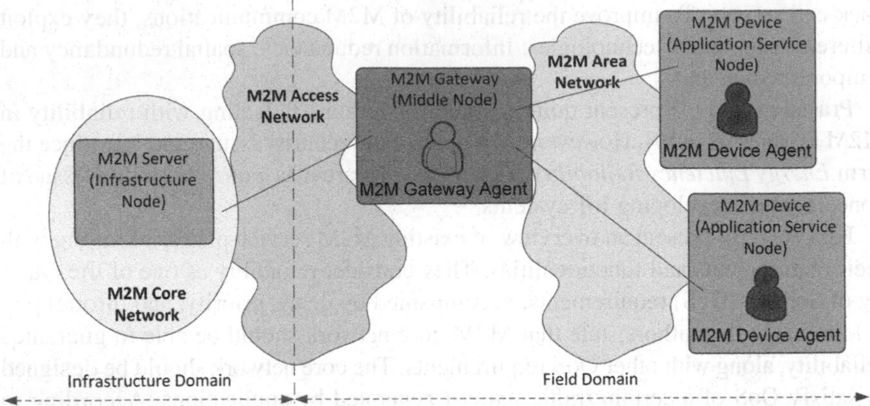

Fig. 1 M2M field and infrastructure domain

compliance with oneM2M functional architecture [11]. It consists of two main domains: field domain, and infrastructure domain. We focus on field domain which is modeled as a multi-agent system placed in a network of nodes, as shown in Fig. 1. Devices and gateway are represented by nodes that host agents which execute services and communicate.

The model is formalized by using a quadruple $\{A, S, N, C\}$ where A represents a set of agents co-operating and communicating in the environment defined by S and N. S denotes a set of processing nodes in which agents perform dedicated services, and N is a network defined as an undirected graph that connects processing nodes and allows agent communication and mobility. C represents a set of context data handled by the agents. This context data is derived from Rich Presence Information (RPI) for machines which contains information about device's location, load, battery level, charging type and availability [10].

There are two types of agents within set A: M2M Device Agent which is defined as *presentity* and M2M Gateway Agent which acts as a mediator between end devices and applications is defined as *watcher*. Context information C of each M2M entity is defined as $C = \{ctx_1, ctx_2, ..., ctx_j, ..., ctx_nctx\}$ where ctx_j represents one RPI element. Context information ctx_j is defined as $ctx_j = \{c_{state}, c_{ta}, c_{dc}, c_w, c_p\}$ where c_{state} is the state of the entity that can have four values: *on*, *off*, *sleep*, *hibernate*. c_{ta} is the absolute time when entity will change state, c_{dc} is the time period which entity spends in the current state, c_w is a set of *watchers* that the entity will inform about the change in state, while c_p is a set of *presentities* that the *watcher* receives information from.

The functionality of agents within a set A is defined by a set of elementary services $ES = \{es_1, es_2, ..., es_j, ..., es_n\}$. Each of these services can be provided by a single as well as by multiple competing or collaborating agents. A set of services that we have defined are: es_1 executing task (e.g. measuring temperature), es_2 sending measurement, es_3 sending context information, es_4 adjusting operating mode, es_5 collecting measurement, es_6 collecting context information, es_7 creating measurement report and es_8 sending measurement report to server. Services es_1, es_2 and es_3 and es_4 are supported in *presentity* nodes, while activities es_5, es_6, es_7, and es_8 are supported in *watcher* nodes [10].

In common operation mode M2M devices are in low-energy state (hibernate or sleep) or in on state when they carry out their tasks. In one operating cycle, the nodes wake up from hibernate state, carry out their measurement, exchange RPI with gateway, send their measurement and go back to low-energy state. In the implementation of our model on devices which run on Libelium Waspmote platform, we used hibernate for low-energy state since it consumes smallest amounts of energy. In previously conducted simulations the M2M devices had variable operating cycle duration (time between two wake-ups), and different times at which they started their operations. By exchanging RPI and using synchronization mechanism, the M2M devices adjusted their operating cycles and grouped together so that M2M gateway could receive their data and go to hibernate mode instead of always being on and waiting for messages.

In this work we do not use the full functionality of the model since devices only need to send sensor measurements to gateway. Gateway has unlimited power supply, so it does not need to enter low-energy modes in which it would be nonoperational. We focus on reliability mechanisms which are executed within elementary service es_4. In the next section we introduce two reliability levels, and a mechanism which decides under which circumstances should which level be used.

4 Reliability for Machine-to-Machine Communication

As mentioned in Sect. 2, reliability should be implemented throughout all IoT layers. Our model presented in Sect. 3 includes communication on link layer, and this upgrade for enhancing reliability is performed precisely on that layer.

The main motive for the work proposed in this paper is the idea that there is a need for different reliability levels while transmitting sensor data from M2M device to M2M gateway based on transmission intervals. The reason why we introduced this mechanism is experience from a real-world environment, a weather station deployed at our faculty. Sensors connected to it monitor temperature, humidity, percentage of oxygen in the air, concentrations of carbon monoxide and carbon dioxide, wind speed and direction, and the amount of precipitation (rain) over a certain period of time. Since all those sensors could not be connected to one M2M device, we connected them to two devices. Gas sensors, temperature sensor and humidity sensor are connected to one device, which will be referred to as device 1 from now on. Wind speed sensor, wind direction sensor and pluviometer (for monitoring the amount of precipitation) are connected to the other device which will be referred to as device 2 from now on. Both devices are battery powered. Collected data is intended for a monitoring application, i.e. for showing current measurements and historical values on web and mobile application. The initial measurement interval was one minute. However, since gas sensors use significantly more energy than wind sensors and pluviometer, and since they need certain amount of time for heating before collecting measurements, their interval was set to 20 min with the intention not to drain the battery too quickly. During regular operation, some measurements were occasionally lost. For the first device which collected data about wind and rain, the loss was not so significant due to small interval for reporting measurements. On the other hand, when measurements of the other device get lost, the loss is more significant. In some cases, the measurements have not been received for a couple of hours in which temperature values changed noticeably. Therefore, we propose a mechanism which enables reliable transmission of data sent by devices which transmit in larger intervals.

Our approach suggests two levels of reliability—without retransmission and sending acknowledgments, and with retransmission and sending acknowledgments for the received data. M2M device agent decides on the level of reliability based on measurement and transmission intervals. Activity diagram of M2M devices is shown in Fig. 2. In the implementation of our model M2M devices on Libelium Waspmote

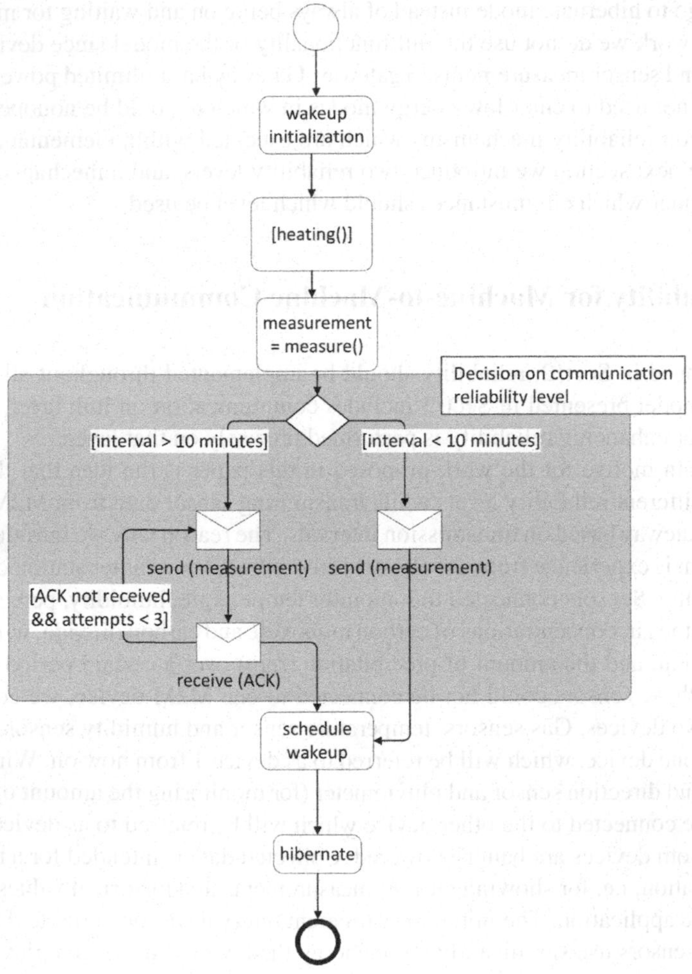

Fig. 2 M2M device agent activity diagram

platform are used, with XBee modules for communication. By using XBee, it is possible to define whether messages should be sent with or without acknowledgments.

After wake-up, the device performs measurements. In case of gas sensors, heating time is needed before reading the values. After the values from sensors are read, the transmission process begins. In the case when interval for the next measurement and transmission is less than 10 min, the measurement is sent only once (lower reliability level). When transmission interval is more than 10 min, the measurement is sent with certain amount of time reserved afterwards for waiting the acknowledgment (higher reliability level). The message can be resent 3 times after the first sending, which is a feature enabled in the technology used for communication, XBee [3]. After sending the measurement, and in some cases after waiting for the acknowledgment and retransmission, the device enters low-energy mode (hibernate). The influence of this approach on number of lost packages and on energy consumption will be shown in the next section.

5 Influence of Agent-Based Reliability Approach on Number of Lost Packages and on Energy Consumption

Approach proposed in Sect. 4 has an impact on reliability and on energy consumption. If the acknowledgment is sent by the gateway and if the device retransmits the message when no acknowledgment is received, the reliability will be higher. However, energy consumption for the aforementioned case will be higher, which will result in shorter lifetime of the device. On the other hand, if each message is sent only once, the reliability will be lower since there is a lower possibility that the message will be received. Nonetheless, there is a benefit in sending message only once, and that is lower energy consumption.

Table 1 shows the number of transmitted and received packages within 24 h collected over a period of 3 months for days in which device operation was regular, i.e. without maintenance interruptions or other unexpected interruptions, for both reliability levels and for device 1. Table 2 shows the same values as Table 1, but for device 2. Reliability is calculated as the number of messages received at destination divided by the number of messages sent from source [5].

Table 1 Communication reliability for device 1 (with larger transmission interval)

	Received messages in 24 h	Sent messages in 24 h	Reliability (%)
Transmission with lower reliability level	43	72	59.72
Transmission with higher reliability level	60	72	83.33

Table 2 Communication reliability for device 2 (with smaller transmission interval)

	Received messages in 24 h	Sent messages in 24 h	Reliability (%)
Transmission with lower reliability level	1153	1440	80.07
Transmission with higher reliability level	1234	1440	85.69

Table 3 Energy consumption for device 1 (with larger transmission interval) during one operating cycle

	Wake-up initialization	Preheating	Sensing	Send	[Receive]	Hibernate	Total
Power consumption (mW)	70.13	197.53	197.53	258,95	262.68	0.000461	
Duration (ms)—lower reliability	80	40 000	10	255.3	–	1 200 000	
Duration (ms)—higher reliability	80	40 000	10	510.6	417.6	1 200 000	
Energy consumption (μWh)—lower reliability	1.56	2194.76	0.55	18.36	–	0.15	**2215.38**
Energy consumption (μWh)—higher reliability	1.56	2194.76	0.55	36.73	30.47	0.15	**2264.22**

According to results shown in Tables 1 and 2, using a higher reliability level presents greater benefit to device 1. Reliability in that case is 23.61 % better than in the case when lower reliability level was used. As for the device 2, benefits of using higher reliability level are not so considerable, i.e. reliability only improves for 5.62 %. And even with lower reliability level, the reliability is quite high (80 %).

Table 3 shows energy consumption in one operating cycle for each task executed on device 1 for both reliability levels. Energy consumption of the tasks was calculated in our previous work [12], apart from sensing tasks which spend different amounts of energy because different sensors are used in this experiment. To obtain energy consumption for the sensors connected to our weather station, values from specifications were used [7, 8]. In our earlier work [12], the values from specifications were proven to be trustworthy.

For the case when lower level of communication reliability was used, the message was sent only once. On the other hand, for the case when retransmission was used, maximum number of sent messages was 4 (first message and 3 chances for retransmission). In this calculation, we assume that the message was averagely sent 2 times because, as it can be seen from Table 1, about 50 % of messages reach their destination even from the first time. So, for the transmission consumption we assume that energy for sending is averagely twice higher than when message was sent with lower reliability level, and we also include energy consumption for the reception of

Table 4 Energy consumption for device 2 (with smaller transmission interval) during one operating cycle

	Wake-up initialization	Sensing	Send	[Receive]	Hibernate	Total
Power consumption (mW)	70.13	2.59	258.95	262.68	0.000461	
Duration (ms)—lower reliability	80	10	255.3	–	60 000	
duration (ms)—higher reliability	80	10	382.95	313.2	60 000	
Energy consumption (μWh)—lower reliability	1.56	7.19×10^{-3}	18.36	–	7.68×10^{-3}	**19.93**
Energy consumption (μWh)—higher reliability	1.56	7.19×10^{-3}	27.55	22.85	7.68×10^{-3}	**51.97**

the acknowledgment. Table 4 shows the same values as Table 3, but for device 2. For that device, we assume that the message was averagely sent 1.5 times because more messages than for device 1 reach their destination from the first attempt.

From the data shown in Tables 3 and 4, it is obvious that introduction of higher reliability level has a higher impact on energy consumption of a device which connects anemometer, wind vane and pluviometer. Energy consumption in such case is 160 % higher than when using lower reliability level. The reason for that is small energy consumption of sensing activities, and a larger impact on overall consumption by communication activities, which make 92 % of total energy consumption in the case where lower reliability level is used. On the other hand, on device 1 energy consumption when using higher reliability level is only around 2.2 % higher than for the cases when lower reliability level is used. The reason for that is the fact that communication does not consume that much energy during one operating cycle. The highest amount is accounted for sensor preheating and sensing (around 99 %), and transmitting a message once or twice does not make the overall consumption significantly higher.

When analyzing results from previous tables, it is obvious that using higher reliability level makes more sense for device 1 because reliability in that case is significantly better, and the influence on energy consumption is negligible. As for device 2, the use of higher reliability level would not be reasonable because using higher reliability level improves reliability only slightly, but significantly increases energy consumption of the device.

6 Conclusion

Reliability in communication is an important parameter which needs to be taken into account when developing M2M communication systems. In our scenario, where M2M devices are battery powered, reliability has to be in balance with energy efficiency because enhancing reliability often increases energy consumption. In this paper it is shown that enhancing reliability levels has different influence on energy efficiency. For the device which reports measurements every 20 min, it was proved suitable to use higher reliability level because it enhanced communication reliability, and did not increase energy consumption much. On the other hand, for the device which reported measurements every minute, enhancing reliability level did not improve communication reliability much, but energy consumption did increase significantly. When making decisions about which reliability mechanisms to use, analysis like the one described in this paper should be performed.

In the executed experiment, reliability levels were set before putting the device in operation. In order to fully enable the agent-based mechanism proposed in Sect. 4, the possibility to change reliability levels on XBee modules dynamically, during operation, should be tested.

In our model communication is deployed on link layer. Applications are defined by specifying sensors from which measurements need to be obtained and by indicating intervals, without using application protocols. In future work we plan to expand our model so that protocols from higher level within protocol stack can be used for starting services and obtaining sensor measurements, such as CoAP and MQTT.

Acknowledgments This work has been supported by Croatian Science Foundation under the projects 8065 (Human-centric Communications in Smart Networks) and 8813 (Managing Trust and Coordinating Interactions in Smart Networks of People, Machines and Organizations).

References

1. Al-Fuqaha, A., Guizani, M., Mohammadi, M., Aledhari, M., Ayyash, M.: Internet of things: a survey on enabling technologies, protocols, and applications. IEEE Commun. Surv. Tutorials **17**(4), 2347–2376 (2015)
2. Chen, K.C., Lien, S.Y.: Machine-to-machine communications: technologies and challenges. Ad Hoc Netw. **18**, 3–23 (2014)
3. Digi International Inc.: XBee®/XBee-PRO® RF Modules. Product manual v1.xex-802.15.4 protocol (2009). https://www.sparkfun.com/datasheets/Wireless/Zigbee/XBee-Datasheet.pdf
4. Gubbi, J., Buyya, R., Marusic, S., Palaniswami, M.: Internet of things (iot): a vision, architectural elements, and future directions. Future Gener. Comput. Syst. **29**(7), 1645–1660 (2013)
5. Jereb, L.: Network reliability: models, measures and analysis. In: Proceedings of the 6th IFIP Workshop on Performance Modelling and Evaluation of ATM Networks, pp. 1–10 (1998)
6. Kim, J., Lee, J., Kim, J., Yun, J.: M2M service platforms: survey, issues, and enabling technologies. IEEE Commun. Surv. Tutorials **16**(1), 61–76 (2014)
7. Libelium Comunicaciones Distribuidas S.L.: Agriculture 2.0. Technical guide (2015). http://www.libelium.com/downloads/documentation/agriculture_sensor_board_2.0.pdf

8. Libelium Comunicaciones Distribuidas S.L.: Gases 2.0. Technical guide (2015). http://www.libelium.com/downloads/documentation/gases_sensor_board_2.0.pdf
9. Lu, R., Li, X., Liang, X., Shen, X., Lin, X.: GRS: the green, reliability, and security of emerging machine to machine communications. IEEE Commun. Mag. **49**(4), 28–35 (2011)
10. Maracic, H., Miskovic, T., Kusek, M., Lovrek, I.: Context-aware multi-agent system in machine-to-machine communication. Proc. Comput. Sci. **35**, 241–250 (2014). Knowledge-Based and Intelligent Information & Engineering Systems 18th Annual Conference, KES-2014 Gdynia, Poland, September 2014 Proceedings
11. oneM2M: M2M Functional Architecture. Technical specification (2015). http://www.onem2m.org/images/files/deliverables/TS-0001-Functional_Architecture-V1_6_1.pdf
12. Skocir, P., Zrncic, S., Katusic, D., Kusek, M., Jezic, G.: Energy consumption model for devices in machine-to-machine system. In: 2015 13th International Conference on Telecommunications (ConTEL), pp. 1–8 (2015)
13. Skocir, P., Maracic, H., Kusek, M., Jezic, G.: Data filtering in context-aware multi-agent system for machine-to-machine communication. In: Agent and Multi-Agent Systems: Technologies and Applications: 9th KES International Conference, KES-AMSTA 2015 Sorrento, Italy, June 2015, Proceedings, pp. 41–51 (2015)
14. Sundar Prasad, S., Kumar, C.: An energy efficient and reliable internet of things. In: 2012 International Conference on Communication, Information Computing Technology (ICCICT), pp. 1–4 (2012)

Part II
Agent-Based Modeling and Simulation

Herding Algorithm in a Large Scale Multi-agent Simulation

Richard Cimler, Ondrej Doležal, Jitka Kühnová and Jakub Pavlík

Abstract This research is focused on creation of a herding algorithm suitable for a large map area which will be used in an agent-based simulation of an ancient Celtic society development. Algorithm is designed in order to find suitable place for grazing of animals in a satisfactory time on a map composed of more than 700 000 cells. Parameters of the algorithm are adjusted due to the results of a statistical research. Simulation is created in the AnyLogic multimethod simulation modeling tool. Virtualized server is used for experiments because of a complexity of the simulation.

Keywords Simulation · Agent-based · Model · Herding · Model optimization

1 Introduction

Computer simulations are used for research purposes for decades. With the increasing computational power also complexity of the simulations increases. There are many simulation tools which can run even on the personal computers in order to obtain desired result data. On the other hand for the large scale simulations it is necessary to provide powerful computers and optimize simulations algorithms.

Multiagent simulations are often used for modeling a real world situations. There are many simulations from different areas of social, environment or medical research. Set the appropriate level of model abstraction is crucial during design of the model.

R. Cimler (✉) · O. Doležal · J. Pavlík
Faculty of Informatics and Management, Department of Information Technologies,
University of Hradec Králové, Rokitanskeho 62, 500 03 Hradec Králové, Czech Republic
e-mail: richard.cimler@uhk.cz
URL: http://www.uhk.cz

J. Kühnová
Faculty of Science, University of Hradec Králové, Hradec Králové, Czech Republic

R. Cimler
Center for Basic and Applied Research (CZAV), University of Hradec Králové,
Hradec Králové, Czech Republic

G. Jezic et al. (eds.), *Agent and Multi-Agent Systems: Technology and Applications*, Smart Innovation, Systems and Technologies 58,
DOI 10.1007/978-3-319-39883-9_7

Our simulation deals with a problem of animal herding on a vast area. Level of abstraction is set due to the time step which is equal to one day. Searching problem in a vast area is described in this paper.

Different herding algorithms can be found at [2, 4, 14]. Framework focused on the sheep herding is introduced in [5]. Time units in described simulations are very small compared to the time units in our simulation. Because of the needs of archaeological research several decades has to be simulated. Due to this length of the simulation our herding algorithm has one step equal to one day. Level of abstraction in our simulation is higher. We do not need to simulate movement of animals during the day as is typical for herding algorithms thus we proposed our herding algorithm for this level of abstraction and size of a map.

Values of the model parameters are often not fixed but in the given range with a discrete step size. During the model testing and creating the experiments number of parameters combinations grows rapidly. It is also necessary to repeat model runs with the same parameters in order to obtain sufficient amount of data to ensure reliable results for a statistical analysis. Thus there could be hundreds, thousands or more experiments runs. For these experiments virtualized high performance computer has been used.

Paper is divided into five sections. In the following, Sect. 2, the model and a herding problem is described. Section 3 contains detailed description of the simulation. Herding algorithm is shown in the Sect. 4. Experiments and results are in the Sect. 5 of the paper

2 Problem Description

Presented model is a replication and an extension of a model previously created in a NetLogo [3, 7, 8, 11, 12]. The general objective of our research is to develop the complex agent-based simulation of an ancient Celtic society population, food production and animal husbandry. Simulation, described in this paper, is focused only on one part of the model—herding of animals. There are several parts of the simulation in the original model such as simulation of population, agriculture, nitrogen cycle or animal herding. It is intended to replicate original model in the AnyLogic [1] which brings possibilities for creating more accurate model. AnyLogic is advanced computer simulation tool that supports all the most common simulation methodologies such as system dynamic, agent based modeling and discrete events.

Some of our model's basic principles are the same for NetLogo and AnyLogic but there are also very essential differences such as change of the time steps from one year to one day and also each animal is simulated separately. Change of the time step brings new challenges. One of them is described in this paper—realistic and effective herding of animals. In a such detailed simulation it is necessary to simulate movement of herds between pastures. In previous simulation, where one step was equal to one year, total size of a grazed land was counted only once a year. This amount was based only on a number and a type of animals in a herd. In the current simulation grazing of each animal is simulated for each day. It is not simulated how animals are moving during the day but an animal moves only once a day and during its movement it is simulated on which area was animal grazing on in a current day. Animals are amassed in herds which move to a new location if a certain amount of animals from the herd is starving—which means area around the herd has been grazed.

3 Model Description

Certain parts of the ODD protocol [10] are used for description of this model. Due to the size of this paper it is not possible to use all parts of the protocol.

3.1 *Model Purpose*

Purpose of a model is to create a model of an animal herding which will be a part of complex Celtic oppidum simulation. In general meaning "oppidum" was a fortified settlements situated on high ground, mostly used by Celts in central and western Europe around one hundred years B.C. Aim of the simulation is to test proposed algorithm and find dependencies between input parameters and simulation results. The model was developed based on the requirements resulting from previous archaeological research. Size of maps, geographic properties, types of animals, number of animals, all these parameters have been the subject of many archaeological researches of given settlement, and for reasons of preserving the authenticity, all these properties remain unchanged.

3.2 *Entities, State Variables, and Scales*

Each animal is modeled as an agent. Second (static) agent type is a land cell. Animals grouped in a herd are grazing on a land during a grazing period of a year (181 days). If there is not enough grass in the grazing area, herd change its location to some more suitable area.

Animal Variables

- *Basic parameters*: AnimalSpecies (cow, sheep/goat, horse), Sex (male/female), Age, Location (x, y coordinates), Id (animal identification)
- *Grass eaten*: refers to an amount of grass which has been already eaten this year
- *Herd*: information to which herd animal belongs to
- *Shepherd*: reference to a leader of the herd. Leader is also an animal and simulates decision of the herd shepherd. It was not necessary to model new agent type—

shepherd. Every new year is chosen new herd leader for each herd. Because after the grazing period of a year leader may die.

- *GrazeType*: collection of possible land types where can animal graze.
- *HungerRatio*: Leader variable, ratio of a grass amount which herd should eat and actually had eaten. This parameter is used to determine how big was the lack of grass.
- *HungerCounter*: Leader variable, how many days in a row herd had the lack of grass and therefore they had to move to next pasture. When the value exceeds the limit, it indicates that there is not enough food around, and the herd has to move to another part of the map.

There are also static parameters same for all animals:

- *Herd size*: Number of the animals in one herd. One herd contains animals of one species (cow, horses, goat/sheep or pigs).
- *Herd area*: Refers to a square area around the middle of the herd where are animals allowed to graze. This size is also part of the experiment. Too small area may result in a frequent movement of whole herd.
- *Grass needed*: Refers to an amount of grass which is needed for each animal every day.

Land

There are several layers of GIS data which are loaded during the initialization of the simulation. Layers are: Forest type, wetness, slope, distance from streams, distance from oppidum and suitability for planting crops. Background map for the simulation is divided into the discrete grid with size 908×778 cells. Real size of a map is 9080×7780 m one cell refers to 10×10 m (100 cells are 1 ha). Only certain part of the map can be used for the agricultural and other work based on its distance from the oppidum. Size of the usable land is smaller than size of the entire grid.

- Size of entire map = 706 424 cells ~ 7064 ha
- Size of usable land = 526 969 cells ~ 5299 ha
- Size of oppidum = 4 250 cells ~ 42 ha

During the initialization parameters cells are filled with the GIS values. There are 8 types of cells in the simulation based on this data.

- *Wasteland*: Cells which are not include in the usable land.
- *Oppidum*: Location of the oppidum. This land cannot be used for any actions.
- *Grassland*: Lands with a grass. Based on its quality and an amount it is more or less suitable for animals grazing. Type of the grassland is stored in a parameter grassType.
- *Wheat*: Land which is currently used for a wheat farming.
- *Pulses*: Land which is currently used for a pulses farming.
- *Fallow*: After harvesting of a crop, field becomes fallow. If the cultivation strategy is intensive, field is used again next year. If cultivation strategy is extensive, field will be used at least after 3 years.

- *Forest*: Used for pigs grazing and source of wood.
- *Water*: No actions can be performed on this cell.

Land Parameters

- grassType
  Pasture—Land where animals can graze. Each animal type has its preferences for attributes of the land such as wetness, slope, and the distance from the water source.
  Meadow—Land which is used for collecting hay for animal feeding during winter season. Grass from this land has better quality and a higher yield because this land is not used for grazing.
  Fallow—It is possible to graze here but an amount of grass on this land is low.
  Unreachable Animals are not able to reach this land due to its slope, wetness or distance from streams.
- amountGrass—refers to current amount of grass on a land
- mapDistanceOppidum, mapDistanceStreams, mapCrops, mapSlope, map Wetness—layers loaded from the GIS data. MapCrops layer refers to suitability of the land for planting crops.

3.3 Time Steps

In previous two models one time step (tick) referred to one month in the first model and one year in the second model [11]. In the current model one tick refers to one day. Usage of powerful virtual PC enabled to create such detailed simulation with several complex computations which are conducted every step. Smaller time steps such as hours or minutes could not be implemented into our model. There are two main reasons. First reason is limited computational resources. Even with virtualized solution which is currently used one simulation run with a tick equal to one day takes approximately up to 30 min. For the experiments it is necessary to run at least 20 but preferable 100 repetitions with the same setup in order to eliminate effect of random events in the model. Because of the parameter variation there could be hundreds of the setups. With these values model cannot be 24 times (in case tick = hour) or 1440 times (in case tick = minute) detailed. Second reason is a sufficient level of an abstraction of the model. We dont have enough information to create such detailed simulation where life in an oppidum is simulated for each hour or more detailed. One simulation step equal to one day has been chosen because it is known approximately in days how long agriculture and other works took. It is also possible to compute how much grass has been eaten by the animals each day and thus when it was necessary for herd to change the pastures. In a previous simulations with a ticks equal to one year it was not possible to count bottlenecks of works—that means situation with not enough workforce to maintain all of the necessary works. Postponement of non-urgent works during the bottlenecks was also not possible to simulate. In previous

Table 1 Animal scenarios

Animals per person	Cow	Oxen	Calf	Horse	Sheep/Goat
S1	0.2	0.75	0.2	0.075	0.1
S2	0.4	0.15	0.4	0.15	0.2

simulation it is also not possible to simulate grazing of the animals and movement of the herd. On the other hand with lower level of the abstraction there was smaller prone to errors in previous models.

3.4 *Initialization*

Map Data for creating an environment of the model are loaded from several files. Each file represents one layer of a map. AnyLogic tool includes a module for working with GIS maps, but unfortunately it does not support a GIS layers and a specific format used for our files. It was therefore necessary to create a function that retrieves data from files and prepare them into a form that allows us to create an entire map in the simulation.

Population Population is loaded from the file. For the purpose of testing the algorithm population in the simulation is static.

Number of animals Number of animals is also fixed in this simulation. There are two scenarios (S1, S2) of the herd size in the complex simulation. Human population in the simulation grows based on the algorithms described in [3, 8]. Number of animals is counted due to the size of the population every year. Number of animals per one person is in the Table 1. Experiments in this paper are based on a fixed population corresponding to the fortieth year of the population simulation and scenario 2. Movement of the herds containing totally 318 cows is simulated.

4 Herding Algorithm

Herd stays in a selected location as long as there is enough grass for grazing. If their current pasture runs out of grass, they have to move to another one. Shepherd chooses one of the nearest possible grazing areas. If there are no suitable grazing areas in the surrounding herd moves to different location on the map. It is not possible to search the entire map (526 969 cells) in order to find most suitable location for grazing. There are several herds in the simulation and computational workload would be too high. When it is necessary to move the herd, few cells are randomly selected as a candidates. From these candidates the cell with the biggest amount of grass is selected. Algorithm is shown on the Fig. 1. Blocks of code are described in the following text. Experiments are at the last part of the paper.

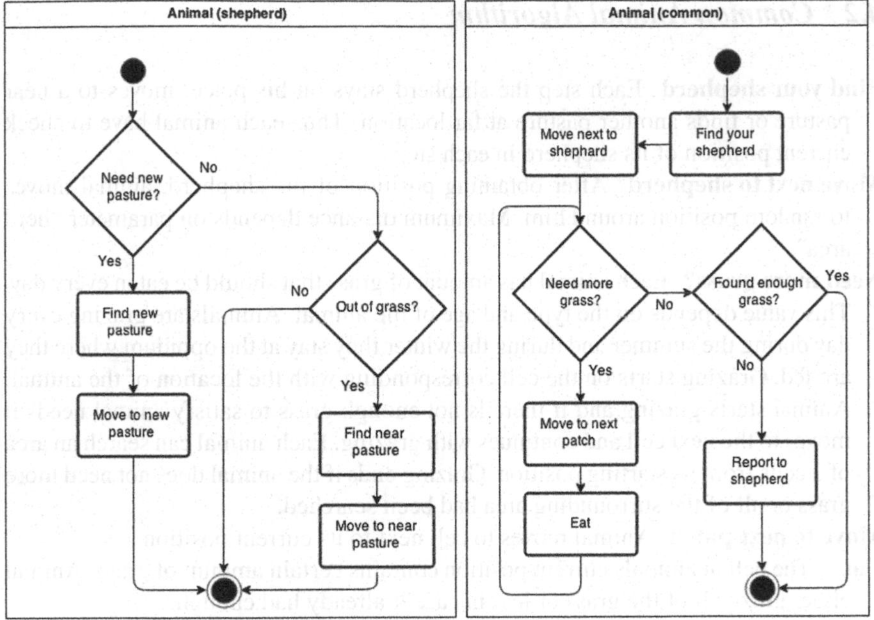

Fig. 1 Herding algorithm flowchart

4.1 Shepherd Algorithm

Need new pasture? Condition is fulfilled if certain number of days in a row, based on the parameter 'hungerCounter', herd had a lack of grass. In a such case ten cells are searched in order to find more suitable location for grazing. This action is called in experiment "Far movement".

Find new pasture Shepherd checks ten randomly selected cells with type pasture all around the map and choose the one with the greatest amount of grass.

Move to new pasture Shepherd moves to a location chosen in block "Find new pasture".

Out of grass? For each herd it is registered how much kilograms of grass has been grazed each day. If this amount is smaller than expected amount of grass, which should be eaten each day, it is necessary to move the herd to the next grazing area. This action is called in experiment "Near movement". It is also registered that this day was lack of food. Number of days with lack of grass is evaluated in the block "Need new pasture?".

Find near pasture Shepherd inspects pastures in the direct surrounding to his current pasture and selects the one with the largest amount of grass. Number of inspected places depends on a parameter "AreaSize" and is counted as: $(\frac{4}{5} \cdot \text{areaSize})$

Move to near pasture Shepherd moves to a location chosen in a previous step.

4.2 Common Animal Algorithm

Find your shepherd Each step the shepherd stays on his place, moves to a near pasture or finds another pasture at far location. Thus each animal have to check current position of its shepherd in each step.

Move next to shepherd After obtaining position of the shepherd, animal moves to random position around him. Maximum distance depends on parameter "herd area".

Need more grass? Each animal has amount of grass that should be eaten every day. This value depends on the type and age of the animal. Animals are grazing every day during the summer and during the winter they stay at the oppidum where they are fed. Grazing starts on the cell corresponding with the location of the animal. Animal starts grazing and if there is not enough grass to satisfy animal needs it moves to the next cell and continues with grazing. Each animal can search an area of 3 cells from its starting position. Grazing ends if the animal does not need more grass or all of the surrounding area had been searched.

Move to next patch Animal moves to cell next to its current position.

Eat The cell at animals current position contains certain amount of grass. Animal either graze all of the grass or less in case it already had enough.

Found enough grass? Has animal already grazed as much grass as it should?

Report to shepherd In case animal has not found enough grass in a step "Need more grass?" missing amount of grass is registered by the herd boss. Herd boss has information about missing amount of grass from all animals in the herd. Sum of the missing grass is used in making the decision if herd should move to another grazing location in block "Out of grass?".

5 Experiments

In the experiments different settings of five algorithm parameters has been tested. Parameter values can be seen in the Table 2. There were 108 combinations of input parameters. Each combination has been tested 5 times. Each run simulated 40 years.

Table 2 Experiment parameters

Parameter	From	To	Step
Pasture size	20000	40000	20000
Hunger ratio	0.1	0.5	0.2
Hunger counter	3	4	1
Area size	10	50	20
Herd size	10	50	20

Number of cows was 318 according to the size of the population and S1 scenario (See Table 1). Needed amount of grass for each cow is 14 kg per day. Optimal number of cells for all cows is approximately 40 000. Experiments were conducted on a virtual PC with 24 processors Westmere E56xx/L56xx/X56xx and 32 GB RAM. There were 20 repetitions of each experiment.

Description of the simulation parameters:

- *Pasture size*: the total size of pastures available for grazing. (Cells)
- *Hunger ratio*: the maximum allowable value for the ratio between grass that has been actually grazed and total requirement for one day in one herd.
- *Hunger counter*: how many days in a row herd can have the lack of grass.
- *Area size*: size of the area on which graze one herd.
- *Herd size*: number of animals in one herd.

5.1 *Results*

We would like to find which parameters influence results like Time of the simulation, Starving of cows and so. The result values are:

- *Time*: Refers to real time in milliseconds that one simulation last.
- *Starving*: The difference between what was actually grazed and what should be grazed by all animals in kilograms.
- *Grass left*: Refers to an amount of grass that was left on a grassland after one year.
- *Near movement*: Herd is moving to one of the surrounding areas.
- *Far movement*: Herd is moving to brand new location.

Maximal and minimal values of main three result variables can be seen at Tables 3 and 4. These results were inconclusive thus more sophisticated statistical methods has been used.

Table 3 Experiment results (Pasture size 2000)

Result	Value	Hunger ratio	Hunger counter	Area size	Herd size
Minimal run time	444326	0.5	3	30	30
Maximal run time	751397	0.1	3	50	30
Minimal starving	7234611	0.5	4	50	10
Maximal starving	10012189	0.5	4	10	50
Minimal grass left	12067	0.1	4	50	10
Maximal grass left	387466	0.5	4	10	50

Table 4 Experiment results (Pasture size 40000)

Result	Value	Hunger ratio	Hunger counter	Area size	Herd size
Minimal run time	323459	0.5	4	10	50
Maximal run time	699359	0.1	4	30	50
Minimal starving	1255654	0.1	3	50	10
Maximal starving	6699633	0.5	4	10	50
Minimal grass left	1779795	0.1	4	10	30
Maximal grass left	2052225	0.5	4	50	10

Table 5 Estimated parameters of linear regression models—basic

	Starving	Time	Grass left
Intercept	1.454×10^7***	8.160×10^5***	-1.778×10^6***
Pasture Size	-3.190×10^2***	-2.899***	9.027×10^1***
Hunger Ration	2.165×10^6***	-2.583×10^5***	1.666×10^5***
Hunger Counter	2.455×10^3	-1.556×10^4	5.347×10^2
Area Size	-2.766×10^4***	1.815×10^1	-1.982×10^2
Herd Size	3.153×10^4***	-8.416×10^2*	-97.61
R^2	0.9557	0.1508	0.9936
p-value	$<2.2 \times 10^{-16}$	$<2.2 \times 10^{-16}$	$<2.2 \times 10^{-16}$

Linear regression model were used for finding dependencies of these result variables. As is said in [6], a linear regression model is given by a formula

$$Y = \beta_0 + \beta_1 X_1 + \beta_2 X_2 + \cdots + \beta_n X_n + \varepsilon,$$

where X_i are independent variables, β_i are unknown parameters, $i = 0, \ldots, n$, and ε is a random variable of errors. Our goal is to find if there is a statistically significant relation between independent variables (as Hunger ratio, Area size, ...) and dependent variable Y (in our case Starving, Time or Grass left). We will use statistical software R and the function `lm()` (See [13]). Our findings are summarized in Tables 5 and 6.

Firstly, we tried to find statistically significant relation between one of dependent variables (Starving, Time and Grass left) and independent variables (Pasture Size, Hunger Ration, Hunger counter, Area Size and Herd Size). Secondly, we tried to explain the dependent variable only with two variables—Far movement and Near movement. In each case, we tested the significance of estimated parameters and the significance of a model. If parameters were significant, we put '*' (if it was on the 0.05 significance level), '**' (on 0.01 level) or '***' (on 0.001 level). R^2 in tables means *coefficient of determination* and it tells us how much model fits the data ($R^2 \in \langle 0; 1 \rangle$). If R^2 is close to 1, data are almost perfectly explained by model. The last

Table 6 Estimated parameters of Linear regression models—basic

	Starving	Time	Grass left
Intercept	4.520×10^6***	5.551×10^5***	1.337×10^6***
Far movement	-2.467×10^2***	2.892	6.853×10^1***
Near movement	9.731×10^1***	2.769×10^{-1}	-2.800×10^1***
R^2	0.337	0.06051	0.4006
p-value	$<2.2 \times 10^{-16}$	$<5.267 \times 10^{-8}$	$<2.2 \times 10^{-16}$

information in our tables is so called p-value, in this case it is p-value of the F-test in linear regression model (see [9]). The model is significant if this p-value is lesser than 0.05.

One can see that each model is significant. From coefficients of determination is seen that Starving and Grass left is highly dependent on variables, but Time is more loose. If we want to minimize Starving we have to maximize Pasture Size and Area size but minimize Hunger Ration and Herd Size. For minimization of Time we have to maximize Hunger Ration and Herd Size and for maximization of Grass left we have to minimize Hunger Ration and maximize Area size. Which is quite obvious, but now we have verified that model corresponds to reality.

For the verification of these results we have prepared the last experiment. Parameters of this experiment were set to 40000 for Pasture size, 0.1 for Hunger ratio, 50 for Area size and 10 for Herd size. Then we have investigated the starving of cows with these settings. After 40 years of simulation (7240 grazing days) average number of days when cow was starving (did not eat more than 50 % necessary grass) was only 2 days a year.

Different settings of parameters in a block "Find new pasture?" (number of searched cells) has been tested. There was no significant change of results for 10, 60 and 120 searched cells. Same results were also obtained for a test where a number of animal moves in a block "Need more grass?" has been increased to 5 or 10.

6 Conclusion

Aim of this research was to create an algorithm of herding process for the simulation of Celtic oppidum development. Simulation had to be very detailed due to needs of archaeological research. Previous model created in NetLogo has been replicated in AnyLogic and level of abstraction has been lowered. Due to archaeologists' requirements, in the current simulation one simulation step represents one day and one cell of the map represents $10 \times 10\,m^2$ area. Size of the area is more than 7 000 ha thus the map consists of more than 700 000 cells. Search all cells to find a suitable place for herding would be very demanding on computational resources thus we propose an algorithm for herding on such big map. Algorithm has been created in order to find in a short time suitable locations for grazing of animals. Based on results from statistical research algorithm parameters were adjusted due requirements of the simulation. In our simulation has been parameters chosen as follows: Areasize 30, Herdsize 10, Hunger counter 4, Hunger ration 0.2, in order to minimize starving and to find compromise between run time and algorithm effectiveness. Suitable settings of parameters depends on the simulation requirements as been discussed in Sect. 5.1. Result of our work is proposed algorithm suitable for a vast map which will be used in complex simulation of Celtic oppidum development. Algorithm can be used also in other similar simulations where is needed to search in a vast map.

Acknowledgments The research described in the paper was supported by grant GACR-405/12/0926 Social modeling as a tool for understanding Celtic society and cultural changes at the end of the Iron Age and UHK specific research project.

References

1. Anylogic. http://www.anylogic.com/ (2015). Accessed 03 Dec 2016
2. Bennett, B., Trafankowski, M.: A comparative investigation of herding algorithms. In: Procedings of Symposium on Understanding and Modelling Collective Phenomena (UMoCoP), pp. 33–38 (2012)
3. Danielisová, A., Olševičová, K., Cimler, R., Machálek, T.: Understanding the iron age economy: Sustainability of agricultural practices under stable population growth. In: Agent-based Modeling and Simulation in Archaeology, pp. 183–216. Springer (2015)
4. Dijkstra, J., van Otterlo, M.: Herding sheep (2014)
5. Dorssers, F., van Otterlo, M.: Sheeplog: creating a prolog framework to herd sheep (2014)
6. Faraway, J.J.: Linear Models With R, 2nd edn., Taylor and Francis
7. Machálek, T., Cimler, R., Olševičová, K., Danielisová, A.: Fuzzy methods in land use modeling for archaeology. In: Proceedings of Mathematical Methods in Economics (2013)
8. Machálek, T., Olševičová, K., Cimler, R.: Modelling population dynamics for archaeological simulations. In: Mathematical Methods in Economy, pp. 536–539 (2012)
9. Mathworks.com: Interpret linear regression results—matlab & simulink (2015). http://www.mathworks.com/help/stats/understanding-linear-regression-outputs.html
10. Müller, B., Bohn, F., Dreßler, G., Groeneveld, J., Klassert, C., Martin, R., Schlüter, M., Schulze, J., Weise, H., Schwarz, N.: Describing human decisions in agent-based models-odd+ d, an extension of the odd protocol. Environ. Modell. Softw. **48**, 37–48 (2013)
11. Olševičová, K., Cimler, R.: Agent-based model of carrying capacity of celtic settlement agglomeration. Glob. J. Technol. **3** (2013)
12. Olševičová, K., Cimler, R., Machálek, T.: Agent-based model of celtic population growth: Netlogo and python. In: Advanced Methods for Computational Collective Intelligence, pp. 135–143. Springer (2013)
13. Statmethods.net: Quick-r: Multiple regression (2015). http://www.statmethods.net/stats/regression.html
14. Strömbom, D., Mann, R.P., Wilson, A.M., Hailes, S., Morton, A.J., Sumpter, D.J., King, A.J.: Solving the shepherding problem: heuristics for herding autonomous, interacting agents. J. R. Soc. Interface **11**(100), 20140719 (2014)

I-Fuzzy Core for Cooperative Games with Vague Coalitions

Elena Mielcová

Abstract The main aim of this article is to discuss the construction of the core of a transferable utility cooperative game, when possible coalitions of agents are vague—in this case expressed as I-fuzzy coalitions using I-fuzzy setting. In general, the theory of I-fuzzy sets (originally introduced as intuitionistic fuzzy sets) is considered to be an extension of fuzzy set theory, where the degree of non-membership denoting the non-belongingness to a set is explicitly specified along with the degree of membership of belongingness to the set. The indecisiveness part of I-fuzzy sets implies vague definition of the core, and the necessity of dividing definition of a core into two parts—into the possible and the essential core.

Keywords I-fuzzy sets · Cooperative game · Preimputation · Imputation · Core of I-fuzzy cooperative game

1 Introduction

In complex multiagent systems, objective of agents is to successfully carry out their tasks. Agents are expected to interact—that means cooperate—with other agents. An obvious problem is that of reaching agreements in a group of cooperating agents. One of the standard descriptions of the multiagent problem of cooperation is given by the theory of cooperative games [6].

In general, agents in multiagent systems are considered to be to some extent autonomous—capable of making decisions in order to satisfy their designed objectives [8]. Agents—in game theory called players—of a model game are cooperating in order to increase a mutual profit, and therefore each agent's profit. Considering that the profit can be distributed (transferred) among players with respect to some

E. Mielcová (✉)
Department of Informatics and Mathematics, School of Business Administration in Karviná, Silesian University in Opava, Univerzitní náměstí 1934/3, 733 40 Karviná, Czech Republic
e-mail: mielcova@opf.slu.cz

G. Jezic et al. (eds.), *Agent and Multi-Agent Systems: Technology and Applications*, Smart Innovation, Systems and Technologies 58,
DOI 10.1007/978-3-319-39883-9_8

coalition agreement, these games are called also transferable utility games. Cooperative game theory studies the mechanism of possible profit distribution, and provides several solution concepts. The concept of core is considered to be the basic possible solution concept.

The main aim of this article is to discuss the construction core of the transferable utility cooperative game when possible created coalitions of agents are vague—in this case expressed as I-fuzzy coalitions using I-fuzzy setting. Similar research was done on transferable utility coalition games with fuzzy coalitions [7], however this concept was not yet discussed for the games with I-fuzzy coalitions. In general, the theory of I-fuzzy sets (originally introduced as intuitionistic fuzzy sets) is considered to be an extension of fuzzy set theory, where the degree of non-membership denoting the non-belongingness to a set is explicitly specified along with the degree of membership of belongingness to the set [1]. The I-fuzzy set theory is used to describe game theory concepts in order to more realistically describe real-world concepts [5].

The next text is organized as follows; the preliminaries cover basic terms from the theory of I-fuzzy sets, as well as the theory of transferable utility games with basic definitions of an imputation and a core. Section 3 introduces a concept of coalition games with I-fuzzy coalition and discusses a construction of an imputation and a core of such games. The conclusion followed by the list of references ends the text.

2 Preliminaries

2.1 I-Fuzzy Sets

Atanassov (1986) [1] extended Zadeh's idea of fuzzy sets [9] into the theory of intuitionistic fuzzy sets by incorporating the idea of indecisiveness; in the Atanassov's intuitionistic fuzzy set theory, the membership degree as well as non-membership degree are assigned to each of their elements. Moreover, the sum of the membership degree and non-membership degree is not necessarily one. Dubois et al. [4] discussed terminological difficulties concerning term "an intuitionistic fuzzy set" and proposed use of different term; therefore throughout this text, the term "I-fuzzy" will be used instead of the term "intuitionistic fuzzy".

Formally, let a set $X = \{x_1, x_2, \ldots, x_n\}$ be fixed. Then an I-fuzzy set is defined as a set of triples $A = \{\langle x_i, \mu_A(x_i), \nu_A(x_i)\rangle;\ x_i \in X\}$, where functions $\mu_A : X \to [0,1]$ and $\nu_A : X \to [0,1]$ define the degree of membership and the degree of non-membership of the element $x_i \in X$ to set A. The condition $0 \le \mu_A(x_i) + \nu_A(x_i) \le 1$ holds for all $x_i \in X$, $i = 1, 2, \ldots, n$.

2.2 Imputation and Core of Transferable Utility Games

A cooperative game is in general considered to be a pair (N, v), where N is a set of players, and v is a characteristic function of a game.

Definition 1 A cooperative game is a pair (N, v), where $N = \{1, 2 \ldots n\}$ is a set of n players, and $v : 2^N \rightarrow R$ is a mapping defined on subsets of N with the property $v(\emptyset) = 0$. The function v is called a characteristic function of a game.

For a set of players $N = \{1, 2 \ldots n\}$, the expression 2^N denotes the collection of all subsets of N. Any nonempty subset of N is called a coalition. Thus, the characteristic function v connects each coalition $K \subset N$ with a real number $v(K) \in R$ representing total profit of coalition K, while $v(\emptyset) = 0$. Any cooperative game is usually denoted as (N, v), or simply only by its characteristic function v.

For any pair of disjoint coalitions $K, L \subset N$, $K \cap L = \emptyset$, a coalition game (N, v) is called superadditive if $v(K \cup L) \geq v(K) + v(L)$; subadditive if $v(K \cup L) \leq v(K) + v(L)$; and additive if $v(K \cup L) = v(K) + v(L)$. A cooperative game (N, v) is convex if for every pair of coalitions $K, L \subset N$: $v(K \cup L) + v(K \cap L) \geq v(K) + v(L)$. Convexity implies superadditivity, inverse implication is not valid. Players will have incentive to cooperate if a game is convex or superadditive.

Expected distribution of profit among players is represented by a real-valued payoff vector $x = (x_i)_{i \in N} \in R^n$, where x_i is a received payoff of player i. There are several reasonable conditions which should be fulfilled for each payoff distribution. The first condition states that a payoff vector should be efficient—that means a payoff vector should represent a distribution of maximal possible profit—in case of superadditive games the profit gained by grand coalition.

Definition 2 A preimputation is any x from the preimputation set $\mathcal{I}^*(v)$ of a cooperative game (N, v) defined as

$$\mathcal{I}^*(v) = \{x \in R^n; \sum_{i \in N} x_i = v(N)\}. \tag{1}$$

Example 1 Let (N, v) be a cooperative game with $N = \{1, 2, 3\}$ and a characteristic function v such that $v(\emptyset) = 0$, $v(1) = v(2) = v(3) = 5$, $v(1, 2) = v(1, 3) = v(2, 3) = 15$, $v(1, 2, 3) = 30$. The preimputation $x = (x_1, x_2, x_3)$ of this game fulfills condition $x_1 + x_2 + x_3 = 30$. Examples of preimputation: triplets $(0, 0, 30)$, $(0, 15, 15)$, $(5, 5, 20)$ or $(10, 10, 10)$.

A payoff vector $x \in R^n$ is individually rational if for each $i \in N$ there is $x_i \geq v(\{i\})$. An individually rational preimputation is called an imputation:

Definition 3 An imputation is an element from an imputation set $\mathcal{I}(v)$ of a cooperative game (N, v) defined as

$$\mathcal{I}(v) = \{x \in R^n; \sum_{i \in N} x_i = v(N), x_i \geq v(\{i\}) \quad \forall i \in N\} \tag{2}$$

Example 2 Let (N, v) be the same cooperative game as in Example 1 with $N = \{1, 2, 3\}$ and a characteristic function v such that $v(\emptyset) = 0$, $v(1) = v(2) = v(3) = 5$, $v(1, 2) = v(1, 3) = v(2, 3) = 15$, $v(1, 2, 3) = 30$. The imputation $x = (x_1, x_2, x_3)$ of this game fulfills conditions $x_1 + x_2 + x_3 = 20$; $x_1 \geq 4$; $x_2 \geq 4$; and $x_3 \geq 4$. Preimputations $(0, 0, 30)$ and $(0, 15, 15)$ are not imputations, while preimputations $(5, 5, 20)$ and $(10, 10, 10)$ are imputations.

A core represent a set of stable imputations—that means imputations under which no coalition has incentive to leave the grand coalition and receive a larger payoff.

Definition 4 Let the game (N, v) be superadditive. Then the core of (N, v) is the set of payoff vectors

$$C(v) = \{x \in R^n; \sum_{i \in N} x_i = v(N), \sum_{i \in L} x_i \geq v(L) \quad \forall L \subset N\}. \tag{3}$$

Example 3 Let (N, v) be a cooperative game with $N = \{1, 2\}$ and a characteristic function v such that $v(\emptyset) = 0$, $v(1) = 5$, $v(2) = 10$, and $v(1, 2) = 20$. The core $x = \{x_1, x_2\}$ of this game fulfills conditions

$$x_1 + x_2 = 20,$$

and

$$x_1 \geq 5, \quad x_2 \geq 10.$$

In this case the core is nonempty set containing line segment $x_2 = 20 - x_1$ for $x_1 \in [5, 10]$ (see also Fig. 1).

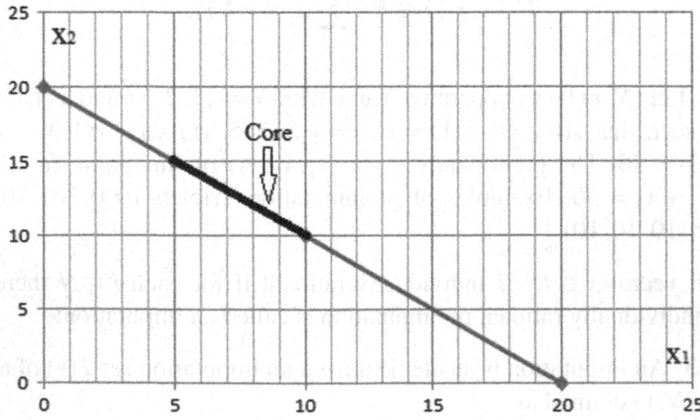

Fig. 1 The core of the game from Example 3. *Source: Own calculations*

3 Cooperative Games with I-Fuzzy Coalitions

In a fuzzy cooperative game the players may choose to partially participate in a coalition. Generalization of this assumption into I-fuzzy sets involves also the possible choice of the level of non-participation in a coalition. The example of such type of a I-fuzzy game can be, for example a voting of political party in a voting body, or shareholder voting at cooperative meeting.

A fuzzy coalition consists of a group of participating players along with their participation level [2, 3]. Analogically, I-fuzzy coalition should consist of participation as well as non-participation level for each player. Formally, let $N = \{1, 2 \dots n\}$ be a set of n players. An I-fuzzy coalition $\tilde{C}$ is given by a pair of vectors $\tilde{C} = \langle \mu^C, \nu^C \rangle$ with coordinates $\mu^C = (\mu_1^C, \mu_2^C, \dots \mu_n^C)$ and $\nu^C = (\nu_1^C, \nu_2^C, \dots \nu_n^C)$ such that $0 \leq \mu_i^C + \nu_i^C \leq 1$ for all $i \in N$. Then the ith coordinate of vector μ^C gives a level of membership (participation) of player i in an I-fuzzy coalition $\tilde{C}$; the ith coordinate of vector ν^C gives a level of nonmembership of player i in an I-fuzzy coalition $\tilde{C}$.

Any crisp coalition $S \subset N$ can be expressed as I-fuzzy coalition $\tilde{C}^S$ such that $\tilde{C}^S = \langle \mu^S, \nu^S \rangle$ for which $\mu_i^S = 1$ for all $i \in S$ and $\mu_i^S = 0$ for all $i \notin S$ while $\nu_i^S = 1$ for all $i \notin S$ and $\nu_i^S = 0$ for all $i \in S$. Analogically, the empty coalition can be expressed as $\tilde{C}^\emptyset = \langle \mu^\emptyset, \nu^\emptyset \rangle$ for which $\mu_i^\emptyset = 0$ and $\nu_i^\emptyset = 1$ for all i, while the grand coalition is of the form $\tilde{C}^N = \langle \mu^N, \nu^N \rangle$ for which $\mu_i^N = 1$ and $\nu^N t_i = 0$ for all i. Let $\tilde{C}^i$ denote the I-fuzzy relation corresponding to the crisp coalition $S = \{i\}$. Let $L(\tilde{U})$ be the class of all I-fuzzy subsets of the set $\tilde{U}$ (this set can be both crisp or I-fuzzy). That means, that any I-fuzzy subset $\tilde{S}$ of a set $\tilde{U}$ satisfies $\mu_i^S \leq \mu_i^U$ and $\nu_i^S \geq \nu_i^U$. Accordingly, $L(N)$ is the class of all I-fuzzy subsets of the set N; this set is infinite, because possible levels of memberships and nonmemberships levels are potentially values from $[0, 1]$.

The characteristic function of an I-fuzzy game should specify the worth of each I-fuzzy coalition. The definition of cooperative game with I-fuzzy coalitions is derived from the definition of cooperative game (Definition 1):

Definition 5 An I-fuzzy cooperative game is a pair (N, ν) where $N = \{1, 2 \dots n\}$ is a set of n players $\nu : L(N) \to R$ is a mapping defined on subsets of N with the property $\nu(\tilde{C}^\emptyset) = 0$ The function ν is called a characteristic function of an I-fuzzy game.

The characteristic function ν connects each I-fuzzy coalition $\tilde{C}$ with a real number $\nu(\tilde{C}) \in R$ representing total profit of coalition $\tilde{C}$; we assume $\nu(\tilde{C}^\emptyset) = 0$. An I-fuzzy cooperative game can by denoted as (N, ν), or simply only by characteristic function ν.

Example 4 Let (N, ν) is an I-fuzzy cooperative game with three players N=$\{1, 2, 3\}$, and a characteristic function

$$\nu(\tilde{C}) = \sum_{1=1}^{3} 2(\mu_i^C - \nu_i^C + 1).$$

Calculate $\nu(\tilde{C}^\emptyset)$, $\nu(\tilde{C}^N)$ and $\nu(\tilde{L})$ where $\tilde{L} = \langle (0.3, 0.5, 1), (0.1, 0.2, 0) \rangle$.

Solution: In the case of three players the empty coalition is expressed as $\tilde{C}^{\emptyset} = \langle(0,0,0),(1,1,1)\rangle$, and the grand coalition is of the form $\tilde{C}^N = \langle(1,1,1),(0,0,0)\rangle$. Then:

$$v(\tilde{C}^{\emptyset}) = 2((0-1+1)+(0-1+1)+(0-1+1)) = 0$$
$$v(\tilde{C}^N) = 2((1-0+1)+(1-0+1)+(1-0+1)) = 12$$
$$v(\tilde{L}) = 2((0.3-0.1+1)+(0.5-0.2+1)+(1-0+1)) = 9$$

The superadditivity and convexity in cooperative games with I-fuzzy coalitions are defined as follows:

Definition 6 An I-fuzzy cooperative game (v, N) is said to be superadditive if

$$v(\tilde{K} \cup \tilde{L}) \geq v(\tilde{K}) + v(\tilde{L}) \text{ for all } \tilde{K}, \tilde{L} \in L(N) \text{ such that } \tilde{K} \cap \tilde{L} = \emptyset. \tag{4}$$

Definition 7 An I-fuzzy cooperative game (v, N) is said to be convex if

$$v(\tilde{K} \cup \tilde{L}) + v(\tilde{K} \cap \tilde{L}) \geq v(\tilde{K}) + v(\tilde{L}) \text{ for all } \tilde{K}, \tilde{L} \in L(N). \tag{5}$$

As in the crisp case, any convex game is superadditive.

Example 5 Let (N, v) is an I-fuzzy cooperative game with two players $N = \{1, 2\}$, and a characteristic function v. Players can create only coalitions $\emptyset = \tilde{L}_{\emptyset} = \langle(0,0),(1,1)\rangle$, $\tilde{L}_1 = \langle(1,0),(0,1)\rangle$, $\tilde{L}_2 = \langle(0,1),(1,0)\rangle$, $\tilde{L}_3 = \langle(1,1),(0,0)\rangle$, $\tilde{L}_4 = \langle(0.3,1),(0.5,0)\rangle$, and $\tilde{L}_5 = \langle(0.3,0),(0.5,1)\rangle$. Respective values of a characteristic function are:

$$v(\tilde{L}_{\emptyset}) = 0 \quad v(\tilde{L}_1) = 2 \quad v(\tilde{L}_2) = 2 \quad v(\tilde{L}_3) = 12 \quad v(\tilde{L}_4) = 4 \quad v(\tilde{L}_5) = 5.$$

Show that this I-fuzzy game is superadditive and convex.
Solution: It is sufficient to show that the game is convex. We will use relations for union and intersection of two I-fuzzy sets [1]:

$$A \cap B = \{\langle x_i, \min(\mu_A(x_i), \mu_B(x_i)), \max(\nu_A(x_i), \nu_B(x_i))\rangle; x_i \in X\};$$

$$A \cup B = \{\langle x_i, \max(\mu_A(x_i), \mu_B(x_i)), \min(\nu_A(x_i), \nu_B(x_i))\rangle; x_i \in X\}.$$

Results of unions and intersections all combinations of two I-fuzzy sets are given Table 1. Left and right hand side of Eq. 5 are given in last two columns of Table 1.

For all combinations relation $v(\tilde{A} \cap \tilde{B}) + v(\tilde{A} \cup \tilde{B}) \geq v(\tilde{A}) + v(\tilde{B})$ holds, therefore this game is convex and superadditive.

Table 1 Results of unions and intersections all combinations of two I-fuzzy sets from Example 5

$\tilde{A}$	$\tilde{B}$	$v(\tilde{A} \cap \tilde{B})$	$v(\tilde{A} \cup \tilde{B})$	$v(\tilde{A} \cap \tilde{B}) + v(\tilde{A} \cup \tilde{B})$	$v(\tilde{A}) + v(\tilde{B})$
$\tilde{L}_\emptyset$	$\tilde{L}_1$	$v(\langle(0,0),(1,1)\rangle = 0$	$v(\langle(1,0),(0,1)\rangle = 2$	2	2
$\tilde{L}_\emptyset$	$\tilde{L}_2$	$v(\langle(0,0),(1,1)\rangle = 0$	$v(\langle(0,1),(1,0)\rangle = 2$	2	2
$\tilde{L}_\emptyset$	$\tilde{L}_3$	$v(\langle(0,0),(1,1)\rangle = 0$	$v(\langle(1,1),(0,0)\rangle = 12$	12	12
$\tilde{L}_\emptyset$	$\tilde{L}_4$	$v(\langle(0,0),(1,1)\rangle = 0$	$v(\langle(0.3,1),(0.5,0)\rangle = 4$	4	4
$\tilde{L}_\emptyset$	$\tilde{L}_5$	$v(\langle(0,0),(1,1)\rangle = 0$	$v(\langle(0.3,0),(0.5,1)\rangle = 5$	5	5
$\tilde{L}_1$	$\tilde{L}_2$	$v(\langle(0,0),(1,1)\rangle = 0$	$v(\langle(1,1),(0,0)\rangle = 12$	12	4
$\tilde{L}_1$	$\tilde{L}_3$	$v(\langle(1,0),(0,1)\rangle = 2$	$v(\langle(1,1),(0,0)\rangle = 12$	14	14
$\tilde{L}_1$	$\tilde{L}_4$	$v(\langle(0.3,0),(0.5,1)\rangle = 5$	$v(\langle(1,1),(0,0)\rangle = 12$	17	6
$\tilde{L}_1$	$\tilde{L}_5$	$v(\langle(0.3,0),(0.5,1)\rangle = 5$	$v(\langle(1,0),(0,1)\rangle = 2$	7	7
$\tilde{L}_2$	$\tilde{L}_3$	$v(\langle(0,1),(1,0)\rangle = 2$	$v(\langle(1,1),(0,0)\rangle = 12$	14	14
$\tilde{L}_2$	$\tilde{L}_4$	$v(\langle(0,1),(1,0)\rangle = 2$	$v(\langle(1,1),(0,0)\rangle = 12$	14	6
$\tilde{L}_2$	$\tilde{L}_5$	$v(\langle(0,0),(1,1)\rangle = 0$	$v(\langle(1,1),(0,0)\rangle = 12$	12	7
$\tilde{L}_3$	$\tilde{L}_4$	$v(\langle(0.3,1),(0.5,0)\rangle = 4$	$v(\langle(1,1),(0,0)\rangle = 12$	16	16
$\tilde{L}_3$	$\tilde{L}_5$	$v(\langle(0.3,0),(0.5,1)\rangle = 5$	$v(\langle(1,1),(0,0)\rangle = 12$	17	17
$\tilde{L}_4$	$\tilde{L}_5$	$v(\langle(0.3,0),(0.5,1)\rangle = 5$	$v(\langle(0.3,1),(0.5,0)\rangle = 4$	9	9

Source: Own calculations

3.1 *Preimputation and Imputation for Cooperative Games with I-Fuzzy Coalitions*

As in the case of crisp cooperative games, the expected distribution of profit among players is represented by a real-valued payoff vector $x = (x_i)_{i \in N} \in R^n$. A preimputation should be an efficient payoff vector—that means a vector representing distribution of maximal possible profit. The definition of preimputation and imputation of cooperative game with I-fuzzy coalitions is derived from the respective definitions for these terms on crisp cooperative games (Definitions 2 and 3):

Definition 8 A preimputation is any x from the preimputation set $\mathcal{I}^*(v)$ of an I-fuzzy cooperative game (N, v) defined as

$$\mathcal{I}^*(v) = \{x \in R^n; \sum_{i \in N} x_i = v(\tilde{C}^N)\}. \tag{6}$$

A payoff vector $x \in R^n$ is individually rational if for each $i \in N$ there is $x_i \geq v(\tilde{C}^i)$. An individually rational preimputation is called an imputation:

Definition 9 An imputation is an element from an imputation set $\mathcal{I}(v)$ of a TU cooperative game $(N, \tilde{v})$ defined as

$$\mathcal{I}(\tilde{v}) = \{x \in R^n; \sum_{i \in N} x_i = \tilde{v}(\tilde{C}^N)), \; x_i \geq v(\tilde{C}^i) \quad \forall i \in N\} \tag{7}$$

Example 6 Let $(N, \tilde{v})$ is the same I-fuzzy cooperative game as in Example 4 with three players $N = \{1, 2, 3\}$, and a characteristic function

$$\tilde{v}(\tilde{C}) = \sum_{1=1}^{3} 2(\mu_i^C - \nu_i^C + 1).$$

Give an example of preimputation and an imputation of such a set.
Solution: Value of characteristic function for the grand coalition is $\tilde{v}(\tilde{C}^N) = 12$, and for the crisp coalition composed of one player is $\tilde{v}(\tilde{C}^i) = 4$ for any $i = 1, 2, 3$.

- The preimputation of this game should fulfill a condition $x_1 + x_2 + x_3 = 12$; there exist infinitive amount of triples $([x_1, x_2, x_3])$ in a preimputation of this game, for example $(4, 4, 4)$, $(1, 1, 10)$ or $(0, 0, 12)$.
- The imputation of this game should fulfill condition for preimputation, as well as additional conditions $x_1 \geq 4$, $x_2 \geq 4$, and $x_3 \geq 4$. Therefore, the imputation consist of the unique point $(4, 4, 4)$.

3.2 *Core for Cooperative Games with I-Fuzzy Coalitions*

The core $C(v)$ of an I-fuzzy game v, similarly as in the fuzzy case [7], consists of imputations stable against any possible deviation—that means imputations under which no coalition has incentive to leave the grand coalition and receive a larger payoff. In an I-fuzzy case both membership and nonmembership function should be taken into account. That means, that all imputations are divided into three groups: imputations, which are in a core, imputations, which are not in a core, and imputations we have no information about their core membership:

Definition 10 Let the game (N, v) is a cooperative game with N players and an I-fuzzy coalitions. Then the essential core $C^+(v)$ of the game (N, v) is defined as:

$$C^+(v) = \{x \in R^n; \sum_{i \in N} x_i = v(\tilde{C}^N), \sum_{i \in \tilde{L}} \mu_i x_i \geq v(\tilde{L}) \quad \forall \tilde{L} \subset L(N)\}. \tag{8}$$

Definition 11 Let the game (N, v) is a cooperative game with N players and an I-fuzzy coalitions. Then the possible core $C^-(v)$ of the game (N, v) is defined as:

$$C^-(v) = \{x \in R^n; \sum_{i \in N} x_i = v(\tilde{C}^N), \sum_{i \in \tilde{L}} (1 - \nu_i) x_i \geq v(\tilde{L}) \quad \forall \tilde{L} \subset L(N)\}. \tag{9}$$

Clearly, the set $C^-(v)$ covers the set $C^{uncert}(v)$ of imputations we have no information about: $C^{uncert}(v) = C^-(v) - C^+(v)$. Imputations from $(\mathcal{I} - C^-(v))$ are not in a core; moreover $C^+(v) \subset C^-(v) \subset \mathcal{I}$. In the case of classical fuzzy sets $C^+(v) = C^-(v)$ and the result corresponds with definition of core as discussed in [7].

Example 7 Let (N, v) be an I-fuzzy cooperative game with two players $N = \{1, 2\}$, and a characteristic function v. Players can create only coalitions $\emptyset$, $\tilde{L}_1 = \langle(1,0),(0,1)\rangle$, $\tilde{L}_2 = \langle(0,1),(1,0)\rangle$, $\tilde{L}_3 = \langle(1,1),(0,0)\rangle$, $\tilde{L}_4 = \langle(0.3,0.8),(0.5,0.2)\rangle$, $\tilde{L}_5 = \langle(0.2,0.8),(0.5,0.2)\rangle$, and $\tilde{L}_6 = \langle(0.1,0.5),(0.8,0.5)\rangle$. Respective values of a characteristic function are:

$$v(\tilde{C}^{\emptyset}) = 0 \;\; v(\tilde{L}_1) = 2 \;\; v(\tilde{L}_2) = 2 \;\; v(\tilde{L}_3) = 12 \;\; v(\tilde{L}_4) = 8.4 \;\; v(\tilde{L}_5) = 8.1 \;\; v(\tilde{L}_6) = 4.5.$$

Write down conditions for $C^-(v)$ and $C^+(v)$.
Solution: Conditions for imputation are based on values of characteristic function for the empty coalition, the grand coalition, and crisp one-player coalitions. In this case $v(\tilde{C}^1) = v(\tilde{L}_1) = 2$, $v(\tilde{C}^2) = v(\tilde{L}_2) = 2$, and $v(\tilde{C}^N) = v(\tilde{L}_3) = 12$. Therefore conditions for imputation $x = (x_1, x_2)$ are:

$$x_1 + x_2 = 12, \;\; x_1 \geq 2, \;\; x_2 \geq 2.$$

Thus, the imputation in this case is the part of a line $x_1 + x_2 = 12$ for $x_1 \in [2, 10]$, and $x_2 \in [2, 10]$. Imputation is depicted in Fig. 2.

Similarly, additional conditions for calculation of the essential core C^+ for this game are based on the relation $\sum_{i \in \tilde{L}} \mu_i x_i \geq v(\tilde{L})$:

- for $\tilde{L}_4$: $0.3x_1 + 0.8x_2 \geq 8.4$,
- for $\tilde{L}_5$: $0.2x_1 + 0.8x_2 \geq 8.1$,
- for $\tilde{L}_6$: $0.1x_1 + 0.5x_2 \geq 4.5$.

Incorporating these conditions into the condition of imputation gives the result that the C^+ core consists of the part of a line $x_1 + x_2 = 12$ for $x_1 \in [2, 2.4]$, and $x_2 \in [9.6, 10]$.

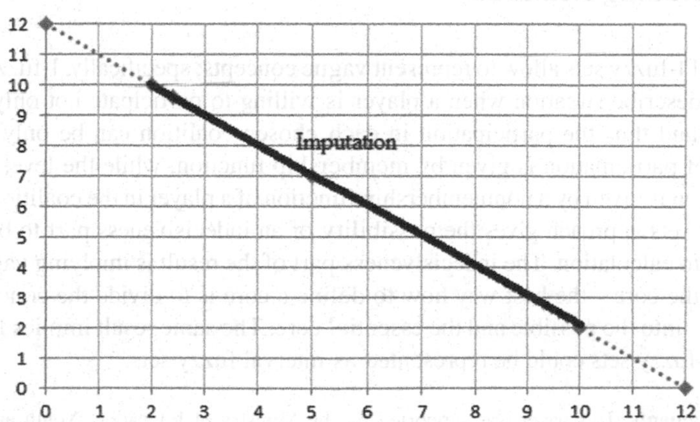

Fig. 2 The imputation of the game from Example 7. *Source: Own calculations*

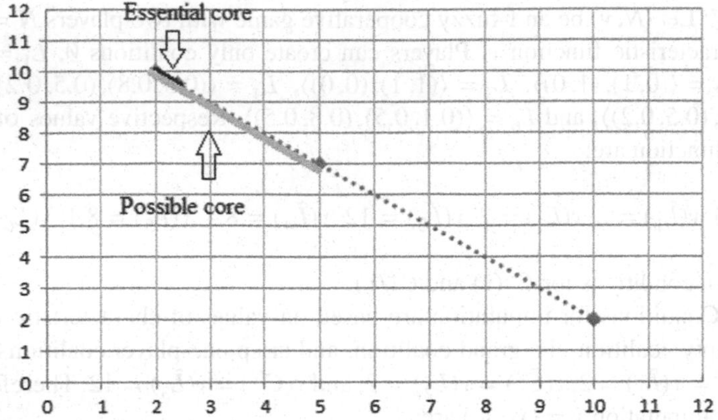

Fig. 3 The core of the game from Example 7. *Source: Own calculations*

Additional conditions for calculation of the possible core C^- for this game are based on the relation $\sum_{i\in\tilde{L}}(1-v_i)x_i \geq v(\tilde{L})$:

- for $\tilde{L}_4$: $0.5x_1 + 0.8x_2 \geq 8.4$,
- for $\tilde{L}_5$: $0.5x_1 + 0.8x_2 \geq 8.1$,
- for $\tilde{L}_6$: $0.2x_1 + 0.5x_2 \geq 4.5$.

Incorporating these conditions into the condition of imputation gives the result that the C^- core consists of the part of a line $x_1 + x_2 = 12$ for $x_1 \in [2, 5]$, and $x_2 \in [7, 10]$. Both sets C^+, and C^- are illustrated in Fig. 3.

4 Concluding Remarks

Concept of I-fuzzy sets allow to represent vague concepts; specifically, I-fuzzy coalitions can describe situation when a player is willing to participate not only in one coalition, and thus the participation in each chosen coalition can be only partial; the level of participation is given by membership function, while the level of nonparticipation is given by a nonmembership function of a player in the coalition. Thus, an I-fuzzy sets approach gives the possibility of an indecisiveness part to be incorporated into calculation. The indecisiveness part of the result is implying vague definition of the core—the best way how to define a core is to divide the concept into two parts—into the possible and the essential core. The same result implies from the fact that I-fuzzy sets could be represented as interval fuzzy sets.

Acknowledgments This paper was supported by the Ministry of Education, Youth and Sports within the Institutional Support for Long-term Development of a Research Organization in 2016.

References

1. Atanassov, K.T.: Intuitionistic fuzzy sets. Fuzzy Sets Syst. **20**(1), 87–96 (1986)
2. Aubin, J.P.: Cooperative fuzzy games. Math. Oper. Res. **6**(1), 1–13 (1981)
3. Azrieli, Y., Lehrer, E.: On some families of cooperative fuzzy games. Int. J. Game Theor. **36**(1), 1–15 (2007)
4. Dubois, D., Gottwald, S., Hajek, P., Kacprzyk, J., Prade, H.: Terminological difficulties in fuzzy set theory—the case of "intuitionistic fuzzy sets". Fuzzy Sets Syst. **156**(3), 485–491 (2005)
5. Li, D.-F.: Decision and Game Theory in Management with Intuitionistic Fuzzy Sets. Studies in Fuzziness and Soft computing. Springer, Heidelberg (2014)
6. Shoham, Y., Leyton-Brown, K.: Multiagent Systems: Algorithmic, Game-Theoretic, and Logical Foundations. Cambridge University Press, Cambridge (2009)
7. Tijs, S., Branzei, R., Ishihara, S., Muto, S.: On cores and stable sets for fuzzy games. Fuzzy Sets Syst **146**(2), 285–296 (2004)
8. Wooldridge, M.J.: An Introduction to Multiagent Systems. J. Wiley & Sons, New York (2002)
9. Zadeh, L.A.: Fuzzy sets. Inf. Control **8**(3), 338–353 (1965)

Formalizing Data to Agent Model Mapping Using MOF: Application to a Model of Residential Mobility in Marrakesh

Ahmed Laatabi, Nicolas Marilleau, Tri Nguyen-Huu, Hassan Hbid and Mohamed Ait Babram

Abstract Modeling and simulating the world with agent-based models is a one of the key disciplines that emerge today in the computing area, with the development of power calculation machines and the availability of huge amount of data. Many methodologies have been established to guide the elaboration of different models, but few ones have focused on linking data to model. In this paper, we give a formalized mapping between data and multi-agent components (DAMap: Data to Agent Mapping), as a first step in the process of standardizing the development of a simulation model from raw data. Then we apply it to an household decision-making process in the city of Marrakesh.

Keywords Multi-agent models · Meta-modeling · Mapping · Data analysis · Social simulation · Residential mobility · MOF · UML

1 Introduction

In recent years, multi-agent modeling has become one of the most popular techniques in modeling complex systems, especially in social sciences [1] where they help to build artificial societies and understand the interactions, dynamics and emergent phenomena at spatial and temporal scales. But developing a multi-agent model is not always an easy task because of the complexity introduced by the studied phenomenon; the heterogeneity and the number of types of agents and the data structure could lead to a very complex model. Emphasizing data analysis and processing may help to simplify the model development and conception. But according to our knowledge, existent methodologies like Agent Unified Modeling Language [2] and Gaia

A. Laatabi (✉) · H. Hbid · M.A. Babram
Cadi Ayyad University, Marrakesh, Morocco
e-mail: laatabi44@gmail.com

N. Marilleau · T. Nguyen-Huu · H. Hbid · M.A. Babram
UMI 209 UMMISCO IRD, Bondy, France

G. Jezic et al. (eds.), *Agent and Multi-Agent Systems: Technology and Applications*, Smart Innovation, Systems and Technologies 58,
DOI 10.1007/978-3-319-39883-9_9

Methodology [3] do not focus on how an agent model can be built based on available data, as well as they do not show where and how collected data is localized and used in the model.

In this paper, we extend description and designing methods like the ODD protocol (Overview, Design Concepts, and Details) to drive the process of structuring data and mapping this structure to the multi-agent model components. Such a methodology aims at addressing questions raised in [4] about the reliability of the model, its distance from the reality, its validity regarding the data and the corresponding case study. It will also help to understand how the model is linked to reality, and how it can be developed based on empirical data. Thus, we propose the first version of a new formalized graphic method, as a part of a general methodology based on ODD, which will be completed with other formal tools in future works, in order to build a developer guide until the last part of model structure and code generation. As an application of the method, we propose an example of a case study in social sciences, namely residential mobility and household decision-making, based on survey data and multi-agent systems.

The paper is organized as follows: in the next section, we discuss the building of models based on both theory and data, and discuss the gap existing between data and model. In the following section, we present the Meta Object Facility and the necessity of meta-modeling and describing models for development purposes. Next, we propose a mapping method and a meta-model formalism. Then, we apply these concepts on a simple model of residential mobility implemented on a simulation platform. Finally, we conclude with perspectives and future work.

2 Theory, Data and Model

One difficulty of developing and using models is the calibration and validation of the microscopic sub-models [5]. The modeler has to keep in mind the real system, and try to link it with his model that will be calibrated and validated using data. Many models can be developed for the same phenomenon based on the theory. These models may vary in term of complexity, as well as in their abstraction level where they can go from particular to stylized one [6], and only collected data can specify which one is better to use, even if it is difficult to decide empirically when the data vary considerably [7]. Empirical data, stemming essentially from field survey and GIS databases, represents the real world and must then play a fundamental role in the elaboration of the model to facilitate the comprehension, validation and interpretation of the results. The developed model based just on theory could be far from the real system, because modelers tend to ignore the context that the model is built for. In [8] it is mentioned that when using theoretical knowledge, it is important to remember that the model needs to fit in the application context.

The real world represented by empirical data can be explained by different models implied by different theories, and the question is to investigate their relative relevance [9]. Authors in [10] say that models can be compared to maps that cannot represent the whole aspects of the study area being mapped, but only those seen relevant by the geographer. The modeler has to select which features to inject and represent in his model, and this has to be done during the development itself by including data structure in the model building process. Existing methodologies especially in the multi-agent paradigm are not based on empirical data when designing and developing the model, which loses its connection with the data source [11].

The process of developing a model based on both data and theory as shown in Fig. 1 seems to be more efficient and reliable, even if it could give rise to another difficulty in the case of the absence of information about data, or due to the lack of suitable methods for adapting models to empirical data [5]. The need of a methodology that formalizes the relation between theory, model and empirical data appears to be important, as it will make possible the switch between theories or datasets while using the same model structure. When the modeler knows how data is used in a model, he could specify what part of the model has to be modified whenever data has changed, or he can transform and structure data to make it valid for the model, without affecting theories and results. Following such methodology could make the developed model more accurate to data, relevant to theory and easy to understand and replicate or adapt to other data or context. In the next section, we emphasize the benefits of formalizing the data to model linking and present our meta-modeling technique.

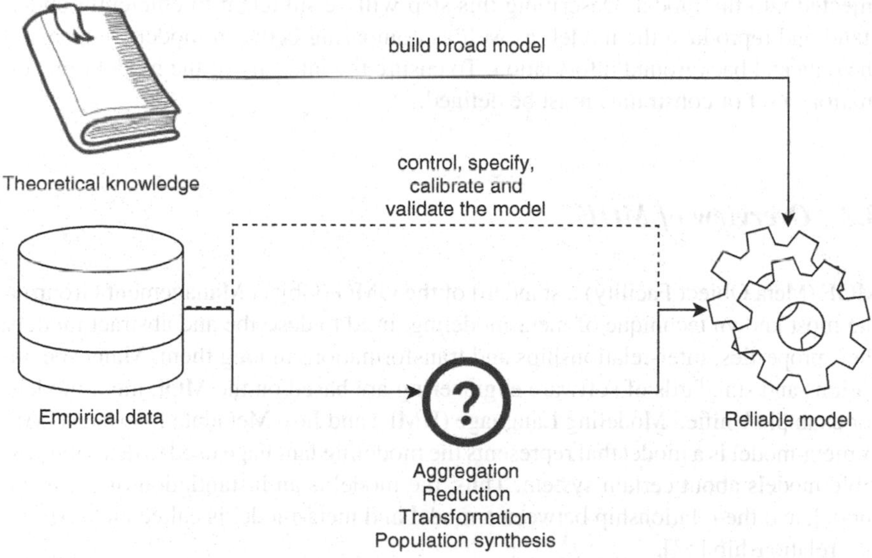

Fig. 1 Building a model from empirical data and theoretical knowledge

3 Formalizing Data to Model Mapping with MOF

3.1 Purposes of Linking Data to Agent Model

How data was transformed to build the entities of the agent model? What are the transformations that were made over data? How the state variables are structured? Generally, raw data cannot be put directly in the model; a set of transformations should be done before processing to data modeling and implementation. To understand these operations and bridge the gap between data and model, a simple formalism is needed to describe the steps followed to accomplish the migration. We note that we are not integrating ontologies or creating a new formal language, but just formalizing and describing the ambiguous link between data and model.

Model descriptions allow users to achieve different purposes of their models. Müller et al. [12] have listed and described these purposes such as in-depth comprehension, development and model replication. The same authors argue that following a model description can improve model development, especially when the model description is elaborated in parallel to the model design. However, developed methods and descriptions have not included the pre-implementation stage, consisting of data preprocessing and extraction of relations, to allow building a data-driven model based on empirical data. These protocols do not tell how sub-models and dynamics were extracted (if it is the case) from data, and how data records are loaded and injected into the model. Describing this step will be sufficient to efficiently understand and reproduce the model, as well as comparing between models in terms of the data and background information. To ensure the integrity of the model transformation, a set of constraints must be defined.

3.2 Overview of MOF

MOF (Meta Object Facility) a standard of the OMG (Object Management Group) is the most known technique of meta-modeling, used to describe and abstract models, their properties, inter-relationships and transformations among them. Many specifications and standards of software engineering are based on the MOF meta-models, for example Unified Modeling Language (UML) and Java Metadata Interface (JMI). A meta-model is a model that represents the modeling language used to describe possible models about certain system. Thus, the model is an instantiation of the meta-model, and the relationship between a model and meta-model is called an "*instance of*" relationship [13].

MOF is an architecture composed by four layers. The top layer M3 provides a meta-meta-model, which is a language and a set of rules that are used to build meta-models of the below layer M2. These M2-layer meta-models describe components of the M1-layer containing M1-models that are used to design elements of the bottom

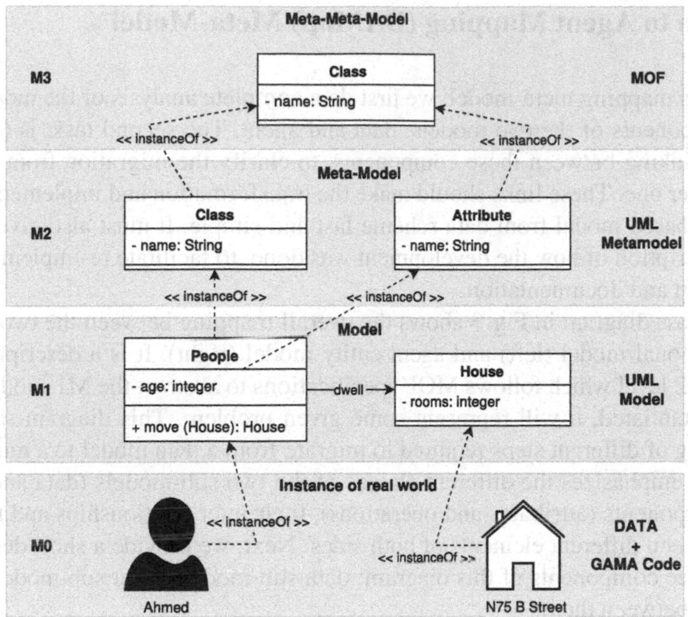

Fig. 2 Illustration of the MOF layered architecture

layer M0 or data layer, where the real world is modeled and described. Thus, MOF is an auto-recursive meta-modeling technique, as every level can be described by the up-level and so on (MOF is described by the MOF itself). The schema in Fig. 2 illustrates the layered architecture of MOF in our case, where the real world is stated by households and dwellings, which are represented by data records and GAMA code.

MOF is a subset of UML that uses class diagrams to serve many purposes, for example to define the migration from a model to another, as in [14] where it was introduced along with the rules language OCL (Object Constraint Language), to link a temporal Entity-Relationship model with a temporal multidimensional model. Gogolla and Lindow [15] used MOF meta-modeling to create a formal connection between conceptual and implementation data models. Using MOF to design models and OCL to express constraints about them in terms of preconditions and postconditions [16], is the perfect way to build a precise protocol allowing to make the link between data and model. We propose here an MOF diagram to formalize the mapping between the relation model (database classes in UML) and the agent model, and as in other works, we use the OCL language to apply some constraints that will verify and guarantee the correct functioning of the system.

4 Data to Agent Mapping (DAMap) Meta-Model

To build a mapping meta-model, we first do a complete analysis of the most important components of the two models: data and agent. The second task, is to find an optimal linking between these components, to clarify the migration from one side to the other one. These links should make the transformation and implementation of an agent-based model from data scheme fast and simple. It must also give a structural description of how the development was done, to facilitate re-implementation, replication and documentation.

The class diagram in Fig. 3 shows the overall mapping between the two models: data relational model (left) and agent entity model (right). It is a description situated in M2 level which follows MOF specifications to abstract the M1-model. Thus, when instantiated, it will represent some given problem. This diagram shows the processing of different steps required to migrate from a data model to a multi-agent model. It emphasizes the different classes of the two sub-models (data and agent), their components (attributes and operations), their inter relationships and the mapping between different elements of both sides. Next, we provide a short description of the three components of this diagram: data sub-model, agent sub-model and the mapping between them.

Data Sub-Model: *RelationEntity* can represent a household/dwelling which may belong to a set of households/district that we model by *Package*. Every *RelationEntity* can be described by many attributes *Attribute* (dwelling: type, #rooms, ...) and many operations *Operation* (household: grow, move, ...). Each class may be linked to other classes by some relation *AssociationLink* (household H lives in dwelling D). An *Operation* manipulates and modifies attributes that can be subject of a *Transformation* (converting categorical variables from strings to numbers) or a discovering pattern (correlation between household size and dwelling surface), which is modeled here by *DataPattern*.

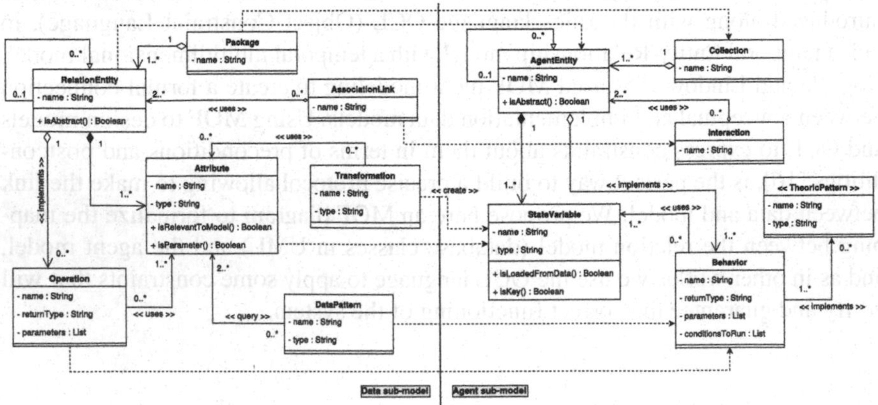

Fig. 3 M2-Model: DAMap meta-model

Agent Sub-Model: The *AgentEntity* class models the basic element of the multi-agent system, namely the agent. These agents can represent households and dwellings that interact through *Interaction* and coalesce in homogeneous groups *Collection*. Agents are described by attributes *StateVariable*, and may have actions *Behavior* that could be extracted from theory represented by *TheoricPattern*. Through these behaviors executed by the agents and interactions between them, global structures emerge from individual actions, for example the form of a city is a result of decision moves taken by households.

Data to MAS linking: Dashed lines show the mapping between the two sub-models, where we can see for example that a *Behavior* comes from *Operation* or *DataPattern* in the data side. In the same way, an *Attribute* of the *RelationEntity* is mapped to a *StateVariable* directly or after a *Transformation* that creates a new state variable, which does not exist as an attribute in the original data. These mapping links resume the migration from data to an agent-based model.

OCL Constraints: We present here a non exhaustive list of some constraints that we can apply on the meta-model, to assure its integrity and implementation without errors. We note that we prefer OCL language to other techniques such as modal logic, since it is already integrated in the UML standard.

```
-- two attributes of the same class cannot have the same name
context RelationEntity inv: self.attribute -> forAll (
          a1, a2 | a1 <> a2 implies a1.name <> a2.name)
-- each operation must act at least on one attribute
context Operation inv: self.attribute -> size() >= 1
```

5 Case Study: Residential Mobility in Marrakesh

5.1 Residential Mobility and DAMap Model

In this paper, we use the previous methodology to a case study of the process of residential mobility and household decision-making. We made a survey to collect data about the choice and preferences of households in Marrakesh. Collected data provide information on both households (number of persons, income, etc.) and dwellings (size, location, cost, nearby facilities, etc.). Our purpose is to build a discrete-time model that represents the residential mobility dynamics at the city scale. Decision rules will be derived from collected data: at each time step, those rules will apply to determine which households will decide to move, and which dwellings will be their settling location. After some descriptive analysis of collected data, we elaborated a relational class model of level M1, representing different classes of our conceptual model. As our final aim was to simulate residential mobility in Marrakesh, we had to define the structure of the multi-agent system that will be a skeleton of the running simulation. The choice of multi-agent paradigm is due to its possibility of modeling decisions at the individual microscopic level, which is what we need to represent the interactions between households and housing units in the context of the city. A first mapping between the two models (data and MAS) is needed to understand this transition, and how to go from a data survey to a multi-agent model. We use the previously presented meta-model (M2-layer) to build the mapping M1-model that shows the complete process, explaining how agents and their state variables are constructed from components of class diagrams (classes and attributes).

The model that we provide here, is a basic mapping showing different relationships and transformations between the elements of the data model, namely classes, and the components of agent model (agents). This mapping diagram is depicted in Fig. 4. Ordinal relationships are drawn in a solid line, while mapping links are represented by dashed lines. For example, class *Household* in the data part (left side), is an agent *Household* (right side) with five attributes mapped from data. One of these attributes (*transports*) was built from two attributes (*trans_pub*, *trans_taxi*) with a simple transformation of summation (*sum*), to show the number of people using public transports in each household. Another example is the agent *Choice* that contains an attribute *work*, built as a *mean* of five attributes from the data class *Choice*, to express a score of the perception of workers about the distance from the house to their workplaces. All other components can be interpreted with the same way. Attributes of the agent part are chosen by the modeler as required by the final user, or by outputs of the simulation. For example, in our case we chose only the attributes that are most

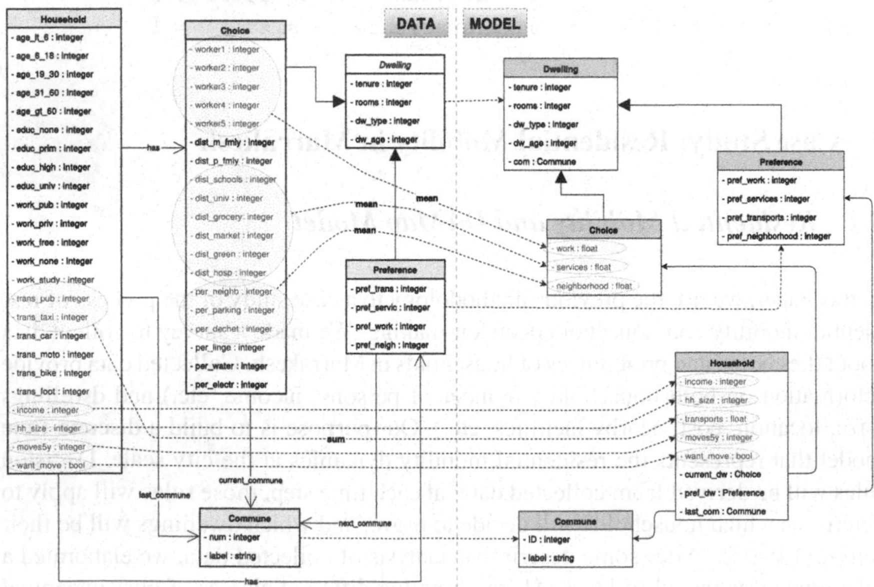

Fig. 4 M1-Model: DAMap model

important and relevant to our questions, meaning those that we want to link to the answers and to the results of the simulation. These attributes should not be numerous in order to keep the model simple and easy to understand.

5.2 *Model Implementation and Simulation*

To implement the model above, we use GAMA platform, a free and open-source development environment for modeling and building spatially agent-based simulations [17]. Using GAMA special language (GAML), we code the five entities from the M1-model (Household, Dwelling, Choice, Preference, Commune) with their state variables extracted from data, and behaviors that we learn from our knowledge and literature about residential mobility, but also from analyzing and processing data to detect patterns and relationships between variables. Figure 5 shows an example of programming code, expressing the modeled operations by the meta-model above. We can see for example the structure of the agents and the different mechanisms of the model such as declaration and inheritance. Transformations between data and mapping operations are executed in the initialization of the simulation (init procedure), where data is loaded to the platform directly from a CSV file. Outputs of the simulation are defined by the modeler as required by the study and its goals, but learning the

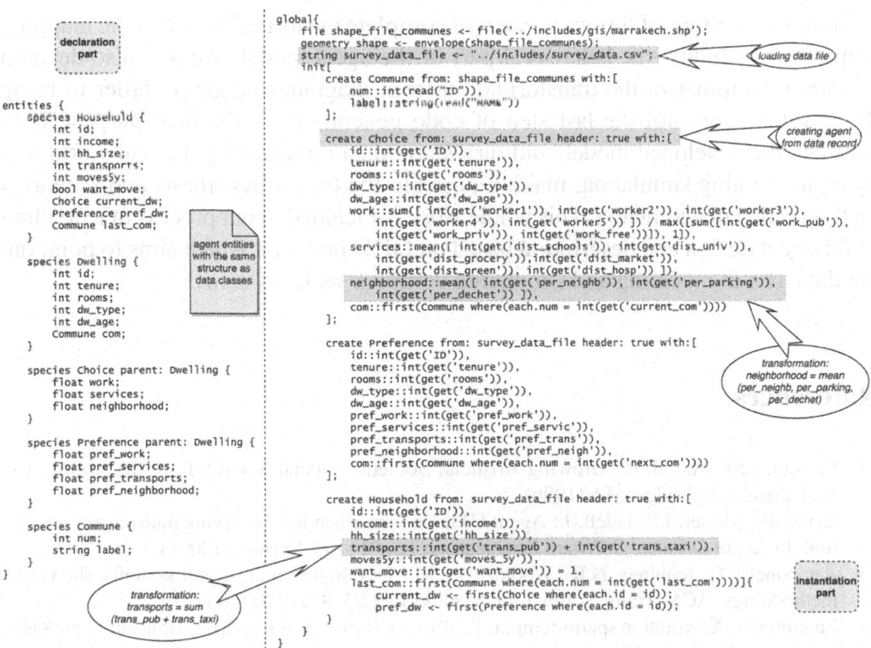

Fig. 5 Overview of the model classes under GAMA

inputs and transformations they undergo would make defining these output indicators more simple, relevant and easy to interpret, as the main purpose of modeling starts after finishing the implementation, by analyzing results to address the questions and problems that the model was formulated for [18]. As a result of this implemented model, we can study and analyze the relationship of residential mobility with the different socio-economic and demographic characteristics of households, as well as dwelling and environmental factors.

6 Conclusion and Perspectives

In this work, we treated the problem of developing models without basing on empirical data. This process is considered a big handicap for modeling paradigm, and can sometimes give useless and ambiguous models that are difficult to replicate and link to the real system. To overcome this lack of consistency, we proposed a formalized approach to connect a survey data model to an agent-based model, which would largely help understanding and documenting the model itself, but also the link between the two sub-models (data and agent). To implement the meta-models, we used Meta Object Facility and Unified Modeling Language as tools of making diagrams. We noticed that following our method guided rapidly to a simple and clean model, where we used to waste more time to develop complex and ambiguous models.

In future versions of this work, we will complete the formalization of the mapping method by including the behavioral part of the agent model. We will also develop a script to automatize the transformation from diagram to code, in order to bring the development until the last step of code generation, as the final purpose is to simulate the developed model, and turn static data representing the real world to a dynamic running simulation, making possible the test of hypothesis and scenarios on the studied phenomenon. This whole work is included in our process of extending multi-agent design and description based on ODD protocol, which aims to point out the data axis in modeling and simulating complex social systems.

References

1. Epstein, J.M., Axtell, R.: Growing Artificial Societies: Social Science from the Bottom Up. MIT Press, Cambridge, MA (1996)
2. Bauer, B., Müller, J.P., Odell, J.: Agent UML: a formalism for specifying multi-agent interaction. In Agent-Oriented Software Engineering, pp. 91–103. Springer (2001)
3. Zambonelli, F., Jennings, N.R., Wooldridge, M.: Developing multi-agent systems: the GAIA methodology. ACM Trans. Softw. Eng. Methodol. **12/3**, 317–370 (2003)
4. Lammoglia, A.: volution spatio-temporelle d'une desserte de transport flexible simul en SMA. In: Cybergeo: Revue Europenne de gographie/European Journal of geography, UMR 8504 Gographie-cits, Document, p. 555 (2011)

5. Brockfeld, E., Kuhne, R.D., Wagner, P.: Calibration and validation of microscopic traffic flow models. Trans. Res. Rec. **1876**, 62–70 (2004)
6. Banos, A.: Pour des pratiques de modlisation et de simulation libres en Gographie et SHS (Doctoral dissertation, Universit Paris 1 Panthon Sorbonne) (2013)
7. Helbing, D., Balietti, S.: Agent-based modeling. In: Social Self-Organization. pp. 25–70 Springer, Berlin (2012)
8. Siebers, P. O., Aickelin, U.: Introduction to multi-agent simulation. Comput. Res. Repository. abs/0803.3905 (2008)
9. Livet, P., Müller, J.-P., Phan, D., Sanders, L.: Ontology, a mediator for agent-based modeling in social science. J. Artifi. Soc. Soc. Simul. **13**(1) (2010)
10. Giere, R.: Using Models to Represent Reality. Model-based Reasoning in Scientific Discovery, Kluwer, Dordrecht (1999)
11. Bykovsky, V.K.: Data-driven modeling of complex systems. In: Unifying Themes in Complex Systems, pp. 34–41. Springer, Berlin (2008)
12. Müller, B., Balbi, S., Buchmann, C.M., de Sousa, L., Dressler, G., Groeneveld, J., Klassert, C.J., Le, Q.B., Millington, J.D.A., Nolzen, H., Parker, D.C., Polhill, J.G., Schlüter, M., Schulze, J., Schwarz, N., Sun, Z., Taillandier, P., Weise, H.: Standardised and transparent model descriptions for agent-based models: current status and prospects. Envrion. Model. Softw. **55**, 156–163 (2014)
13. Overbeek, J.: Meta Object Facility (MOF)—investigation of the state of the art, Master's thesis, University of Twente (2006)
14. Pons, C., Neil, C.G.: Formalizing the model transformation using metamodeling techniques. In: Proceedings of Argentine Symposium on Software Engineering, Jornadas Argentinas de Informtica e Investigacin Operativa (2004)
15. Gogolla, M., Lindow, A.: Transforming data models with UML. In: Omelayenko, B., Klein, M. (eds.) Knowledge Transformation for the Semantic Web, pp. 18–33. IOS Press, Amsterdam, The Netherlands (2003)
16. Loecher, S., Ocke, S.: A metamodel-based OCL-compiler for UML and MOF. Electron. Notes Theor. Comput. Sci. **102**, 43–61 (2004)
17. GAMA Platform website. http://gama-platform.org
18. Railsback, S.F., Volker, G.: Agent-based and individual-based modeling: a practical introduction. Princeton University Press (2011)

A Communication and Tracking Ontology for Mobile Systems in the Event of a Large Scale Disaster

Mohd Khairul Azmi Hassan and Yun-Heh Chen-Burger

Abstract Communication and tracking capabilities during and immediately after a large-scale natural disaster are one of the most important components of speedy response and recovery. In that, it discovers affected people and connects them with their families, friends, and communities with first responders and/or their support computational systems. Capabilities of current mobile technologies can be expanded to become effective large-scale disaster tool aid. To facilitate effective communication and coordination across different parties and domains, ontologies are becoming crucial in providing assistance during natural disasters, especially where affected locations are remote, affected population is large and centralized coordination is poor. Although there are several existing competing methodologies with regard to as how an ontology may be built, there is not a single right way to build an ontology. Furthermore, there is not a (de facto standard) Disaster Relief Ontology, although separated related ontologies may be combined to create an initial version. This article discusses our on-going development of an ontology for a Communication and Tracking System (CTS), based on existing related ontologies, that is aimed to be used by mobile applications to support disaster relief at the real-time. For future work, this ontology will be used to provide a multi-disciplinary knowledge foundation in a distributed multi-agent based environment, where mobile devices, rescue workers and their organizations are modelled and functioned as distributed and collaborative agents to support each other in the event of a large-scale natural disaster.

Keywords Ontology · Mobile application · Semantic web · Linked data · Communication · Natural disaster · Earthquake · Disaster relief · Intelligent systems · Decision support systems

M.K.A. Hassan (✉) · Y.-H. Chen-Burger
Department of Computer Science, Heriot-Watt University, Edinburgh, UK
e-mail: mh42@hw.ac.uk

Y.-H. Chen-Burger
e-mail: y.j.chenburger@hw.ac.uk

G. Jezic et al. (eds.), *Agent and Multi-Agent Systems: Technology and Applications*, Smart Innovation, Systems and Technologies 58,
DOI 10.1007/978-3-319-39883-9_10

1 Introduction

In recent years, earthquake disasters have caused terrible losses, fatalities and missing people. Sichuan, China earthquake in 2008 killed at least 69,000 people, injured more than 374,000 [1], and left about 4.8 million homeless [2]. While, Haiti earthquake in January 2010, causing over 200,000 fatalities, 300,000 injuries and leaving over 1 million people homeless [3]. Another big impact earthquake is Pacific Ocean earthquake and subsequent tsunami in Japan in March 2011 that cost the Japanese economy more than $300 billion and caused unprecedented loss to the Japanese people, their environment, and the global economy [4].

A few of application system has been developed in a web and mobile application to fulfill tracking missing people, disaster management and emergency response. SEA-EAT blog, Nepal Earthquake Missing People on Facebook, Nepal Earthquake Missing People Website, PeopleFinder and others are amongst of well-known application developed for this purpose especially in very large-scale disaster. Some of the application are using a semantic web and linked data as a solution to integrate the information between multiple of agencies for top management to make a decision and also for reporting purposes but there are still gaps to be filled especially in a communication and tracking people during and after earthquake using an ontology in a mobile application.

Mobile phone nowadays is one of the top multi-functional devices which capable to be a broad range of application and able to serve people in communication. This is proved by the numbers growth from 4.3 billion users in 2015 and expected to become 5.07 billion users in 2019 [5]. We hope, with a linked data and ontology implemented in mobile application, smart device can assist agencies and community to track people who need a help or trapped in the prone area, rescuer can easily find victims nearby their current location and the important thing is survivors can easily find the nearest shelter or agencies who can help them.

This article will discuss on the ontology development in linked data and semantic web technologies [6] that can be used especially in mobile phone application to make people connecting with others during the large-scale disaster. We aimed to improve the existing current and stable ontologies to fulfill the communication and tracking people using a mobile phone as an instrument. Therefore, step by step on the development of Communication and Tracking Ontology (CTO) will be discussed and the remainder of this paper is structured as follows. In Sect. 2 will provide a few of related work and motivation in ontology development, Sect. 3 describes the process of how the CTO was developed. The merging and building CTO in Sect. 4, while Sect. 5 gives an overall conclusion and future work.

2 Motivation and Related Work

The main motivation for developing a CTS is to provide effective communication and assist coordination among victims, communities, rescuers and organizations who are involved directly or indirectly during and just after a large-scale disaster, thereby providing speedier recovery and relief to the victims as possible. With a mobile phone supporting internet connection on hand, people/organization may use their smart devices to know about how to help people especially nearby their location without knowing where the exact location, address, who's their family to contact and also they don't know even victims background information such as their name or blood type. The stakeholders (government agencies) of this domain find it increasingly difficult to coordinate and respond to emergency situations. The result of this has increased the number of deaths, delay in access to basic needs and slower recovery time [7]. There are many types of the disasters exist in this world such as epidemic, eruption, fire, flood, forest fire, hailstorm, heat wave, hurricane, ice storm, lahars, landslides, limnic eruption, maelstrom, mudslide, sinkholes, storm, thunderstorm, tornado, tsunami, typhoon, volcano, wildfire and terrorist bombing [8–13]. Among them, an earthquake is one of the most poorly-understood disasters that may happen with no warning sign and nobody will know for sure, including geologist scientist concerning where and when it may happen [14]. The CTO is very important to enable the systems, find and collect the information from many resources on the net such as the name of victims, their places, their relative and their current coordinate. Therefore, the objectives to build up the CTO are:

1. To reduce fatalities and missing people in case of an earthquake.
2. To help personnel to monitor a real-time information about missing people during an earthquake.

From Table 1, the Earthquake Reports from United States Geological Survey (USGS) [15] show to the world that the big magnitude (magnitude > 6.0) impact earthquake happened yearly since 1990. The year 2010 history in Haiti indicated a large number of fatalities with 316,000 people and many of them might not be traced (missing) until today.

Quite early on, a number of researches have shown that the emergency response domain can especially benefit from Semantic Web technology. Ontology has been used and some of the applications have been developed to find missing people and such as PeopleFinder that has been developed from the experience of Katrina hurricane [7]. Another research regarding the use of ontology of a web base application is a blog (SEA-EAT) that has been set up during the Asian tsunami in 2014. It was developed into an information exchange system for missing persons, requests for help and news updates [7]. For example, it had used an ontology as a knowledge base for sharing and extracting information. They have developed an ontology for situation awareness in crisis management to provide a contextual understanding of the post-disaster environment for different users.

Table 1 Number of earthquakes and fatalities from year 2000–2014 by US Geological Survey (USGS)

Largest earthquakes					Deadliest earthquakes			
Year	Date	Magnitude	Fatalities	Region	Date	Magnitude	Fatalities	Region
1990	16-Jul	7.7	1,621	Luzon, Philippine Islands	20-Jun	7.4	50,000	Iran
1991	22-Apr	7.6	75	Costa Rica	19-Oct	6.8	2,000	Northern India
1991	22-Dec	7.6	–	Kuril Islands				
1992	12-Dec	7.8	2,519	Flores Region, Indonesia	12-Dec	7.8	2,519	Flores Region, Indonesia
1993	08-Aug	7.8	–	South of Mariana Islands	29-Sep	6.2	9,748	India
1994	04-Oct	8.3	11	Kuril Islands	06-Jun	6.8	795	Colombia
1995	30-Jul	8	3	Near Coast of Northern Chile	16-Jan	6.9	5,530	Kobe, Japan
1995	09-Oct	8	49	Near Coast of Jalisco Mexico				
1996	17-Feb	8.2	166	Irian Jaya Region Indonesia	03-Feb	6.6	322	Yunnan, China
1997	14-Oct	7.8	–	South of Fiji Islands	10-May	7.3	1,572	Northern Iran
1997	05-Dec	7.8	–	Near East Coast of Kamchatka				
1998	25-Mar	8.1	–	Balleny Islands Region	30-May	6.6	4,000	Afghanistan-Tajikistan Border Region
1999	20-Sep	7.7	2,297	Taiwan	17-Aug	7.6	17,118	Turkey
2000	16-Nov	8	2	New Ireland Region, P.N.G.	04-Jun	7.9	103	Southern Sumatera, Indonesia
2001	23-Jun	8.4	138	Near Coast of Peru	26-Jan	7.7	20,023	India
2002	03-Nov	7.9	–	Central Alaska	25-Mar	6.1	1,000	Hindu Kush Region, Afghanistan
2003	25-Sep	8.3	–	Hokkaido, Japan Region	26-Dec	6.6	31,000	Southeastern Iran
2004	26-Dec	9.1	227,898	Off West Coast of Northern Sumatra	26-Dec	9.1	227,898	Off West Coast of Northern Sumatra
2005	28-Mar	8.6	1,313	Northern Sumatra, Indonesia	08-Oct	7.6	80,361	Pakistan
2006	15-Nov	8.3	–	Kuril Islands	26-May	6.3	5,749	Java, Indonesia
2007	12-Sep	8.5	25	Southern Sumatera, Indonesia	15-Aug	8	514	Near the Coast of Central Peru
2008	12-May	7.9	87,587	Eastern Sichuan, China	12-May	7.9	87,587	Eastern Sichuan, China

(continued)

Table 1 (continued)

Largest earthquakes					Deadliest earthquakes			
Year	Date	Magnitude	Fatalities	Region	Date	Magnitude	Fatalities	Region
2009	29-Sep	8.1	192	Samoa Islands region	30-Sep	7.5	1,117	Southern Sumatra, Indonesia
2010	27-Feb	8.8	507	Offshore Maule, Chile	12-Jan	7	316,000	Haiti
2011	11-Mar	9	20,896	Near the East Coast of Honshu, Japan	11-Mar	9	20,896	Near the East Coast of Honshu, Japan
2012	11-Apr	8.6	–	Off the west coast of northern Sumatra	06-Feb	6.7	113	Negros-Cebu region, Philippines
2013	24-May	8.3	–	Sea of Okhotsk	24-Sep	7.7	825	61 km NNE of Awaran, Pakistan
2014	01-Apr	8.2	6	NW of Iquique, Chile	03-Aug	6.2	729	Near Wenping, China

Efforts to apply linked data and ontology in the field of emergency response are presented in recent researches on various facets. In the context of Weather Ontology, it was discussed in detail about AEMET, the Spanish Public Weather Service in [16] where they are using the ontology to make meteorological data publicly available via their website, as registered by its weather stations, radars, lightning detectors and ozone soundings. They also discussed the reusing of Time Ontology and Location Ontology to make it more suitable to cater for the Weather Ontology itself.

Many reports showed that a lot of people, who needed to be evacuated, had problems finding the nearest evacuation centers that the government and companies had set up for them, thereby receiving necessary assistant in a timely fashion [17]. Therefore, the way of providing information about evacuation centers for those people is an important issue in the future and research in [18]. In this article, they firstly design an Earthquake Evacuation Ontology and secondly, indicate that can computers provide the most suitable evacuation center, by using the ontology based on earthquake victims' behaviors in real-time.

3 Developing a Communication and Tracking Ontology to Assist Disaster Relief Efforts

Tom Gruber defined an ontology as the following "An ontology is an explicit specification of a conceptualization." [19]. It is important to understand what ontology is for. The ontology is to enable knowledge sharing and data consistency and as such an ontology is a specification for making ontological commitments.

There is no one "correct" way or methodology for developing ontologies and we have followed the seven top-level steps from [20] (see Fig. 1) where the main component will be described from 3.1 to 3.7 below.

3.1 Determine Domain and Scope

Naturally, the concepts describing a communication and tracking during a large-scale disaster, type of disaster, weather, places, shelter, who to contact and etc. will figure into CTO. At the same time, it is unlikely that the ontology will include concepts for victims, rescuer, community and families to trace and connecting each other. The ontology will be designed focused to make people's traceable during the large-scale disaster, able to safe and secure after the disaster such as how community or agencies can help peoples in their surrounding (walking distance) and also victims can find the nearest shelter from their current location to get food, water or blanket supplied.

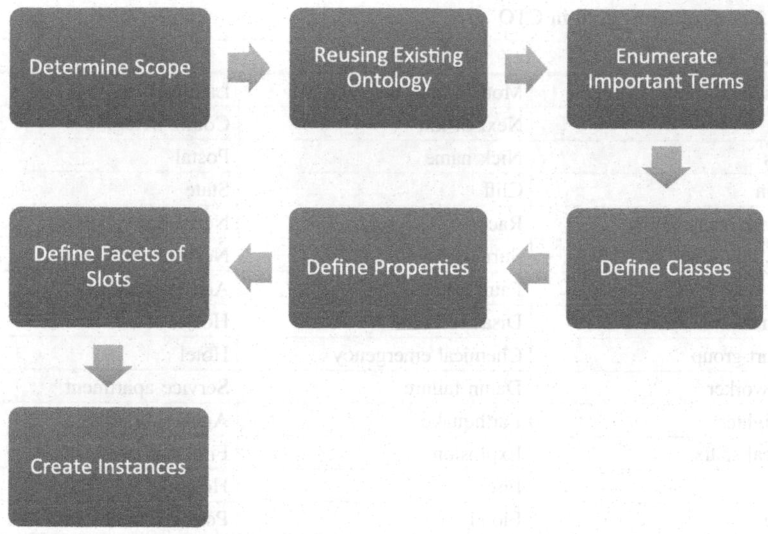

Fig. 1 Seven steps to develop ontology [20]

3.2 *Reusing, Merging and Tailoring Existing Ontologies*

Many ontologies are already available in electronic form and can be imported into an ontology development environment. This method called reuse process where this process is a cost effective way to build ontologies. The CTO was built by means of reuse, following an evolving prototyping life cycle. From 10 main ontologies component in CTO, several of them were reused from existing stable and maintained ontologies such as Friend of a Friend (FOAF) [21–25], weather [25–27], Disaster [25, 28], Time [16, 29], Places [16, 23] and Location [16] ontology. Component from these ontologies has been selected and focused to the communication and tracking area.

3.3 *Enumerate Important Terms*

It is useful to write down a list of all terms that would like either to make statements about or to explain. The example of listed terms are in Table 2 below:

Table 2 List of terms used in CTO

Terms		
Location	Mothers name	Latitude
Agent	Next of kin	Country/Region
Places	Nick name	Postal
Terrain	Cliff	State
Network connectivity	Race	Network available
Time	Surname	Network down
Event	Fault	Accommodation
Weather	Disaster	Hostel
Support group	Chemical emergency	Hotel
Help worker	Damn failure	Service apartment
Fire fighter	Earthquake	Agency building
Medical staffs	Explosion	Fire station
NGO	Fire	Hospital
Police	Flood	Police station
Red Cross	Flash flood	School
Person	Riverine flood	Commercial resources
Date of birth	Landslide	Shopping mall
Blood type	Nuclear power	Shop
Contact details	Radiation	Houses
Peninsula	Tsunami	Apartment
First name	Location	Flat
Gender	City	Landed house
Island	Coordinate	Other building
Religion places	Longitude	Weather
Church	Shelter	Humidity
Mosque	Available capacity	Pressure
Temple	Total capacity	Temperature
Severity status	Support group	Wind
Red	Agencies	Hill
Green	Family	Mount
Yellow	Friends	Valley
Time	Others	
Date	Colleague	
Duration	Neighbors	
Start	Other individual	
End	Military	

Sources from [8, 9, 21, 23, 25, 26, 29, 30, 18, 31, 32, 33]

3.4 Define the Class Hierarchy

There are several possible approaches in developing a class hierarchy as mention in [34] but in this cases, a combination of the top-down and bottom-up approached in development process were used. Figure 2 below shows the classes and hierarchy for CTO.

3.5 Define Properties

For each property in the list, it will determine which classes will be described. The classes alone will not provide enough information to answer the competency

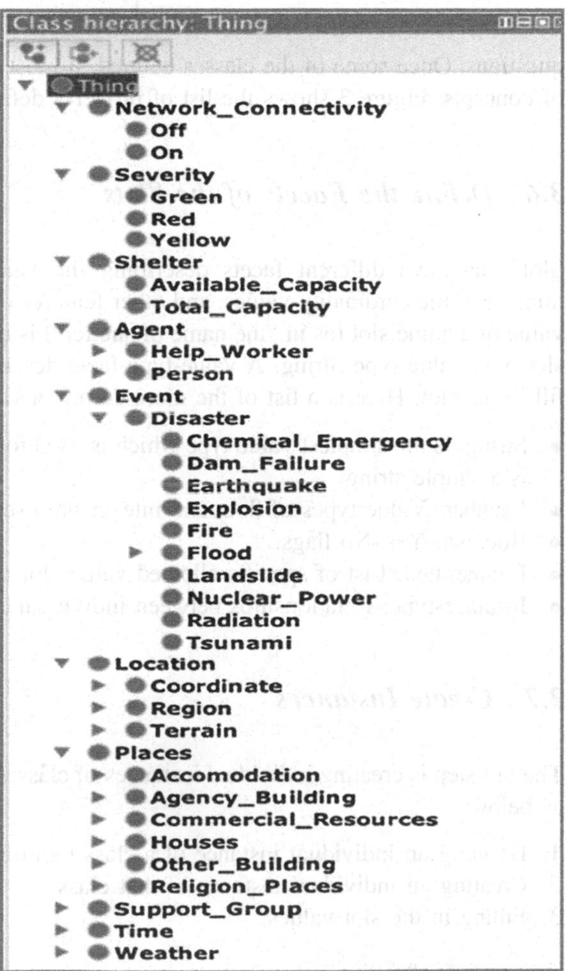

Fig. 2 Classes and hierarchy for CTO

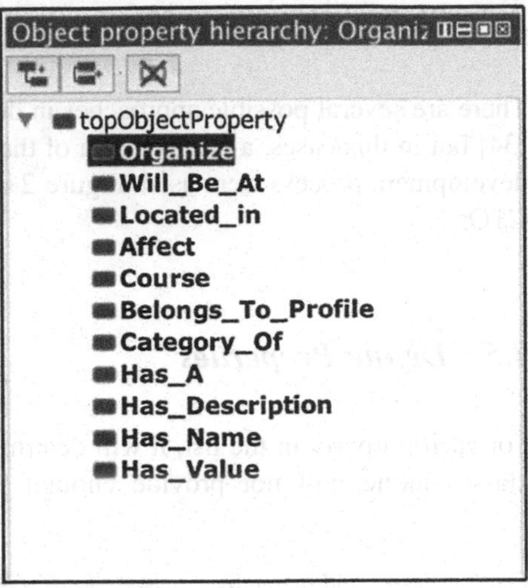

Fig. 3 List of property defined in CTO

questions. Once some of the classes defined, it must describe the internal structure of concepts. Figure 3 shows the list of property defined.

3.6 Define the Facets of the Slots

Slots can have different facets describing the value type, allowed values, the number of the cardinality values, and other features of the values. For example, the value of a name slot (as in "the name of shelter") is one string. That is, a name is a slot with value type String. A value-type facet describes what types of values can fill in the slot. Here is a list of the more common value types in this article:

- String: Is the simplest value type which is used for slots such as name: the value is a simple string
- Number: Value types of float and integer are used.
- Boolean: Yes–No flags.
- Enumerated: List of specific allowed values for the slot.
- Instance-type: Relationships between individuals.

3.7 Create Instances

The last step is creating individual instances of classes in the hierarchy for this CTO as below:

1. Defining an individual instance of a class required
2. Creating an individual instance of that class
3. Filling in the slot values.

The CTO example can create an individual instance Edinburgh-Mosque to represent a specific type of Places. The value of 1,000 is an instance of the class Total-Capacity representing Shelters. This instance has the following slot values defined.

- Total Capacity: 1,000
- Places: Edinburgh Mosque
- Location: City of Edinburgh
- Coordinate: Latitude: 55.9449995, Longitude: −3.1860282
- Event: Earthquake

4 Merging and Building the CTO

Ontologies have become core components of many large applications. Previous research shows a few ontologies for disaster management, emergency response and others have been done. For examples, AEMET Weather, Disaster Management, Management of a Crisis (MOAC), FOAF, etc. was developed and used in web application system for reporting purposes.

This article used the existing of ontologies mentioned above by other researches to addresses the issue on the CTS. To make this CTO suite with any application especially in a mobile application, a research to combine, reuse and create a new ontology must be carried out before testing can be done.

Figure 4 gives an overview of the CTO developed by using seven top-level steps as mention in Sect. 3 above and it has been developed using Protégé v5.0.0. CTO posses 10 main classes in total to fulfill the communication and tracking issues during a large-scale disaster and emergency response. The main classes and their function are described below:

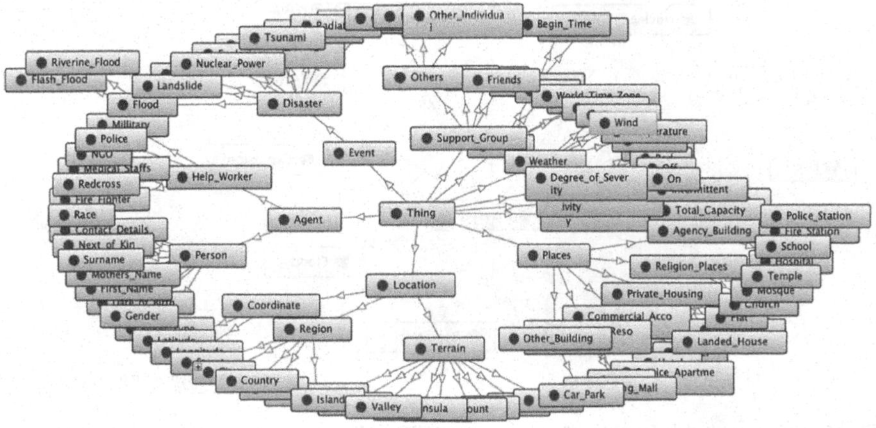

Fig. 4 Details of communication and tracking ontology

1. Events

Event classes are one of the main classes in CTO. It is grouped all type of disaster which base from the research review in [7, 9, 11, 35] and etc. All type of disaster in this subclasses may occur during the earthquake event. From this classes, the information of the type of disaster can be traced from the system. The subclasses for the Disaster Event listed as Earthquake, Radiation, Fire, Tsunami, Chemical Emergency, Landslide, Nuclear power, Flood, Explosion and Damn Failure. Event class is shown in Fig. 5 below.

2. Agents

One of the important classes in CTO is Agent Classes which contain another 2 subclasses called Person and Help Worker. This classes are aims to spread the information about personal information and also the organization who can give a support for the community who maybe in a trap, injury and lost. The personal information can give a better idea to community or help worker to determine and get the background information during and before rescue process. The information about a Person is like Date of Birth, First Name, Surname, Mothers Name, Gender, Race, Contact Details, Blood Type and Next of Kin information. The other subclasses in Agent are Help Worker where the information supplied in this class are about the organization information for rescuers such as Police, Fire Fighter, Red Cross, NGO, Military and Medical Staffs. Refer to Agent Class in Fig. 6 below.

3. Places

Places are a group of locations that consist of formal building information which may help rescuer to know the last places of large-scale disaster victims and

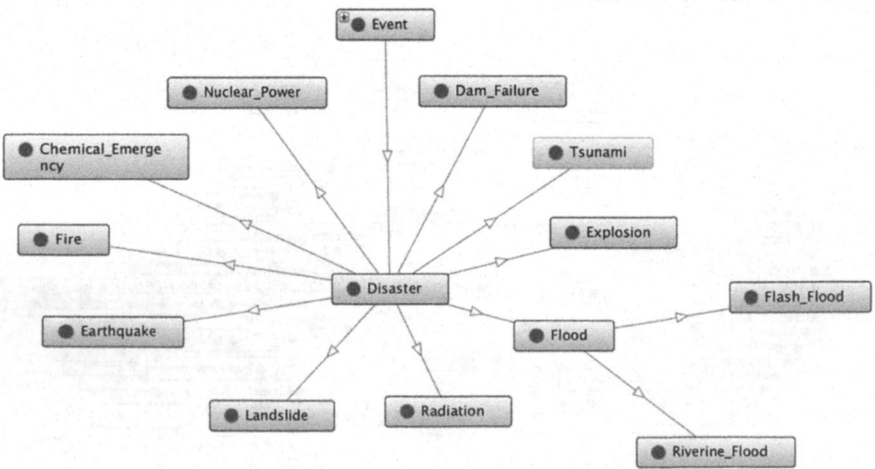

Fig. 5 Event and the branch of disaster in CTO (*directional* links indicate the sub-class relationships)

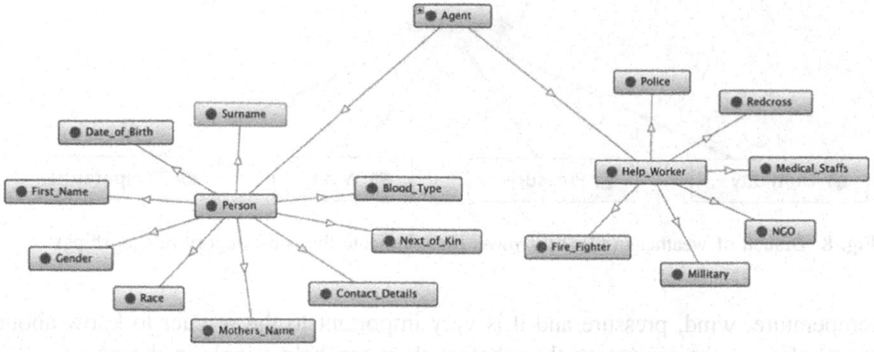

Fig. 6 The branch of person and help worker in agent class (*directional* links indicate the sub-class relationships)

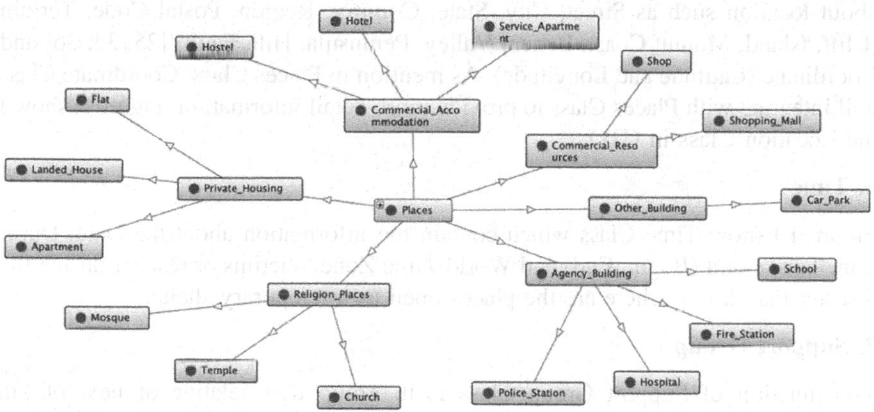

Fig. 7 Branch of places in CTO (*arrowed* links indicate the sub-category relationships)

survivors. A part of this class will integrate with other classes in CTO such as Coordinate in Location Class and Shelter Class which will be described later in point no 5 and no 8 below. Places class contains 6 subclasses in layer 2 and 15 subclasses in layer 3. In layer 2 in Places class have subclasses such as Private Housing, Commercial Accommodation, Commercial Resources, Agency Building, Religion Places and Other Building. Other than that, layer 3 in this class contains a Temple, Church, Mosque, Police Station, Hospital, Fire Station, School, Shopping Mall, Shop building, Hostel, Service Apartment, Hotel, Flat, Landed House and Apartment. It grouped under the layer 2 classes as shown in Fig. 7.

4. Weather

Figure 8 shown the class of Weather where this class will group all the information about weather related to the large-scale disaster. This class will help people such as just after the large-scale disaster happen. It holds an information about humidity,

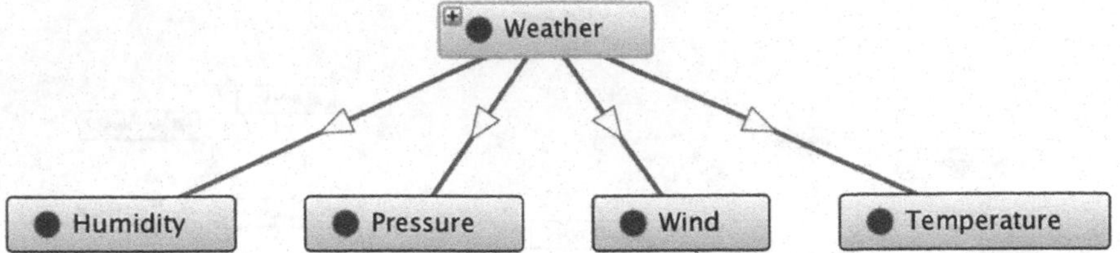

Fig. 8 Branch of weather in CTO (*arrowed* links indicate the sub-category relationships)

temperature, wind, pressure and it is very important to the rescuer to know about their safety or the impact on them before they can help people in their area.

5. Location

Location is a simple class where it was assigned to grouped all the information about location such as Street, City, State, Country, Region, Postal Code, Terrain (Cliff, Island, Mount, Coast, Bench, Valley, Peninsula, Hill, Fault) [25, 33, 36] and Coordinate (Latitude and Longitude). As mention in Places Class, Coordinate Class will integrate with Places Class to provide more detail information. Figure 9 shown the Location Class in CTO.

6. Time

Figure 10 show Time Class which contain the information about the Date, Duration, Time Point (Begin, End) and World Time Zone. Victims or rescuer during the disaster may know where are the places open as a temporary shelter.

7. Support Group

The function of Support Group Class is to group user relative or next of kin information and agencies who can support or help victims during or after the large-scale disaster. This class contains information such as Friends, Agencies,

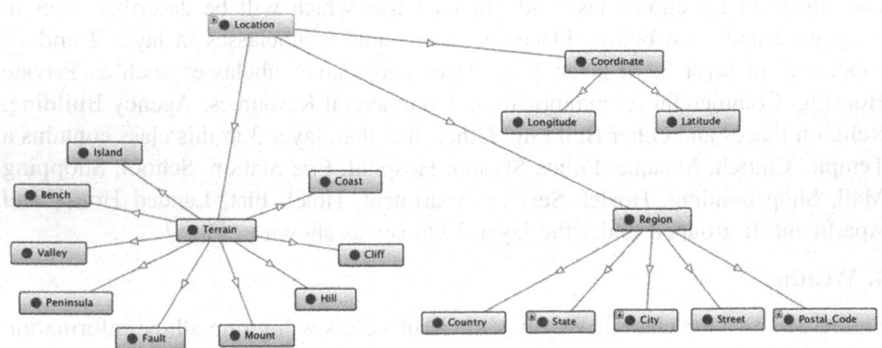

Fig. 9 Branch of location in CTO (*arrowed* links indicate the description relationships)

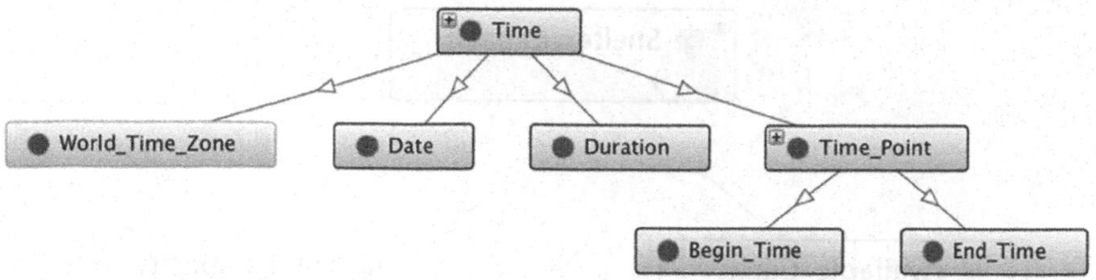

Fig. 10 Branch of time in CTO (*arrowed* links indicate the description relationships)

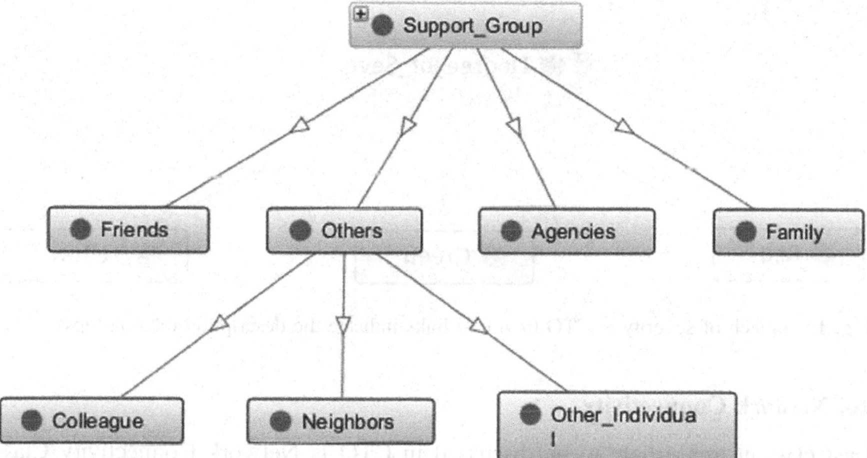

Fig. 11 Branch of support group in CTO (*arrowed* links indicate the sub-category relationships)

Family or Others (Colleague, Neighbors, Other Individual). Figure 11 show the Support Group Class.

8. Shelter Capacity

The Shelter Capacity class will hold the information about the total capacity and current availability shelter to make rescuer or victims will have a better choice to choose which shelter they want to take covered. Figure 12 shown the Class mentioned show the Shelter Class.

9. Degree of Severity

Figure 13 shown the Degree of Severity Class where it will contain color code where it may green, yellow or red depend on the severity of the large-scale disaster. This is important to know what are the action should be taken to follow the International Disaster Relief Organization Standard of Procedure (SOP).

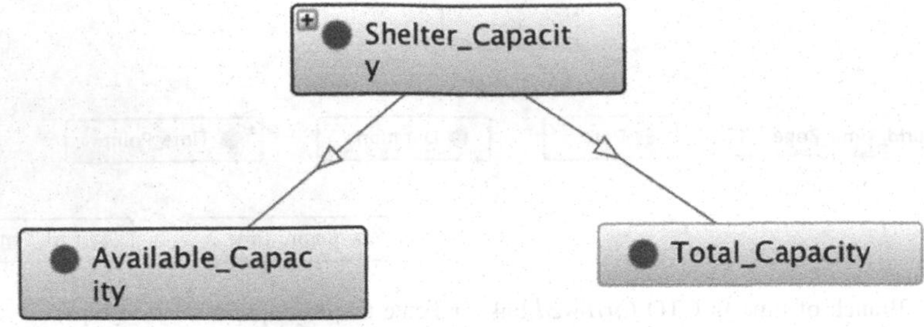

Fig. 12 Branch of shelter in CTO (*arrowed* links indicate description relationships)

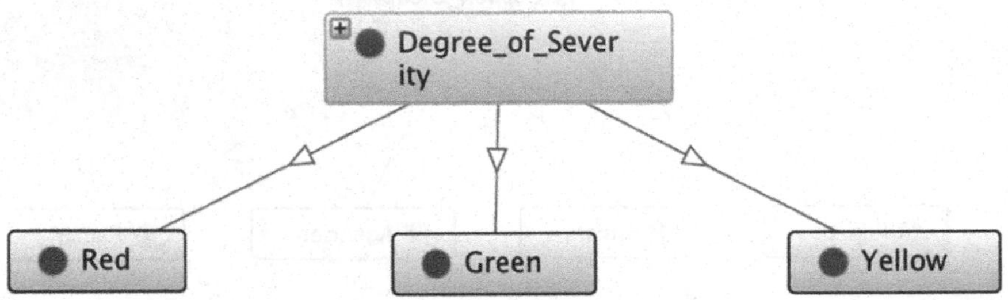

Fig. 13 Branch of severity in CTO (*arrowed* links indicate the description relationships)

10. Network Connectivity

Last class in this article to be discussed in CTO is Network Connectivity Class where the function of this class is to identify either the mobile phone network down, intermittent or shut off because of low battery. Figure 14 show the Network Connectivity Class.

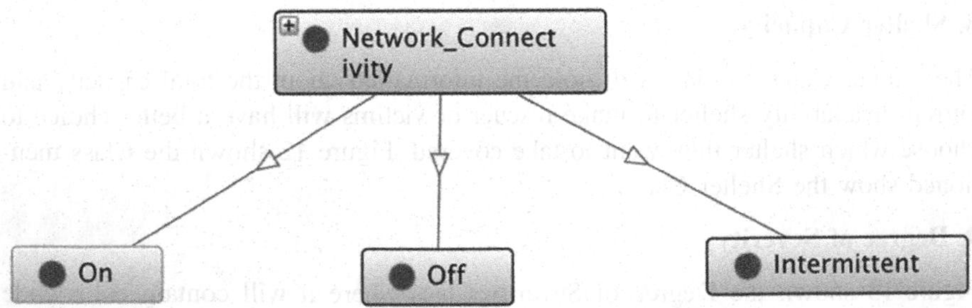

Fig. 14 Branch of network connectivity in CTO (*arrowed* links indicate the description relationships)

5 Conclusion and Future Work

In a large-scale disaster, it is inevitable that many different types of stakeholders (victims, friends and families, disaster relief workers, other support workers and their organisations) are involved. It is common that different types of tasks will need to be carried out between the different types of personnel. It is paramount that all tasks must be well-coordinated and carried out in a timely fashion in order to save lives.

In order to support more effective and timely communication in the event of a large-scale disaster, we have therefore proposed to use ontology as a fundamental technology to pull the information from different relevant domains and store them consistently in one place, thereby promoting information sharing between personnel and interoperability between systems involved. The aim is to facilitate the right types of information to be shared at the right time and in the right format. When successful, the end result of this is faster access to aid, reduce confusion and better coordination during disasters and the ability to save a greater number of lives. In this article, we describe the development of a Communication and Tracking Ontology (CTO) using the seven top-level steps as proposed in [20] to support communication and tracking activities during and after a disaster. The CTO is consisted of ten high-level classes to describe the ten different disaster-relief inter-related sub-domains.

For future research, a distributed multi-agent based Communication and Tracking System (CTS) will be built using an open source framework for mobile devices, i.e. IONIC (Ionic: Advanced HTML5 Hybrid Mobile App Framework). Based on the above ontology, simulations and trials of CTO and CTS involving real users will be carried out based on real-world emergency response scenarios to test the robustness and effectiveness of our proposed CTO and CTS. Separately, we are engaging the emergency response community to get their response on the above ontology and to listen to their recommendations when designing the CTS [37]. In order to ensure that the CTO will fulfill the requirement of CTS, verification and validation with experts will be conducted. The Delphi evaluation technique [38] will use as a method because it is a proven method for gathering data from respondents within their domain of expertise. This technique is also suitable for evaluating group communication which aims to achieve a convergence of opinion on specific real-world issues that suits our requirements well.

References

1. McGill School of Computer Science: http://www.cs.mcgill.ca/~rwest/link-suggestion/wpcd_2008-09_augmented/wp/2/2008_Sichuan_earthquake.htm
2. British Broadcasting Corporation (BBC): http://news.bbc.co.uk/1/hi/world/asia-pacific/7405103.stm
3. Yates, D., Paquette, S.: Emergency knowledge management and social media technologies: a case study of the 2010 haitian earthquake. Int. J. Inf. Manag. **31**(1), 6–13 (2011)

4. About News: http://useconomy.about.com/od/criticalssues/a/Japan-Earthquake.htm
5. The Statistics Portal: http://www.statista.com/statistics/274774/forecast-of-mobile-phone-users-worldwide/
6. Bizer, C., Heath, T., Berners-Lee, T.: Linked data—the story so far. Int. J. Semant. Web Inf. Syst. **5**(3), 1–22 (2009)
7. Ratnam, K.R., Karunaratne, D.D.: Application of Ontologies in Disaster Management (2008)
8. Malizia, A., Onorati, T., Diaz, P., Aedo, I.: SEMA4A: an ontology for emergency notification systems accessibility. Expert Syst. Appl. **37**(4), 3380–3391 (2010)
9. Yan, L.: A survey on communication networks in emergency warning systems. Sci. Comput. **100**(314) (2011)
10. Sakaki, T., Okazaki, M., Matsuo, Y.: Earthquake shakes twitter users: real-time event detection by social sensors. In: Proceedings of the 19th International Conference World wide web, WWW'10, p. 851 (2010)
11. Chou, C.H., Zahedi, F.M., Zhao, H.: Ontology for developing web sites for natural disaster management: methodology and implementation, IEEE Trans. Syst. Man Cybern. Part A Syst. Hum. **41**(1), 50–62 (2011)
12. Associated Press & NORCb Center: Communication During Disaster Response and Recovery, pp. 1–5 (2013)
13. Sharmeen, Z., Martinez-Enriquez, A.M., Aslam, M., Syed, A.Z., Waheed, T.: Multi Agent System Based Interface for Natural Disaster. Lecture Notes in Computer Science, vol. 8610, pp. 299–310 (2014)
14. Planet Science: http://www.planet-science.com/categories/over-11s/natural-world/2011/03/can-we-predict-earthquakes.aspx
15. United States Geological Survey (USGS): http://earthquake.usgs.gov/earthquakes/eqarchives/year/byyear.php
16. Atemezing, G., Corcho, O., Garijo, D., Mora, J., Poveda, M., Rozas, P., Vila-Suero, D., Villazón-Terrazas, B.: Transforming meteorological data into linked data. Semant. Web **1**, 1–5 (2013)
17. Survey Research Center: http://www.surece.co.jp/src/press/backnumber/pdf/press22.pdf
18. Iwanaga, I.S.M., Nguyen, T.M., Kawamura, T., Nakagawa, H., Tahara, Y., Ohsuga, A.: Building an earthquake evacuation ontology from twitter. In: Proceedings of the 2011 IEEE International Conference Granular Computing GrC 2011, pp. 306–311 (2011)
19. Gruber, T.R.: A translation approach to portable ontology specifications. Knowl. Acquis. **5**(2), 199–220 (1993)
20. Noy, N.F., McGuinness, D.L.: Ontology development 101: a guide to creating your first ontology. Stanford Knowledge Systems Laboratory Technical Report KSL-01-05 Stanford Knowledge Systems Laboratory Technical Report SMI-2001-0880, vol. 15, no. 2, pp. 1–25 (2001)
21. Brickley, D., Miller, L.: FOAF Vocabulary Specification, vol. 3 (2010)
22. Hsu, I.-C., Lin, H.-Y., Yang, L.J., Huang, D.-C.: Using linked data for intelligent information retrieval. In: Soft Computing and Intelligent Systems (SCIS), pp. 2172–2177 (2012)
23. Liu, S., Brewster, C., Shaw, D.: A Semantic Framework for Enhancing Information Interoperability in Emergency and Disater Management, pp. 1–20 (2013)
24. Bosch, T., Cyganiak, R., Gregory, A., Wackerow, J.: DDI-RDF discovery vocabulary: a metadata vocabulary for documenting research and survey data. In: Proceedings of the www2013 Work. Linked Data Web (2013)
25. Liu, S., Brewster, C., Shaw, D.: Ontologies for crisis management: a review of state of the art in ontology design and usability. In: ISCRAM, pp. 1–10 (2013)
26. Moran, K., Claypool, K.: Building the NNEW Weather Ontology (2010)
27. Claypool, K., Moran, K.: Ontologies : weather and flight information. In: Integrated Communications, Navigation and Surveillance Conference (ICNS), pp-34 (2012)
28. Limbu, M.: Integration of Crowdsourced Information with Traditional Crisis and Disaster Management Information Using Linked Data (2012)

29. Peralta, D.N., Pinto, H.S., Mamede, N.J., Camp, O., Filipe, J.B.L., Hammoudi, S., Piattini, M.: Reusing a time ontology. Enterp. Inf. Syst. V, pp. 241–248 (2004)
30. Sotoodeh, M.: Ontology-based semantic interoperability in emergency management candidate. Decis. Support Syst. 1–30 (2007)
31. Becker, C., Bizer, C.: DBpedia mobile: a location-enabled linked data browser. In: CEUR Workshop Proceedings, vol. 369 (2008)
32. Xu, Y., Chen, X., Ma, L.: LBS based disaster and emergency management. In: 2010 18th International Conference on Geoinformatics (2010)
33. Lin, Y., Sakamoto, N.: Ontology driven modeling for the knowledge of genetic susceptibility to disease. Kobe J. Med. Sci. **55**(6), 290–303 (2009)
34. Uschold, M., Gruninger, M.: Ontologies: principles, methods and applications. Knowl. Eng. Rev. **11**(2), 93–136 (1996)
35. Babitski, G., Bergweiler, S., Grebner, O., Oberle, D., Paulheim, H., Probst, F.: SoKNOS—using semantic technologies in disaster management software. In: 8th Extended Semantic Web Conference, (ESWC 2011), pp. 183–197 (2011)
36. Smart, P.R., Russell, A., Shadbolt, N.R.: AKTiveSA: supporting civil-military information integration in military operations other than war. In: 2007 International Conference on Integration of Knowledge Intensive Multi-agent Systems KIMAS 2007, pp. 434–439 (2007)
37. Galton, A., Worboys, M.: An ontology of information for emergency management. In: Proceedings of the 8th International ISCRAM Conference, pp. 1–10, (2011)
38. Baker, J., Lovell, K., Harris, N.: How expert are the experts? An exploration of the concept of “expert” within delphi panel techniques. Nurse Res. **14**(1), 59–70 (2006)

Towards an Interaction Protocols Adaptation and Management System for Coordination in Crisis Business Processes

Wassim Chtourou and Lotfi Bouzguenda

Abstract This work addresses the Interaction Protocols (IP) adaptation issue for coordination in the context of Crisis Management Processes (CMP). CMP refers to the *coordination* of different activities performed by involved actors in a dynamic and unstable context. Using interaction protocols (such as contract net, negotiation or vote) is one possible way to deal with this coordination. In order to guarantee an efficient use of IP, we need to adapt them. In our previous work, we proposed an MDA (Model Driven Architecture) framework to support the IP adaptation. More precisely, we proposed the conjointly use of version and context notions to model the IP versions and their contexts (the CIM level). We also proposed an extension of AUML sequence diagram meta-model to specify graphically the contextualized versioned interaction protocols (the PIM level). In this paper, we propose an Interaction Protocols Adaptation and Management System (IPAMS) that supports the previous contributions. This IPAMS ensures the IP adaptation at run time. Finally, we propose a well-known crisis scenario called "Air Crash Management Process" in order to illustrate the proposed solution.

Keywords Interaction protocols · Adaptation · Versions · Context · MDA approach · System architecture

W. Chtourou (✉) · L. Bouzguenda
ISIMS/MIRACL Laboratory, University of Sfax, Route de Tunis, km 10, BP 242, 3021 SakeitEzzit, Sfax, Tunisia
e-mail: chtourouwassim@gmail.com

L. Bouzguenda
e-mail: lotfi.bouzguenda@isimsf.rnu.tn

G. Jezic et al. (eds.), *Agent and Multi-Agent Systems: Technology and Applications*, Smart Innovation, Systems and Technologies 58,
DOI 10.1007/978-3-319-39883-9_11

1 Introduction

Today, faced with the proliferation of crises in all fields such as political crises, natural, socio-economic, the Crisis Management Processes (CMP for short) take considerable importance through the communities engaged in response and management of crises.

The CMP refer to the coordination of several tasks running in different organizations (governmental organizations, humanitarian organizations, hospitals, civil protection, etc.) and in an open, dynamic and unstable context [1]. A fundamental problem for CMP is the coordination between agents which is an important challenge to guarantee an efficient execution of activities.

Using interaction protocols (namely contact nets, negotiation or vote) is one possible way to deal with coordination to rule and structure the communication between partners [2]. In effect, the interaction protocols (IP) based coordination is widely recognized as an efficient mechanism to share resources and coordinate the activities of agents [1]. Several works have been proposed in the literature for instance in [2–4]. These research works consider protocols as entities of first class and address the engineering issue such as specification, validation and implementation of protocols for specifying and developing a Multi-Agent System (MAS) in stable context.

This paper considers also IP as first class entities to deal with coordination of BP in the dynamic context and within an engineering perspective. Thus, the interaction protocols adaptation is needed in order to support the coherent interaction between organizations involved in open, dynamic and unstable context. In the literature, IP adaptation can be studied in the context of the two following distinctive approaches. The first one concerns the management of problems (called exceptions) which can occur under the execution of protocols while the second approach aims at the re-use and the modification out (i.e., at build time) and in progress of execution of the IP modeled. This approach is based on meta-modeling aspect. In this paper, we focus on the IP adaptation according to the second approach. One possible way to deal with this adaptation is the use of versioning technique [5] which captures all the predicable changes of the considered interaction protocol. Versioning technique is an interesting solution to deal with interaction protocols adaptation at build time. Indeed, this technique permits to keep trace of the previous versions of an entity, which supports the re-use of these versions if the same situation arises. Also, it allows the definition for the same entity several versions which can be used in an alternative way. Moreover, the use of the context notion [5] is also interesting in order to describe the conditions use of such IP version.

In our previous work, we have proposed a Model Driven Architecture (MDA) framework for IP adaptation based on version and context notions [6]. More precisely, we have suggested, at Computational Independent Model (CIM) level, a meta-model for supporting the modeling of contextualized versioned IP [5]. At Platform Independent Model (PIM) level, we proposed an extended AUML sequence diagram meta-model to specify graphically the contextualized versioned IP

[6]. In this paper, we focus on the Platform Specific Model (PSM) level of the proposed MDA framework and we propose an Interaction Protocols Adaptation and Management System (IPAMS for short) to implement the proposed models dealing with IP adaptation.

The remainder of this paper is organized as follows. Section 2 introduces our MDA framework that we propose for interaction protocols adaptation. Section 3 presents our proposed Contextualized Versioned Interaction Protocol Meta-Model (CVIP2M) to design IP versions and their contexts. Section 4 details our interaction protocol adaptation and management system architecture, presents our selection method and gives a crisis scenario in order to illustrate the proposed solution. Finally, we conclude the paper and underline the main perspectives.

2 An MDA Framework for Interaction Protocols Adaptation

The purpose of this section is to recall our proposed MDA framework for interaction protocols adaptation. More precisely, this framework supports the design of IP versions and their contexts, specification and implementation. We have chosen to use MDA because it promotes the development of abstract models and it is a standard recommended by the OMG [7] since it supports the engineering perspective. The three abstraction levels proposed by MDA are the following:

- CIM Model: Such model should be independent of any computer system [8]. In our context, the proposed CIM model is the Contextualized Versioned Interaction Protocol Meta-Model (CVIP2M for short). By instantiating this meta-model, the designer creates an IP with its versions. It is noted that our proposition is inspired from [9]. This work [9] aims at supporting the flexibility of business processes by using the versioning technique. We believe that the IP can be viewed as a set of coordinated actions and consequently, it can benefits from this technique.
- PIM Model: Such model should be independent of any technology platform [8]. It represents an enriched and less abstract view of CIM. In our context, the PIM level is divided in two sublevels: PIM1 and PIM2. The first one provides a graphical representation of contextualized versioned interaction protocols with extended AUML sequence diagram meta-model. Regarding, the second sublevel, it provides a formal specification to validate the IP versions by using formal languages: Event-B.
- PSM Model: The consideration of the execution platforms has for objective the management of the dependence between the applications and their platforms of execution [8]. In our context, we propose an Interaction Protocols Adaptation and Management System (IPAMS). Also, we propose to migrate to the XML specification in order to execute the IP by extracting the roles and implement them within software agents.

It is noted that the CIM and PIM models support the IP adaptation at build time. Regarding the latest model (PSM), it supports the adaptation of IP at run time.

In this paper, we only focus on the PSM Model by proposing an Interaction Protocols Adaptation and Management System to deal with the life cycle of Interaction protocols.

3 The Contextualized Versioned Interaction Protocol Meta-Model

The Interaction Protocol Meta Model (IP2M) proposed by [10] is the start point of our contribution. This latter is based on the criteria defined in [5] such as the simplicity (i.e., the basic concepts) and the completeness (i.e., the profile and behavior aspects) of the model. According to this model, the profile aspect is described around the property, category, conditions, input and output parameters classes. Regarding the behavior aspect, it is defined in terms of role, AbstractAction, ConversationAct, data and message classes.

To deal with IP versions modeling, we have adopted the versioning kit proposed by [9] to make some classes of the IP2M versionable. In our context, we think that it is necessary to keep versions for four classes: the Protocol class, the ProtocolProfile class, the ProtocolBehavior class and the Role class. It is indeed interesting to keep changes in history for these classes since these changes correspond to changes in the way that interaction protocol is carried out.

We have integrated the contextual model to describe the conditions use of IP versions. In this model, the contexts are viewed as first class entities and specified through two concepts: context model and context which corresponds to the instance of context model. A context model defines a set of contextual attributes describing the needed knowledge to the definition of context. A contextual attribute belongs to a specific category and takes a well determined value. More precisely, we distinguish four contextual categories: Environment, Service, Strategy and Actor. Moreover, in this model, we distinguish two types of context models: global and local. The first one is devoted to the interaction protocol while the second is devoted to the protocol profile, protocol behavior or role. The global context model/local context model defines a set of contextual attributes describing knowledge necessary for the definition of global context/local context. Regarding the local context of protocol behavior, it is described around a set of objectives. Each one is defined on several contextual attributes. The value of contextual attribute in both global and local contexts is thus determined. Finally, we integrate the contextual model in the VIP2M meta-model in order to describe the contexts of IP versions. More precisely, we associate the global context to the interaction protocol version. Regarding the local context, it is related to the version of protocol entity: protocolprofile, protocolbehavior and role (Fig. 1).

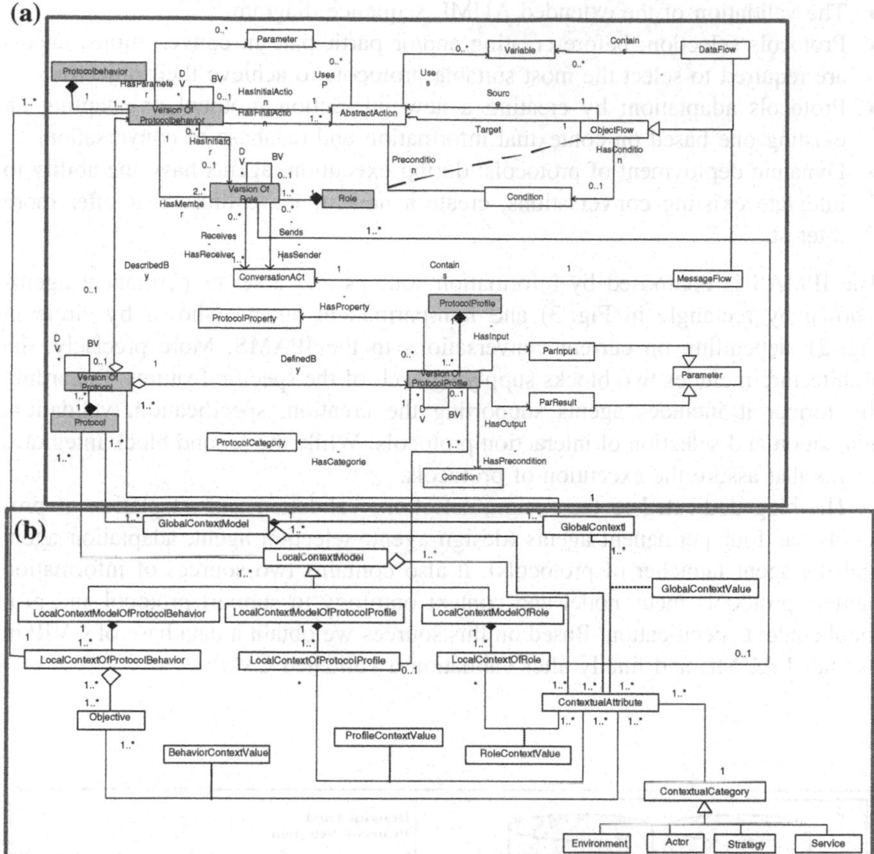

Fig. 1 The CVIP meta-model. **a** VIP2M meta-model. **b** Contextual model

4 An Interaction Protocols Adaptation and Management System, IPAMS

4.1 IPAMS Architecture

The start point of our contribution is the Interaction Protocols Management System proposed by [10]. We have extended it in order to support besides the life cycle of interaction protocols, their adaptation accordance to the solution given above. More precisely, the IPAMS acts as a server and protocols adapter, seen as a space in which participants' agents can connect to create conversations. It should allow:

- The specification of contextualized versioned interaction protocols by providing components dedicated for this purpose. These components use versioned protocols meta-model and contextual ontology.

- The validation of the extended AUML sequence diagram.
- Protocols selection: before creating and/or participate in conversations, agents are required to select the most suitable protocols to achieve their objectives.
- Protocols adaptation: by creating a new interaction protocol or adapting an existing one based on contextual information and database of conversation.
- Dynamic deployment of protocols: during execution, agents have the ability to integrate existing conversations, create a new or leave those that offer more interest.

The IPMAS is composed by information sources and a set of permanent agents (shown by rectangle in Fig. 3) and non-permanent agents (shown by circle in Fig. 2): depending on current conversations in the IPAMS. More precisely, the architecture includes two blocks supplying each of the specific features. Regarding the former it includes agents supporting the creation, specification, validation, adaptation and selection of interaction protocols. While the second block integrates agents that assure the execution of protocols.

The blog dedicated to the design, selection, validation and adaptation of protocols has four permanent agents (design agent, selection agent, adaptation agent and the agent launcher of protocols). It also contains two sources of information namely protocols meta-model and context ontology to support protocol and protocol context specification. Based on this sources we obtain a data base of CVIP in extended AUML and finally after validation a validated CVIP

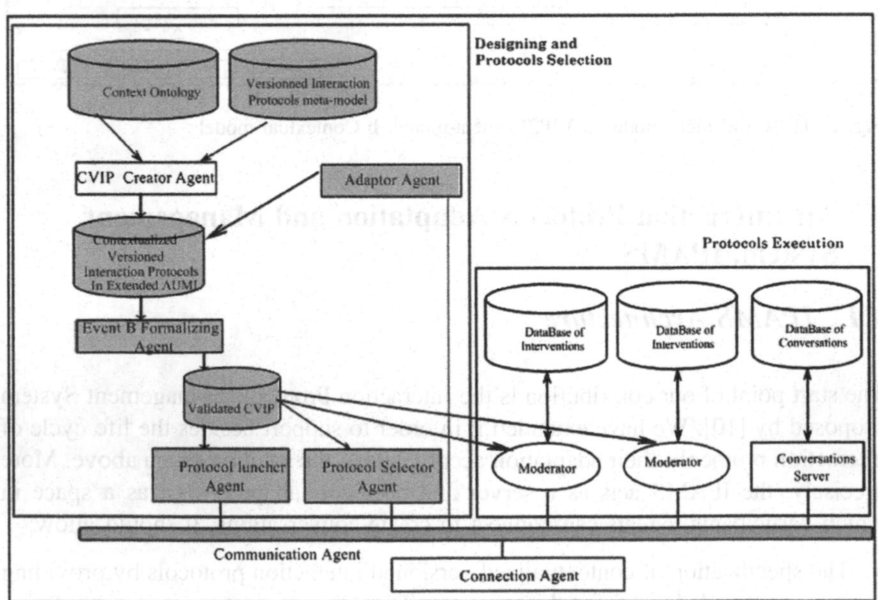

Fig. 2 IPAMS architecture

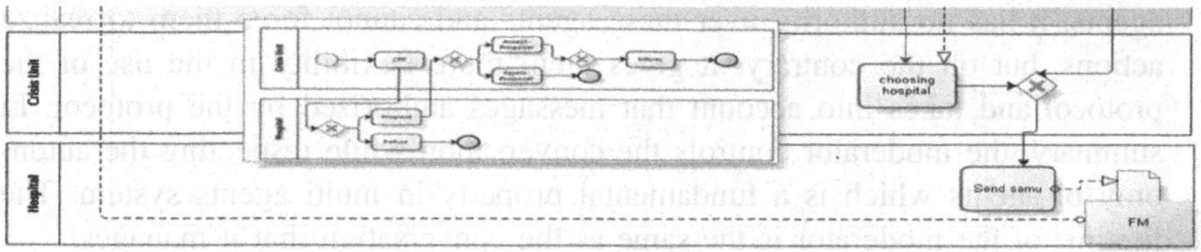

Fig. 3 Extract of air crash process

- CVIP Creator Agent: provides tools (graphical specification interfaces) to designers for the specification of protocols by instantiating both Contextual meta-model and VIP meta-model.
- Event-B Formalizing Agent: it allows validating the created CVIP specified IN extended AUML sequence diagram on event-b formalism.
- Protocol Selection agent: helps agent to choose the most appropriate interaction protocol in order to initiate a new conversation. The choice of the interaction protocol is based on contextual ontology [6].
- Agent launcher protocol: is responsible for creating new conversations. Specifically, this agent provides the launch of the moderator which is a not permanent agent responsible for controlling a conversation. The selection agent also manages the termination of conversations by removing the corresponding moderator agent.
- Adaptor agent: to resolve a blocked interaction protocol in conversation and based on based on contextual information, it proposes a new version of interaction protocol to be used.
- Versioned Protocol Information meta-model: it contains dynamic and static parts of versioned protocols. Regarding the former it means the behavior of the protocol, while the latter describe the profile of protocol.
- Contextual Ontology: describe the contextual information of each protocol. It is used by the selector agent to choose the appropriate interaction protocol.
- Contextualized Versioned Interaction Protocols specified in Extended AUML: contains contextualized versioned interaction protocol specified in extended AUML sequence diagram.
- Validated CVIP: contains contextualized versioned interaction protocol validated with Event-B formalism.

The second block which is dedicated to the implementation of protocols is composed of two types of agents: the agent server conversations and moderator agents. There is only one conversation server. In contrast, the number of moderators is equivalent to the number of running conversations.

- Moderator Agent: allows controlling the execution of interaction protocols. It guarantees the synchronization of the conversations acts realized by involved agents. It has no authority over these agents and cannot force them to realize actions, but on the contrary, it gives them more flexibility in the use of the protocol and takes into account that messages authorized by the protocol. In summary, the moderator controls the conversation while respecting the autonomy of agents which is a fundamental property in multi agents system. The lifetime of the moderator is the same as the conversation that it manages.

- Agent conversation server: store information such as (exchanged conversations acts, the identity of the officers involved in the conversation, the identity of the moderator, the purpose, start date, and the state of the conversation, etc.). The server publishes conversations and makes this information available to registered agents in the IPAMS. This information is stored in the database of conversations and allows to the adaptor agent to adapt and create new interaction protocols.

In addition to these two main blocks, two other components are specified in IPAMS architecture: the agent broadcast messages and communication agent.

The agent broadcast messages enables the interfacing with the outside. More exactly every agent who intends to invoke a service offered by IPAMS must only know the identity of this agent.

Finally, the agent communication supports the internal communication between the agents of the IPAMS and especially between the agent broadcast and the other agents of the architecture. One of its main functions is to solve the conflicts which can appear in the routing of messages, the resolution of the errors or the description of the failures of certain permanent agents of the system.

4.2 Versioned Interaction Protocol Selection

To select the appropriate IP version among several, the idea is to compare the user query described in terms of contextual attributes with the saved contexts of IP versions. This comparison is ensured thanks to the ontology of context designed in OWL [11]. More precisely, the Protocol Agent Selector can start its search based on the global context of IP version. In order to refine the result of this selection, the user can add other contextual attributes describing the locals' contexts of IP versions. Finally, to sort the returned IP version using the protocol matching algorithm, the user can apply the similarity degree calculation between the IP versions according to the predetermined threshold calculated using the similarity calcul algorithm [6].

4.3 *An Illustrative Case Study*

Description

To better illustrate the IP adaptation based coordination for crisis management processes, we give a well-known crisis scenario called "*air crash resolution process*". More precisely, this process is organized around 18 activities: Declaring Plane Disappearance, Building Crisis unit, Designation of Principal Investigator, Training of working groups, Collecting of information's, Reviewing and analysis of information, Establishing of Preliminary Report, Establishing of Final report, Safety recommendation, Locating accident site, Identifying security parameter, Extracting plane debris, Sending fireman team, Extinguishing burned, Evacuating people, Rescuing Injured, Choosing hospital and finally Sending SAMU. Because of space limitation, we present bellow an extract of this process specified in BMPN (Fig. 3). The interested author can refer to [6].

IP Adaptation Based Coordination for Crisis Management Processes

To support the coordination of activities composing the CMP, we propose to integrate communication activities, called also logistic activities. These activities permit to choose an interaction protocol version in accordance with the execution context. In our crisis scenario below, we have suggested to integrate two activities for finding partners to deal with "training of working group" and "choosing hospital". To execute the communication activity (Fig. 4), the requesting agent RA

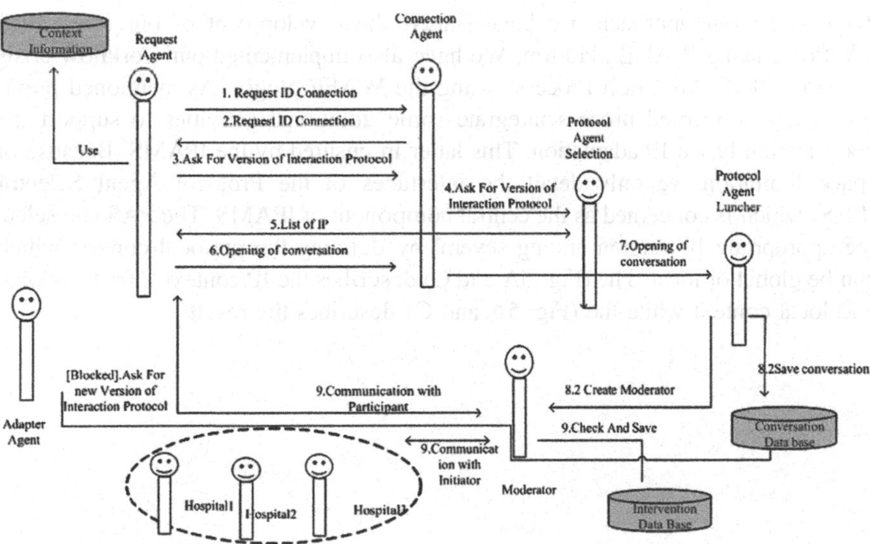

Fig. 4 A running example

(Crisis Unit) connects to the IPAMS to create a new conversation, at first steep and then he chooses a protocol. To create a conversation he sends a request to connection agent to obtain an identifier connection ID. After the abstention of ID connection, the request agent asks the IPAMS to acquire knowledge on the available protocols. The Connection Agent CA forwards this select query to Protocol Agent Selection PAS. The latter handles the request and returns the list of protocols which corresponds to the requirements defined in the demand. Then, the requesting agent decides the protocol to be used and sends a request of opening of conversation for the Connection Agent by specifying the chosen protocol and the parameters necessary for his launch. The Protocol Agent Launcher PAL dynamically creates the Moderator Agent with the type of selected protocol. After creating Moderator agent, PAL officer returns the identity of the moderator to the requesting agent. He also asked the Agent Conservations Server to record information on the new opened conversation such as the identity of the moderator, the type of protocol performed, the identity of participants, the context and purpose of the conversation, etc. The moderating agent begins by notifying the participants of the conversation opening. The participants (hospitals) can join the conversation by playing the roles which they have to hold it. Once a protocol is blocked, the agent adapter proposes a new interaction protocol version based on contextual information.

4.4 Implementation

To validate our approach, we have started the development of our tool called CVIProM using WADE platform. We have also implemented our workflow crisis scenario titled "Air Crash Process" using the WAFE plugin. As mentioned previously, our concerned process integrate some additional activities to support the coordination based IP adaptation. This latter in ensured by the IPAMS. Because of space limitation we only detail the interfaces of the Protocol Agent Selector (PAS) which is concerned as the central component of IPAMS. The PAS can select the appropriate IP version among several by defining the protocol context which can be global or local. The (Fig. 5A and C) describes the IP context form for global and local context while the (Fig. 5A and C) describes the result.

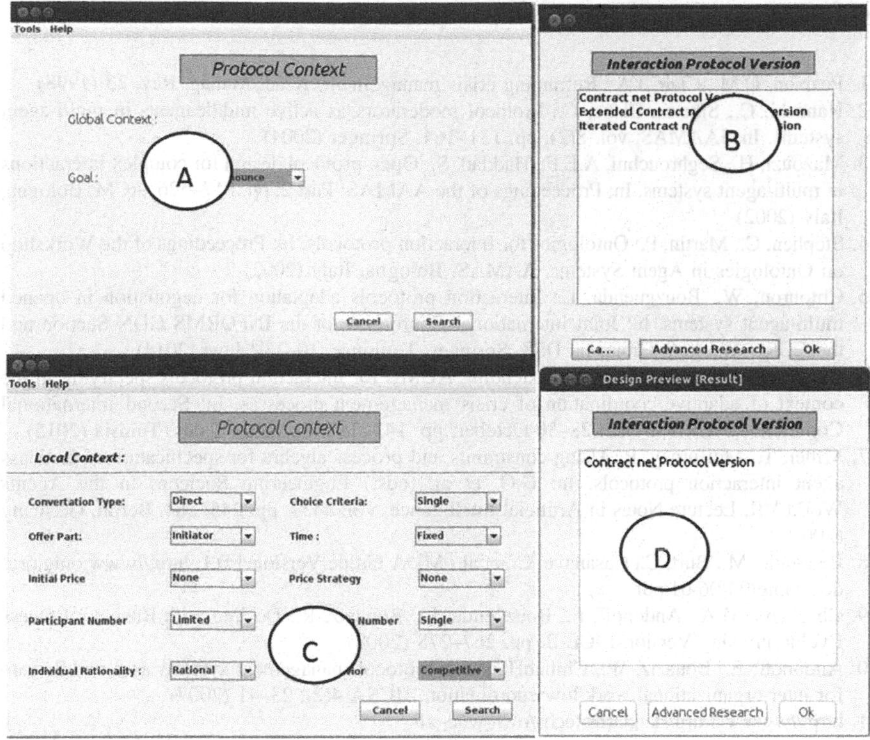

Fig. 5 An overview of CIVProM tool

5 Conclusion

This paper has presented the MDA framework that we have proposed in our previous works. More precisely, it has recalled briefly the CVIP2M meta-model. In this paper, we focused on the IP adaptation at run time. For that, we have proposed an Interaction Protocols Adaptation and Management System (IPAMS). This latter represents the PSM level of the proposed MDA framework. The major benefits of the IPAMS are: (i) it supports the full adaptable protocols development process, (ii) it promotes the reuse and the sharing of adaptable protocols, and therefore and (iii) it offers more flexibility for agents to choose the convenient protocols for a given situation. Finally, we have modeled a concrete crisis scenario called "air crash management process" with BPMN in order to illustrate the proposed solution. As future work, we plan to complete the implementation of the other involved agents.

References

1. Pearson, C.M., Clair, J.A.: Reframing crisis management. Acad. Manag. Rev. **23** (1998)
2. Hanachi, C., Sibertin-Blanc, C.: Protocol moderators as active middleagents in multi-agent systems. In: JAAMAS, vol. 8(2), pp. 131–164. Springer (2004)
3. Mazouzi, H., Seghrouchni, A.E.F., Haddad, S.: Open protocol design for complex interactions in multi-agent systems. In: Proceedings of the AAMAS: Part 2, pp. 517–526. ACM, Bologna, Italy (2002)
4. Stephen, C., Martin, P.: Ontologies for Interaction protocols. In: Proceedings of the Workshop on Ontologies in Agent Systems. AAMAS. Bologna, Italy (2002)
5. Chtourou, W., Bouzguenda, L.: Interaction protocols adaptation for negotiation in opened multi-agent systems. In: Joint International Conference of the INFORMS GDN Section and the EURO Working Group on DSS. Springer, Toulouse, 10–13th June (2014)
6. Chtourou, W., Bouzguenda, L.: Extending AUML for interaction protocols specifying in the context of adaptive coordination of crisis management processes. In: Second International Conference ISCRAM-med, 28–30 October, pp. 143–154. Springer, Tunis, Tunisia (2015)
7. Miller, T., McBurney, P.: Using constraints and process algebra for specification of firstclass agent interaction protocols. In: G.O. et al. (eds.) Engineering Societies in the Agents World VII, Lecture Notes in Artificial Intelligence, vol. 4457. pp. 245–264, Berlin, Germany (2007)
8. Belaunde, M., Burt, C., Casanave, C., et al.: MDA Guide Version 1.0.1. http://www.omg.org/docs/omg/03-06-01.pdf
9. Chaâbane, M.A., Andonoff, E., Bouzgenda, L., Bouaziz, R.: Dealing with Business Process Evolution using Versions. ICE-B, pp. 267–278 (2008)
10. Andonoff, E., Bouaziz, W., ChihabHanachi: Protocol management systems as a middleware for inter-organizational workflow coordination. IJCSA **4**(2), 23–41 (2007)
11. http://www.w3.org/standards/techs/owl#w3c_all

Holonic Multi Agent System for Data Fusion in Vehicle Classification

Ljiljana Šerić, Damir Krstinić, Maja Braović, Ivan Milatić, Aljoša Mirčevski and Darko Stipaničev

Abstract In this paper we describe holonic organization of a multi agent system for automatic vehicle classification in a road toll system. Classification of vehicles in road toll systems is based on physical vehicle features and in this paper we focus on axle counting as the first discriminant feature for class determination. Our system relies on two main sensors—video camera and depth sensor. Video image and depth image processing is performed in several holons. The results from individual holons are fused into the final decision on a number of axles of a passing vehicle. We show that fusion of results from individual holons gives more precise results than individual holons. Holonic organization of the system aids scalability and simplifies inclusion of new sensors and new algorithms.

Keywords Holonic MAS · Vehicle classification · Image processing · Data fusion

1 Introduction

Road toll charging is a measure often used by road managers or government units to improve transport system with additional financing or control congestions [1]. Fees paid by individual cars are calculated on a different basis (journey length, time spent, the weight of the vehicle) but often depend on the type of the vehicle. Vehicle classification is determined by each country regulations. In Croatia there are 5 classes of vehicles for which the price is calculated, as shown in Fig. 1a.

If a human operator is present at a road toll station, he or she visually determines the class of the vehicle and writes the fact into the system. In case of an automatic road toll system (such as ENC or free flow systems) the class of the vehicle must be determined automatically. In automatic road toll the need for automatic vehicle

L. Šerić (✉) · D. Krstinić · M. Braović · I. Milatić · A. Mirčevski · D. Stipaničev
Faculty of Electrical Engineering, Mechanical Engineering and Naval Architecture, University of Split, Rudera Boškovića 32, 21 0000 Split, Croatia
e-mail: ljiljana@fesb.hr
URL: https://www.fesb.unist.hr

G. Jezic et al. (eds.), *Agent and Multi-Agent Systems: Technology and Applications*, Smart Innovation, Systems and Technologies 58,
DOI 10.1007/978-3-319-39883-9_12

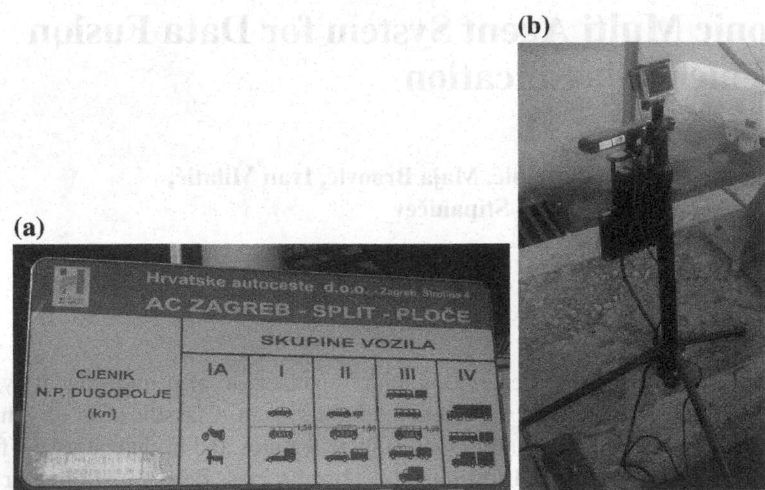

Fig. 1 **a** Vehicle categorization on Croatian Motorways. **b** Installation of a camera and a depth sensor for video stream recording

classification is obvious, but many managers include a classification system in human operated toll station so the operator's work can be tracked. The class of the vehicle is also an important feature for traffic flow analysis often performed by road managers for determining future investments.

In existing systems in Croatia the class of the vehicle is automatically determined by counting vehicle axles and measuring vehicle height. These systems consist of sensors built in pavement for axle counting and optical sensors beside the road for height measurement. However, these sensors have been shown to be rather inefficient and prone to malfunction. The analysis of a ticketing system used by company responsible for toll system maintenance has shown that in the last 7 years there were 93 reports about the malfunctions of sensors in pavement. 12 of those sensors are still not in operation since repairing the sensors demands closing the toll charge lane. This implies that currently operating automatic vehicle classification system is not adequate and there is an obvious need for a better solution. Our solution is a system that is based on easily accessible sensors (video camera and depth sensor) that are placed on the side of the lane and a software system for determining the class of the vehicle by data fusion from several components.

In this paper we present holonic organization of a software system for vehicle axle counting as the first step for automatic vehicle classification. The rest of the paper is organized as follows: first we present related work on signal and image processing for automatic vehicle classification and wheel detection. Further, we present a methodology for software system development. Holonic organization of the system and short description of each holon is given in the next section. Finally we present results of system evaluation on the testing portion of collected video and signal stream from toll charge lane.

2 Related Work

A number of various vehicle classification methods exist, and they mostly vary in different classification algorithms and different sensors used for vehicle detection. In this survey of recent literature revolving around automatic vehicle classification we focus on those classification methods that use stationary camera as a sensor. Fung et al. [2] presented a system for the automatic determination of the number of axles that a vehicle has. They used Gaussian smoothing, Sobel filtering and Hough transform for circle detection in order to identify the wheels of a vehicle in an image. They used image sub-sampling in order to increase image processing speed. Archler and Trivedi [3] presented a novel vehicle detector that uses data received from the rectilinear camera. They convolved the image with difference of Gaussian filterbank, used Principal Component Analysis (PCA) to generate wheel and road model, and used background segmentation to eliminate false detections.

Iwasaki and Kurogi [4] proposed a method for vehicle detection by using shadows beneath them, and they used the distance between the front and rear tires to discover the vehicle size. Gupte et al. [5] presented a system that uses image segmentation to separate vehicles from the background, tracks the obtained regions through the sequence of images, recovers vehicle parameters such as length and width, identifies vehicles by the means of grouping the tracked regions, tracks vehicles and finally classifies them.

Samadi and Kazemi [6] proposed a multi-agent system for vehicle detection. They used eight process agents for vehicle detection in outdoor scenes. Agents that they use are contour agent, wheel agent, etc. They reported that the effectiveness of their system is about 90 %. Some of the vehicle detection systems that do not use stationary camera as the only sensor and/or do not use stationary camera at all are: [7–9].

3 Methodology

Vehicle classification is a task that is easily performed by humans, but in a computer program it is not so straightforward. When performed by machines, classification relies on some measurable parameters of a vehicle. Our idea was to achieve the objective by using various sensors and to develop a process for decision making about the class of the vehicle.

The task of building a system for automatic vehicle classification begins with the selection of appropriate sensors. After survey of literature in search for common solutions and based on our previous experience we decided to use a video camera and a depth sensor for environment sensing and to implement various algorithms for image and signal processing to achieve classifier accuracy that is good enough for practical use. Practical industry use demands accuracy of 99.5 %, but as the first goal we set the objective to reach 90 % accuracy.

The first step of deciding the class of a vehicle on the basis of measurable parameters is counting the vehicle axles. This is the task that is usually done with sensors in the pavement. Those sensors have shown to be the most delicate in the system and that is one reason why this task was selected as the first one. In this paper we will describe the part of the system that performs only this task—counting vehicle axles. The location of the sensors was selected having in mind the task—counting vehicle axles, and physical limitations of the toll lane. Sensors were mounted on the side of the lane on the height of about 50 cm above the ground. On this height sensor does not easily gets dirty and the field of view covers the wheel of every passing vehicle. Installation of a video and a depth sensor on the same pole is shown in Fig. 1b.

We collected several hours of video and depth sensor streams. Algorithms were trained on one portion and tested on the other portion of these streams. Results of the testing are discussed in the results section.

4 Holonic Multi-agent System

An agent is a software entity that acts on behalf of a computer program or a user in a relationship (or agency) and usually makes decisions about actions to take. A software system composed of multiple agents that achieves common objective throughout mutual interaction and communications is called a multi agent system. Multi agent systems are a convenient way for organizing and describing many distributed systems. Agents in multi agent system can be organized in different ways, and one of the ways, previously used by the authors, is **holonic multi agent system** [10].

Holonic multi agent system is a system composed of individual agents that communicate with each other and that can be divided into holons. Each holon can be viewed as a whole multi agent system with its own objective or as a part of the whole system. Holonic multi agent system must undertake process of holonification in which individual agents are assigned to holons that are composing the complete system. In this paper we are presenting holonic multi agent system that is used in vehicle classification system. The system objective is to achieve decision about the number of axles of the passing vehicle on the basis of sensors beside the toll lane. Sensors used in this work are a video camera and a depth sensor. Implementation of only one image or signal processing algorithm cannot achieve desired accuracy because with every approach noise due to imperfect recording conditions (such as variations in light, weather or temperature) influences the results. Our system is composed of several approaches that can be used independently. Each approach is represented with one holon. One holon can work independently on other parts of the system, but has interface on which it emits the results. Interface can be connected to the user, system or data fusion holon. Data fusion holon collects results of each axle counting holons and makes final decision on the presence of the wheel in a stream of data. Holon hierarchy and organization is shown in Fig. 2.

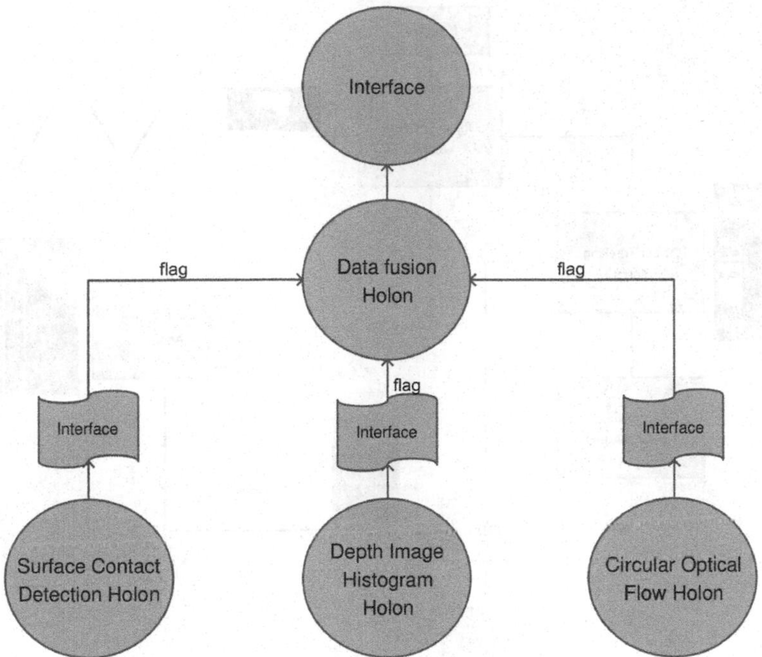

Fig. 2 Holonic organization of the system

4.1 Surface Contact Detection Holon

Detection of wheel contact with surface is a computer vision approximation of contact in pavement sensors. The algorithm relies on two features of the wheel contact: detection of object in an image region by comparing the image frame with background frame and the fact that wheel is coloured black. These features are extracted from the frame image of a video taken in a visible spectrum.

Detection is not performed on the full size image but only on a region (region of interest—ROI) with the most stable background—for region selection we rely on the camera position and in our test streams we selected the region that covers the pillar for the ramp. Position of ROI is shown in Fig. 3.

Background Tracking Agent is responsible for background initialization and update with each frame to compensate variation of light as the time passes. *Object Tracking Agent* makes comparison of current frame with background frame and extracts the pixels that are significantly different compared to background. *Black Tracking Agent* extracts pixels that are darker than threshold that is adapted to average darkness of the ROI to compensate for variations of light in a video stream.*Bottom Tracker Agent* selects the bottom of the object detected by both tracker agents and tracks the development of the bottom of the object inside ROI in time domain. In the case when passing object has touched the surface one of the two typical patterns shown in Fig. 3 will

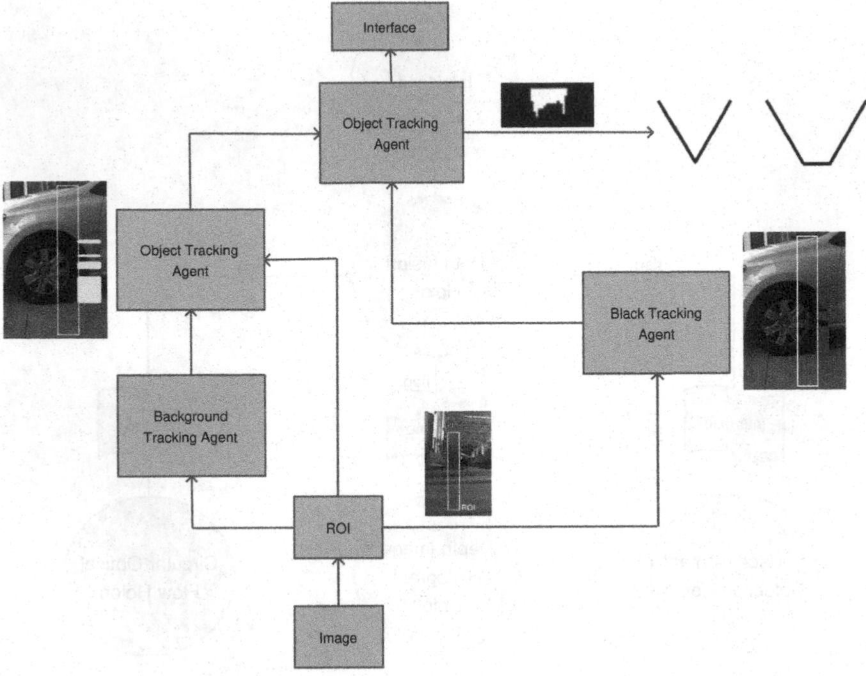

Fig. 3 Surface contact detection holon agents

appear in the *Bottom tracker agent* series. In the case of the detection of the contact, flag that the wheel has passed through ROI is set on the *interface*.

4.2 Depth Image Histogram Holon

Depth image histogram wheels detection is based on *Asus Xtion PRO Live* depth sensor. Grayscale image is acquired with 3D representation of the area in front of the sensor, where each pixel indicates the distance of the object's surface from the sensor.

Depth image is shown in Fig. 4 with marked ROI_L and ROI_R regions of interest. When no obstacle is present in front of the sensor, histograms for both ROI_L and ROI_R are expected to be zero, i.e. all pixels are black, with tolerance for up to 3 %. Depth image histogram holon agents are shown in Fig. 4. *Vehicle Presence Agent* detects presence of the vehicle by analysing both ROI_L and ROI_R. When vehicle enters the scene, the number of black pixels in ROI_R decreases, followed by a decrease in a number of black pixels in ROI_L. When the number of black pixels drops below certain threshold in both regions, *Vehicle Presence Agent* declares that a vehicle is in the scene.

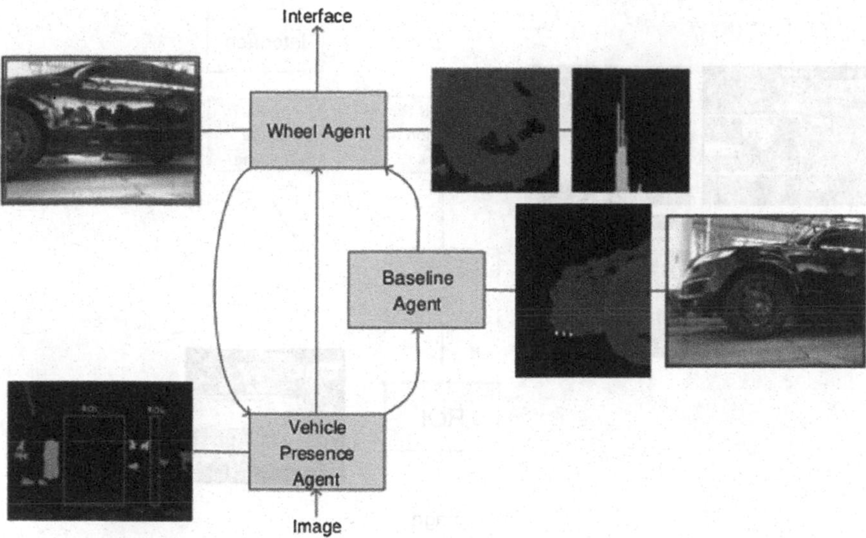

Fig. 4 Depth image

After vehicle detection, agent *Baseline Agent* becomes active. This agent's objective is to detect the baseline of a vehicle. Agent is searching for three points $p(x_i, y_i)$, $i = 1, 2, 3$ in ROI_L which are at least 5 pixels distant from each other, satisfying condition $p(x, y) = 0 \quad \& \quad p(x, y-1) > 0$. When such points are detected, new region $\overline{ROI_L}$ is formed from the lower part of the ROI_L, containing all pixels from the bottom of the ROI to $min\{y_1, y_2, y_3\} - 20$, ensuring that the $\overline{ROI_L}$ contains only pixels representing the wheel and not the upper parts of the vehicle which can confuse *Wheel Agent*. *Wheel Agent* computes histogram for each input frame. When a histogram with a prominent single peak is detected, *Wheel Agent* declares that a wheel is detected and sends information to the interface. The number of wheels for the vehicle is increased as long as *Vehicle Presence Agent* does not detect the end of the vehicle. The end of the vehicle is detected by a decrease of non-zero pixels in ROI_R followed by a decrease of non-zero pixels in ROI_L.

4.3 Circular Optical Flow Detection Holon

The premise behind circular optical flow detection is that while vehicle drives forward/backward, it moves in a linear motion and the wheels of the vehicle rotate. From that premise we can conclude that if we detect the circular motion on the input video, we have detected the wheel.

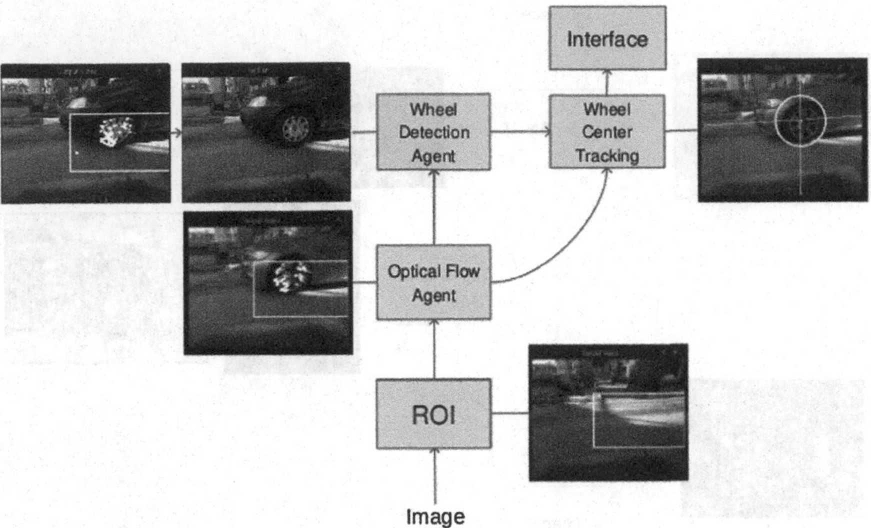

Fig. 5 Circular optical flow detection holon agents

Detection is done only in a part of the video frame where we can expect the wheel—that is in the region of interest (ROI) as shown in Fig. 5. *Optical Flow Agent* detects all motion in ROI by detecting tracking points and calculates optical flow for a sparse feature set using the iterative Lucas-Kanade [11] method with pyramids.

From all detected movements, *Wheel Detection Agent* selects only those that can be a part of the circular movement by examining the distance and the angle of the movement detected. The *Wheel Detection Agent* calculates the center of circular movement by taking into account all possible circles centers from movements detected and calculating the median of those centres. After that *Wheel Center Tracking Agent* tracks both forward and circular movements until the wheel leaves the ROI.

This circular optical flow detection holon has efficiency of 75 % to well over 90 % on tested video sequences, depending on the input image quality and the camera position in regard to the lane.

4.4 Data Fusion

Data fusion holon (in Fig. 6) is the higher tier holon whose objective is to confirm or decline detection of wheels. Its primary goal is to sum all the data from other holons beneath it. *Data Fusion Agent* checks if there is consensus in all holons but if there is a need for a more precise and robust confirmation, two additional agents are called. These additional agents fulfil its secondary goal of acknowledging the wheel detection.

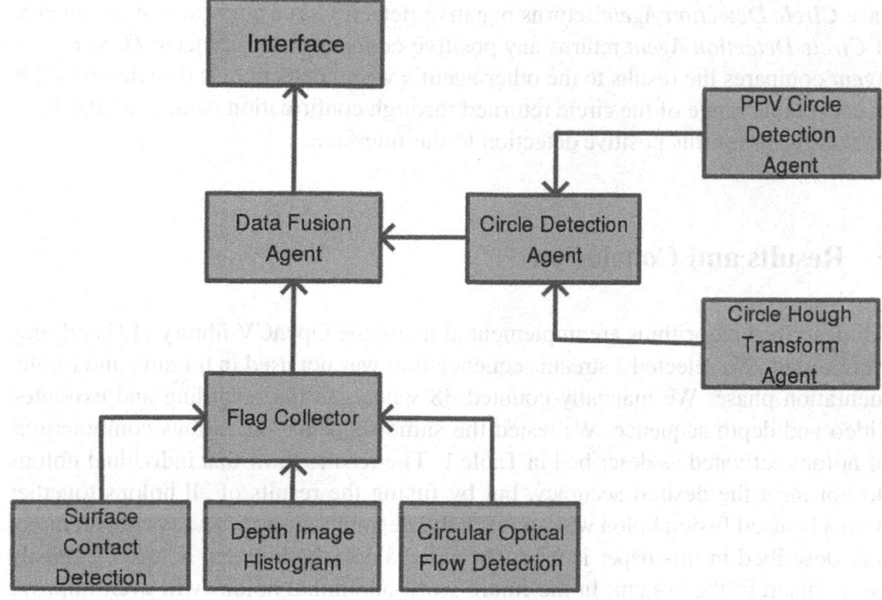

Fig. 6 Data fusion data flow

Data Fusion Agent retrieves all flags from the *Flag Collector Agent*. Flags are computed by data analysis of lower tier holons. After all lower level holons have passed their analysis to *Flag Collector agent*, information is forwarded to *Data Fusion Agent* and it starts its decision making process. If there is any positive response from the *Flag Collector Agent*, data fusion proceeds to the second stage of confirmation. If no confirmation is received, *Data Fusion Agent* returns negative detection to the interface. In case of a positive response, the agent counts all of the responses. The second stage encompasses all the confirmations from the *Flag Collector Agent*. If there is more than one positive response, it means that more than one agent has found a wheel, and they have confirmed each other's analysis. Further on, *Data fusion Agent* returns a positive detection to the interface. If there is only one positive response with no confirmation from other low tier agents through Flag Collector, *Data Fusion Agent* starts its *Circle Detection Agent*. Circle detection agent comprises of two different agents which provide a more robust and a more precise way to confirm a detection. Each of the agents process and return the analysis independently. First agent is PPV circle detection [12] and the second one is Circle Hough Transform. The first agent has a bigger impact factor for the confirmation of the detection because of its better reliability to find circles in a more dense images. The first step is to compare the results of both agents. If the centers of both detected circles are close enough to one another, the circle is taken to be a confirmation parameter. If the first agent finds a circle, but the second agent retrieves none, the circle from the first agent is used as a confirmation parameter. In any other case, *Circle Detection Agent* returns negative detection as a confirmation parameter. If *Circle Detection Agent* returns any positive confirmation parameter, *Data Fusion Agent* compares the results to the other agent's wheel detection. If detected wheel is in acceptable range of the circle returned through confirmation parameter, the *Data Fusion Agent* returns positive detection to the interface.

5 Results and Conclusion

All described algorithms are implemented using the OpenCV library [13] and they were tested. We selected a stream sequence that was not used in training and implementation phase. We manually counted 38 wheels in the recording and extracted video and depth sequence. We tested the same sequence on various combinations of holons activated as described in Table 1. The results show that individual holons do not meet the desired accuracy, but by fusing the results of all holons together with advanced fusion holon we can reach the desired accuracy. The system architecture described in this paper is modular and scalable—additional holons can easily be mounted in the system. In the future work additional holons will even improve the results of the system. Such architecture also facilitates the deployment of the system—software system can easily be deployed to run in parallel on separate processors.

Table 1 Data using different agents on a video with 38 wheels

Holons	Found wheels	False positive	GT percentage (%)
Surface Contact Detection + Depth Image Histogram	29	2	71.05
Surface Contact Detection + Circle Detection	44	16	73.68
Circular Optical Flow Detection + Circle Detection	40	14	68.42
Surface Contact Detection + Circular Optical Flow Detection + Circle Detection	33	2	81.57
Surface Contact Detection + Circular Optical Flow Detection + Depth Image Histogram + Circle Detection	36	2	89.47

Acknowledgments This work was partly supported by the Programme of technological development, research and innovations application of Split-Dalmatia County (in Croatian: Program tehnološki razvoj, istraživanje i primjena inovacija Splitsko-dalmatinske županije) under grant "Automatic vehicle classification based on computer vision and data fusion" (in Croatian: "Automatski klasifikator vozila temeljem računalnog vida i fuzije podataka").

References

1. Eliasson, J.L.M.: Equity effects of congestion pricing quantitative methodology and a case study for Stockholm. Trans. Res. A **40**, pp. 602–620 (2005)
2. Fung, Y.-F., Lee, H., Ercan, M.F.: Image processing application in toll collection. IAENG Int. J. Comput. Sci. **32** (2006)
3. Achler, O., Trivedi, M.M.: Camera based vehicle detection, tracking, and wheel baseline estimation approach. In: The proceedings of the 7th International IEEE Conference on Intelligent Transportation Systems, pp. 743–748 (2004)
4. Iwasaki, Y., Kurogi, Y.: Real-time robust vehicle detection through the same algorithm both day and night. Int. Conf. Wavelet Anal. Pattern Recogn. (ICWAPR) **3**, 1008–1014 (2007)
5. Gupte, S., Masoud, O., Martin, R.F.K., Papanikolopoulos, N.P.: Detection and classification of vehicles. IEEE Trans. Intell. Trans. Syst. **3**, 37–47 (2002)
6. Samadi, S., Kazemi, F.M.: A multi-agent vision-based system for vehicle detection. World Appl. Sci. J. **15**, 1722–1732 (2011)
7. Wang, H., Quan, W., Liu, X., Zhang, S.: A two seismic sensor based approach for moving vehicle detection. In: Procedia—Social and Behavioural Sciences, 13th COTA International Conference of Transportation Professionals (CICTP 2013), vol. 96, pp. 2647–2653 (2013)
8. Chellappa, R., Qian, G., Zheng, Q.: Vehicle detection and tracking using acoustic and video sensors. In: IEEE International Conference on Acoustics, Speech, and Signal Processing, vol. 3 (2004)
9. Steux, B., Laurgeau, C., Salesse, L., Wautier, D.: Fade: a vehicle detection and tracking system featuring monocular color vision and radar data fusion. In: IEEE Intelligent Vehicle Symposium, vol. 2 (2002)
10. Šerić, L., Štula, M., Stipaničev, D.: Engineering of holonic multi agent intelligent forest fire monitoring system. AI Commun. **26**(3), 303–316 (2013)
11. Bouguet, J.-y.: Pyramidal implementation of the Lucas Kanade feature tracker, Intel Corporation, Microprocessor Research Labs (2000)
12. Pan, L., Chu, W.-S., Saragih, J.M., De la Torre, F., Xie, M.: Fast and robust circular object detection with probabilistic pairwise voting. Sig. Process. Lett. IEEE **11**, pp. 639–642 (2011)
13. Bradski, G.: The OpenCV Library Dr. Dobbs J. Softw. Tools (2000)

Modeling and Simulation of Coping Mechanisms and Emotional Behavior During Emergency Situations

Mouna Belhaj, Fahem Kebair and Lamjed Ben Said

Abstract Emotions shape human behaviors particularly during stressful situations. This paper addresses this challenging issue by incorporating coping mechanisms into an emotional agent. Indeed, coping refers to cognitive and behavioral efforts employed by humans to overcome stressful situations. In our proposal, we intend to show the potential of the integration of coping strategies to produce fast and human-like behavioral responses in emergency situations. Particularly, we propose a coping model that reveals the effect of agent emotions on their action selection processes.

Keywords Coping · Emotions · Emergencies · Human behavior

1 Introduction

There are different proposed approaches that intend to reproduce human emotion mechanisms into artificial agents. These approaches have led to the creation of many computational models of emotions in the last years. However, the majority of the proposed models focus on modeling the emotion triggering conditions. Fewer of them study the effect of emotions on agent behaviors [1]. Coping refers to the thoughts and behaviors people use to manage the internal and external demands of stressful events [2]. The coping process is greatly related to emotional mechanisms. In fact, Roseman considers emotion as "a coherent, integrated system of general-purpose coping strategies, guided by appraisal, for responding to situations

M. Belhaj (✉) · F. Kebair · L. Ben Said
SOIE Laboratory, Higer Institute of Management of Tunis, Le Bardo, 2000 Tunis, Tunisia
e-mail: mouna.belhaj@hotmail.com

F. Kebair
e-mail: kebairf@gmail.com

L. Ben Said
e-mail: lamjed.bensaid@isg.rnu.tn

G. Jezic et al. (eds.), *Agent and Multi-Agent Systems: Technology and Applications*, Smart Innovation, Systems and Technologies 58,
DOI 10.1007/978-3-319-39883-9_13

of crisis and opportunity (when specific-purpose motivational systems may be less effective)" [3]. The influence of emotion on behavior is qualitatively different; when emotions are of high intensity, goal directedness diminishes and behavior becomes more organized by action tendencies and readiness [3]. That way, emotions are essential to directly trigger behaviors and evoke mechanisms to cope with stressful situations that are not used in more relaxed emotional conditions [4]. The majority of coping theories and computational models associate coping to negative emotions. However, in [5], the author argues that both negative and positive emotions provide mechanisms to human organisms to cope with life crises. These are situations where there is no time to deliberation that assesses the advantages, disadvantages and consequences of behaviors [5]. In this work, we aim to model and simulate coping mechanisms with negative and positive emotions and to integrate them into an emotional agent model. We particularly aim to study the interplay of emotions and goals and its consequences on behavior in emergency situations. Emotions may induce emotion-specific goals (emotivational goals) or action tendencies, they can also draw attention to, or increase the salience or persistence of goals currently at stake [6]. In order to incorporate coping strategies into the emotional agent, the cognitive mechanisms that choose among goals should be influenced by emotions. In this paper, we adopt the psychological theory of coping proposed by Roseman [3] to model and simulate coping mechanisms in stressful situations. Particularly, we integrate a coping mechanism into our emotional agent model [7, 8]. The agent model makes use of a computational model of emotions we proposed in [9] that is based-on the OCC model [10]. We aim to study the impact of the generated emotions on the agent action selection process. The coping model is implemented within the emotional agent based social simulation of civilians in an emergency situation [7, 8]. The civilian simulator is integrated into the RoboCupRescue (RCR) simulation system [11], which is an agent-based simulator of an emergency situation after an earthquake.

In the remainder of this paper, we first present the theoretical background of the current research work. We then outline related work from the literature. After that, we describe the emotional agent model previously proposed. Then, we provide a new coping model and describe its implementation and experimentations that we have carried on the RCR simulation system. Finally, we draw a conclusion and present perspectives of the current work.

2 Background

2.1 Appraisal and Coping

Humans appraise cognitively their relationship with their environment continuously. The result of the appraisal process is an emotional state. Particularly, the appraisal of a stressful event induces the generation of emotions, then the decision about the convenient behavior to deal with the event called coping [2]. Coping refers to cognitive and behavioral efforts employed by humans to manage the external or internal demands of events with high levels of stress [12]. It determines the response to the appraised significance of events. Therefore, it is an important aspect for creating human-like agents that have to handle different situations depending on their appraisal of the situation [13]. The coping theory proposed by Folkman and Lazarus [2] is the main theory used in computational models of coping. It distinguishes two forms of coping that are the problem-focused coping and the emotion-focused coping [2]. The first form occurs when a human acts on the environment to deal with a stressful situation. It involves the planning of a set of actions that achieve a desired goal and the execution of those actions [14]. However, emotion-focused coping aims at lowering the intensity of strong negative emotions by trying to change the human's interpretation of the causing event. Despite of the strength of this theory, the matching between emotions and particular coping strategies remains not straightforward. Besides, the effect of the coping strategy on the behavior is not presented in the case of emotion-focused coping after the intensity of the negative emotion has been lowered. The distinction between categories of coping is not made in the recent theory of coping proposed by Roseman [3]. Differently, this theory focuses on the behavioral manifestations of emotions in terms of motivation and action readiness which are particularly important for responding to highly emotive circumstances. Besides, Roseman's theory identifies coping strategies as responses characteristic of both negative and positive emotions and that shape agent goals, actions and behaviors. Therefore, we choose to study the human behavioral dynamics during emergencies according to Roseman's theory of coping.

2.2 Emotions and Goals

Roseman presents emotions as having five different components (phenomenological, physiological, expressive, behavioral and emotivational) [3]. In the current work, we aim to study the influence of emotions on agent goals and behaviors. Therefore, we only focus on the emotivational component of an emotion and the associated coping response. The former proposes an emotivational goal that motivates actions. However, the latter suggests particular responses that arise when the emotion is experienced. Table 1 summarizes the emotivational goals that are elicited by emotions and their corresponding coping strategies as defined in [3].

When a significant event takes place, it is assessed to generate a convenient emotion. The latter activates particular goals that enable the human to reestablish or to preserve his well-being challenged by that event [5]. These goals are the emotivational goals defined by [3] (see Table 1). In fact, emotions control the state of achieving or threatening our most important goals. For example, if a person faces an immediate threat while he is pursuing a certain goal, it is vital to alert the organism in order to shift attention to respond to the stressful situation. Therefore, an

Table 1 Emotivational component of an emotion and the corresponding coping strategy

Emotion	Emotivational component (goal)	Coping strategy
Joy	Sustain	Move toward it
Distress	Terminate, get away	Move away from
Hope	Make happen	Prepare to move toward or to stop moving away from it
Fear	Get to safety, prevent	Prepare to move away from or to stop moving toward it
Love/Like	Connect	Move toward other
Dislike	Dissociate	Move away from other
Pride	Recognition, dominance	Move toward self
Shame	Get self out of sight	Move self away
Guilt	Redress	Move against self

emotivational goal arises and triggers a response such as flight in this case [5]. “Emotivational goals” reduce response time by increasing focus on a particular general purpose goal [5] (such as getting to safety in the example). Positive emotions are elicited by events that are consistent with current goals (motives) and are associated with the tendency to preserve the current state. However, negative emotions appear to be associated with a tendency to overcome the situation that elicited them [3].

3 Related Work

Different computational models of emotions exist. However, few of them consider coping mechanisms in their models. Here, we overview approaches to emotion modeling that have included coping mechanisms in their models. In [4], the authors include a motivational component and model the agent reasoning capabilities and behavior according to the BDI (Belief-Desire-Intention) approach. Although different emotions were studied, only three emotions were simulated. This approach explains quite well the different emotional mechanisms. However, the proposed architecture is complex. In fact, it integrates many emotion theories that have different foundations and that may perhaps not converge. Moreover, the experiments presented do not show how emotions affect agent plans and behaviors. In [13], the authors present EMA (Emotion and Adaptation), a computational model of emotions that includes mainly emotion-centered coping strategies. EMA was used to implement emotional agent behaviors in a virtual reality training system. The implemented agent behaviors are believable, according to the study. However, agent mental states are represented by complex structures including appraisal frames and inspired from planning. FAtiMA [15] is an interesting BDI-based approach that implements emotion-focused and problem-focused coping mechanisms. However, its main challenge is that it requires XML definitions of goals importance, emotional reactions and actions tendencies for different characters [14].

The aforementioned models of coping adopt Lazarus's theory of coping. The majority of them focus mainly on emotion-focused coping, giving less attention to the problem-focused coping where actions should be undertaken to cope with emotions. Inspired by existing coping models, we propose our modeling of coping processes. However, unlike these approaches, we adopt Roseman's theory of coping. We aim, to give restrictions on an emotional BDI agent deliberation through emotion-inspired constraints. We intend to model coping strategies with positive and negative emotions. These coping strategies that represent cognitive and behavioral responses to what happens in the environment, enable to model these restrictions since they are associated with emotivational goals that can interrupt, sustain or make happen agent goals.

4 Existing Model

In previous works, we have proposed an emotional BDI agent model [7, 8]. The latter makes use of our computational model of emotions outlined in [9] that is based on the OCC model of emotions [10]. The proposed agent model builds on the BDI paradigm [16] for action selection. Here, we extend the agent model with a coping model to simulate emotion driven behaviors in emergencies. For completeness, we provide a fast overview of the agent model on which we base the current work. The emotional agent model involves three main components (Perception, Appraisal and Behavior modules. The Perception module allows the agent to update its beliefs by processing and categorizing its perceptions into five categories of perceptual data. These categories include Events (Self-related Events SRE, Prospected Events PE and Other-related Events ORE), Actions and Elements (Objects in environment). Each category of perceptions is appraised by an appraiser in the Appraisal module that uses specific dimensions (appraisal variables of the OCC model) to generate emotions. The appraisal of SRE, PE and ORE results on three categories of Goal-based emotions that are Wellbeing emotions, Prospect-based emotions and Empathetic emotions respectively. The appraisal of actions gives rise to Standard-based (or moral) emotions. Finally, the appraisal of objects triggers Attitude-based emotions. The appraisers compute emotion intensities according to our computational model of emotions (Details about the different formula we use to compute emotion intensities along with the implementation results of the emotion generation process within the agent are provided in [7–9]). The output of the appraisal process is an emotional state that affects the agent goals and orients its BDI-based action selection process. The integration of coping mechanisms into the agent deliberation process enables the agent to activate particular goals and actions induced by his emotional state. Actually, we consider that the emotion with the highest intensity is the one considered that affects agent behavior and triggers the corresponding emotivational goal (see Table 1) to adapt to the significant changes in the emergency situation.

5 Coping Behavior During Emergency Situations

In the following, we show how the agent behavior is influenced by his strongest emotion, through coping mechanisms. Our modeling effort of the coping behavior covers the emotions and their associated coping strategies listed in Table 1. We also add the empathetic emotion *SorryFor* (that is not considered in Roseman's theory of coping) because it is necessary to model agent actions and interactions when studying human behavior in emergencies.

5.1 Coping with Goal-Based Emotions

In the context of an emergency situation, we assign to the agent the following general purpose (individual) goals; Escape a risk, Get saved, Find a refuge and Help a person [9]. In emergency situations, safety is the highest priority. Therefore, when the agent is safe his goal would be to preserve this state by finding a refuge ("Find a refuge" is considered as a default purpose). Agent goals priorities change based on the situational context of the agent manifested through its emotions. They are also influenced by the agent ability to cope with the stressful situations he faces (coping potential). In fact, new emotions induce emotivational goals that may result in preserving, activating or suspending agent individual goals. In the following, we show how coping mechanisms with emotions alter agent goals and their behaviors during emergencies. Table 2 summarizes the SRE, PE and ORE an agent may perceive in an emergency situation and the emotions of the OCC model that may result from their appraisal as defined in [7].

Coping with Well-being emotions. Humans may have positive cognitions and emotions even after facing traumatic events. Joy emotion was felt at having survived unharmed [17] and at attaining refugees [18]. Therefore, we assume that the *Joy* emotion may arise from the appraisal of the events *Safe (self)* and *InRefuge (Self)*. However, the *Distress* emotion is elicited when a negative event happens to the agent *(InDanger (self), Injury (self)* or *HealtStateDown (self))*.

Joy: The default goal in an emergency situation is to *"Find a refuge"* in order to preserve the state of safety. The appraisal of a desirable SRE (that is consistent with agent goals) may result in the *Joy* emotion. We relate the desirability of the new event to the importance of the current goal and to the impact of the event on that goal. If *Joy* is the strongest emotion of the agent, then *Joy* will trigger the emotivational goal to *"Sustain"* the current general purpose goal (see Table 1).

Table 2 Perceptual data of an agent, their categories and potential emotions

Perception category	Perception	Emotion	Category
Self-Related Events (ORE)	Safe (self)	Joy	Well-being emotions
	InRefuge (self)		
	Injury (self)	Distress	
	InDanger (self)		
	HealthStateDown (self)		
Prospected Events (PE)	ProspectRescue (self)	Hope	Prospect-based emotions
	Prospect Injury (self)	Fear	
	ProspectDeath (self)		
Other-Related Events (ORE)	Injury (other)	SorryFor (pity)	Empathetic emotion
	InDanger (other)		
	HealthStateDown (other)		
	Death (other)		

Therefore, the agent continues pursuing his current intention until a strongest emotion is generated. We assume that in an emergency situation, agents feeling *Joy* are safe agents (*Safe (self)*) that are able to "Find a refuge" by performing the "Walk" action in the crisis environment. These agents may engage in rescuing activities when they feel *SorryFor* other agents they may perceive. They may also choose the "Rest" action if they are in refuges.

Distress: The *Distress* emotion, that may be triggered from the appraisal of an undesirable SRE (i.e. an event that has a negative impact on the current goal of the agent), gives rise to the emotivational goals to *"Terminate"* the new situation. Therefore, *Distress* is associated with a goal revision strategy where current intentions and actions are no more consistent with the new distressing situation. Thus, when the agent strongest emotion is *Distress*, it enters a new deliberation cycle and a convenient goal is adopted to improve his current state. The events that may be the cause of *Distress* are the *InDanger (self)*, *Injury (self) or HealtStateDown (self)* SRE. In the case of a danger, if the agent is still safe and able to avoid the danger, the *"Escape a risk"* goal may be adopted as current intention. Here, a fast action in the physical space is needed to move away from the danger (*"RandomWalk"* action). The *Distress* emotion is also elicited if the agent, after facing a danger, was not able to escape it. In that case, the agent could be injured or having his health state being worse. Here, the agent will try to *"Get saved"* and to *"Terminate"* the current state by asking for help (*"AskForHelp (Self)"* action).

Coping with Prospect-based emotions. Prospected events are cognitive internal events that may be positive or negative. The appraisal of a positive (respectively negative) prospected event may trigger the emotion *Hope* (respectively *Fear*) (See Table 2).

Hope: If the agent prospects a positive event (*ProspectRescue(self)*), the *Hope* emotion is triggered. When *Hope* is the strongest emotion of the agent, the associated emotivational goal is to try to *"Make happen"* the positive expected situation. Therefore, a convenient individual goal that will enable the agent attain the prospected state is activated and maintained until it is achieved or recognized as unachievable. Thus, if the agent is facing a danger without being injured and is still able to move, his most important goal would be to "Escape a risk" and the action would be to try to escape the danger (*"RandomWalk"* action). Elsewhere, he will try to *"Get saved"*. Thus, he tries to make true his prospects to be rescued by asking for help.

Fear: The appraisal of a negative PE (*ProspectInjury(self)* or *ProspectDeath (self)*) triggers the emotion *Fear*. This emotion is considered as an emergency emotion associated with a threat to one's safety. Therefore, when the agent is fearful, any goal that is not associated with preserving agent safety is *suspended*. A new convenient goal should be *activated*. In fact, the emotivational goal associated with *Fear* is to *"Get to Safety"*. Thus, the agent suspends his current intention and activates one of the goals that permit him to preserve its safety. Therefore, if the agent has the ability to flight, he runs away (the intention is to *"Escape a risk"* through the *"RandomWalk" action*). However, if he is unable to change his state by himself, he tries to get social support by asking for help (*"Get saved"* intention and *"AskForHelp(Self)"* action).

Coping with the Empathetic emotion SorryFor. The *SorryFor* emotion is an empathetic emotion elicited as a result of the appraisal of a negative ORE. We presume that when an agent feels *SorryFor* another agent because of a matter that happened to this other agent, he copes with that negative emotion by trying to provide him with support. In fact, the empathetic emotion *SorryFor* (*pity*) causes the motivation to help. Therefore, pity induces the activation of an emotivational goal that we name *"Support"*. Thus, a safe agent, whose most intense emotion is the *SorryFor* emotion, copes with it by trying to provide others with help. The *"Help a person"* goal is then adopted and the corresponding *"Help(Other)"* action is executed.

5.2 Coping with Standard-Based Emotions

We are actually considering one standard in the emergency situations context that is *"Helping a person in need is a praiseworthy action"*. After having the *SorryFor* emotion triggered by the appraisal of a negative event that happens to others or after perceiving or hearing a person that is asking for help, the agent may either provide the person in need with help or prefer to continue pursuing his current intention. The emotion *Pride* arises if the agent performs the praiseworthy action to help a person in need. However, the emotions *Shame* and *Guilt* arise if the agent does not or was not able to provide the agent in need with help respectively.

Pride: *Pride* arises if the agent performs an action that it considers as consistent with its Standards (norms). The intensity of the *Pride* emotion depends on the potential of the agent to perform the action that maintained the norm from being violated. The coping response associated with *Pride* is to *"Move toward self"*. This can be translated into an increase in the standard importance for the agent or a tendency to help other agents.

Shame and Guilt: These emotions may arise when the agent performs an action that is inconsistent with his Standards. The difference between *Shame* and *Guilt* resides in the nature of the fact that elicited them. *Shame* arises if the agent violates the standard with the inability and incompetence to maintain it. However, *Guilt* is associated with negative power. In fact, it could be able to avoid the standard violation. *Shame* is associated with the emotivational goal *"Get self out of sight"*. Conversely, *Guilt* gives rise to the emotivational goals *"Redress"* and the coping strategy to *"Move against self"*. Thus, if one of these emotions corresponds to the agent strongest emotion, he tries to lower its intensity by beginning other activities.

5.3 Coping with Attitude-Based Emotions

Attitude-based or Attraction emotions are elicited by appraising the alteration of object aspects in the environment. Positively (negatively) attractive objects result on a *Like* (*Dislike*) emotion. The intensity of these emotions decreases if the observed objects become familiar to the agent.

Like: When an agent views an appealing object, it could like it. In the emergency context, we assume that refuges are the most positively attractive objects for an agent. Thus, the fact that an agent is attracted to a Refuge gives rise to the *Like* emotion. The latter elicits the emotivational goal to *"Connect"* to it and the corresponding coping strategy to *"Move toward it"*.

Dislike: The *Dislike* emotion is elicited as a result of observing a negative aspect of an object in the environment. We suppose that a blocked road or a collapsed building is negatively attractive to the agent in the crisis environment and may elicit the *Dislike* emotion. The emotivational goal is then to *"Dissociate"*. The agent copes with the Dislike emotion in a way that decreases attention to the object by trying to *"Move away from it"*.

6 Implementation and Experimentations

We implemented the coping model with Goal-based emotions into our emotional agents. The latter are used to simulate civilian agents in RCR. The perception module of an agent filters the perceptual data that come from RCR simulator in order to extract significant events. The desirability (appraisal variable) of the events is then computed and the corresponding emotions are triggered. The strongest emotion gives rise to an emotivational goal. Then, the agent selects the convenient intention and action. In the following paragraphs, we illustrate this mechanism, by showing the state evolution of an experimented emotional civilian agent in the RCR environment. Figures 1 and 2 display the Goal-based emotions evolution of the agent and provide their related triggering events. This will make sense to the strong relation between events and emotions. The emotivational goals triggered by the strongest emotions of the agent and the actions performed to attain its committed goal are illustrated in Figs. 3 and 4.

We note that the choices of actions performed by the agent stem from his strongest emotions. In fact, at the beginning of the simulation, the agent default intention is to *"Find a refuge"*. The agent tries then to achieve it through the *"Walk"* action (Fig. 4). This goal is maintained by the emotivational goal *'Sustain'* (Fig. 1) induced by the agent most intense emotion *Joy* (Fig. 1). The latter corresponds to being safe (Safe event, Fig. 2). However, facing a danger (*InDanger* event, Fig. 2) gives rise to an internal cognitive PE (*ProspectRescue* event, Fig. 2). The appraisal of these events elicits *Distress* and *Hope* emotions respectively

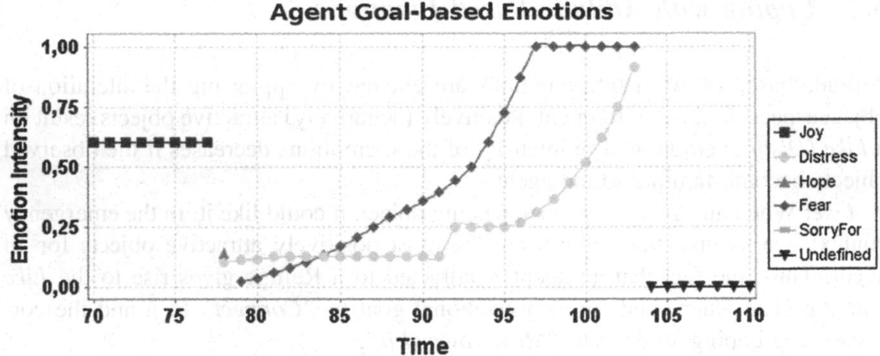

Fig. 1 Evolution of goal-based emotions of an agent

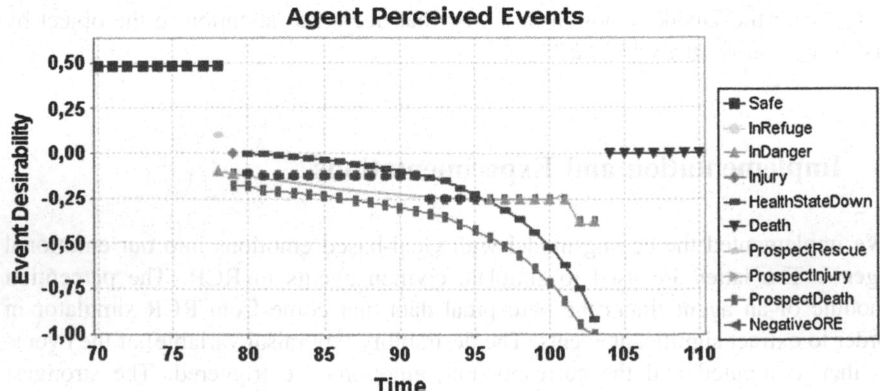

Fig. 2 Evolution of events desirability of an agent

(Fig. 1). The agent strongest emotion elicits the convenient emotivational goal (*"Make happen"* in this case triggered by the *Hope* emotion, see Table 1). The agent is still able to move, thus, the corresponding intention is then to *"Escape a risk"* accomplished by the "RandomWalk" action to try to go away from the danger (Fig. 4). The injury of the agent, the decrease of his health state and the negative PE (*ProspectDeath* event, Fig. 2) trigger the *Distress* and *Fear* emotions. When, *Fear* becomes the most intense emotion, the corresponding emotivational goal *"Get to safety"* is induced. In both cases, the agent is unable to move by his own, therefore, the intention is to *"Get saved"* accomplished by asking for help (Fig. 4). When he is dead (*Death* event, Fig. 2), agent emotion becomes undefined (Fig. 2) and he has no emotivational goal (Fig. 3) and no action (Fig. 4).

In the following, we show the effect of modeling the *SorryFor* empathetic emotion on the behavior of civilian agents in RCR. Experimentations have proved that the agent actions are influenced by empathetic emotions. In fact, the number of civilians that are walking in the crisis environment decreases when considering the *SorryFor* emotion (70 in Fig. 5 vs. 40 in Fig. 6 at time t = 20) since agents stop walking when they are trying to help other agents (*AskForHelpOther* action). Besides, this influences the number of civilians that attain refuges by the end of the simulation (66 vs. 41 agents having a "Rest" action, Fig. 6).

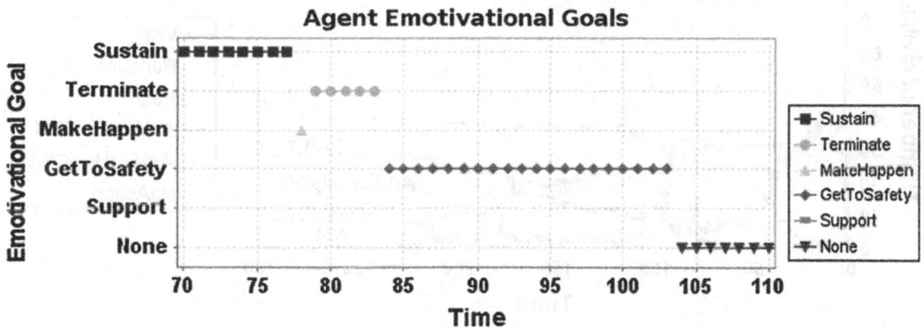

Fig. 3 Evolution of the emotivational goals of an agent

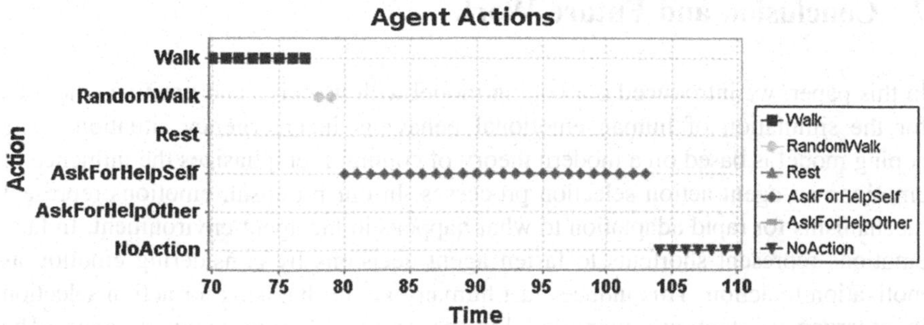

Fig. 4 Evolution of the actions of an agent

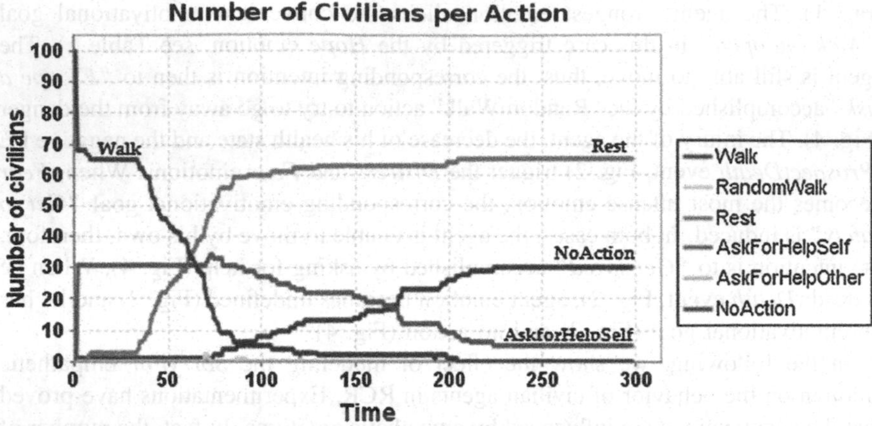

Fig. 5 Evolution of the number of agents per action without considering the *SorryFor* emotion

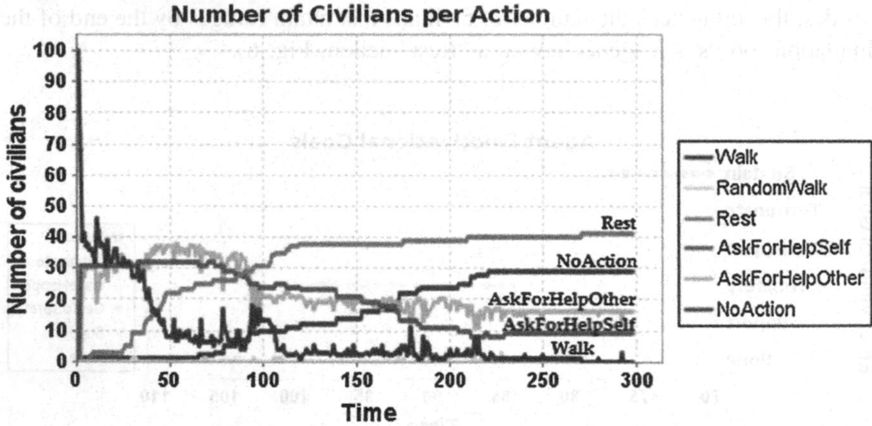

Fig. 6 Evolution of the number of agents per action when the *SorryFor* emotion is considered

7 Conclusion and Future Work

In this paper, we introduced our coping model with negative and positive emotions for the simulation of human emotional behaviors in emergency situations. The coping model is based on a modern theory of coping. It emphasizes the influence of emotions on agent action selection processes. In our proposal, emotions represent mechanisms for rapid adaptation to what happens in the agent environment. In fact, emotions represent shortcuts to fasten agent decisions by considering emotion as motivation to action. This induces fast human-like mechanisms for action selection in emergency situations supported by the notion of emotivational goals. The implementation of the coping model with different emotions shows the impact of the emotional state of the agent on its behavior. In fact, agents are aware of their context, are able to manage their states and to respond to the requirements of the environment and other agents. An aspect that may need a more profound study is the empathy phenomenon which is induced by empathetic emotions. Its importance arises from its potential to produce pro-social behaviors. Finally, personality represents another moderating factor of human behavior that may be considered. It influences emotional experience and action tendencies.

References

1. Steunebrink, B.R., Dastani, M., Meyer, J.J.: A formal model of emotion-based action tendency for intelligent agents. In: Lopes, L., Lau, N., Mariano, P., Rocha, L. (eds.) Progress in Artificial Intelligence. LNCS, vol. 5816, pp. 174–186. Springer, Heidelberg (2009)
2. Folkman, S., Lazarus, R.S.: If it changes it must be a process: study of emotion and coping during three stages of a college examination. J. Pers. Soc. Psychol. **1**, 150–170 (1985)
3. Roseman, I.J.: Appraisal in the emotion system: coherence in strategies for coping emot. Rev. 5, 141–149 (2013)
4. Becker, C., Lessmann, N., Kopp, S., Wachsmuth, I.: Connecting feelings and thoughts modeling the interaction of emotion and cognition in embodied agents. In: 7th International Conference on Cognitive Modeling, pp. 32–37 (2006)
5. Roseman, I.J.: Motivations and emotivations: approach, avoidance, and other tendencies in motivated and emotional behavior. In: Elliot, A.J. (eds.) Handbook of Approach and Avoidance Motivation, pp. 343–366. Psychology Press, New York (2008)
6. Reisenzein, R., Hudlicka, E., Dastani, M., Gratch, J., Hindriks, K., Emiliano, L., Meyer, J.J.: Computational modeling of emotion: toward improving the inter- and intradisciplinary exchange. IEEE Trans. Affect. Comput. **4**, 246–266 (2013)
7. Belhaj, M., Kebair, F., Ben Said, L.: Agent-based modeling and simulation of the emotional and behavioral dynamics of human civilians during emergency situations. In: Müller, J., Weyrich, M., Bazzan, A.C. (eds.) Multiagent System Technologies, LNAI, vol. 8732, pp. 266–281. Springer (2014)
8. Belhaj, M., Kebair, F., Ben Said, L.: An emotional agent model for the simulation of realistic civilian behaviors during emergency situations. In: the 2014 IEEE/WIC/ACM International Joint Conferences on Web Intelligence (WI) and Intelligent Agent Technologies (IAT), vol. 3, pp. 262–269. IEEE Computer Society (2014)
9. Belhaj, M., Kebair, F., Ben Said, L.: A computational model of emotions for the simulation of human emotional dynamics in emergency situations. J. Comput. Theor. Eng. **6**, 227–233 (2014)
10. Ortony, A., Clore, G.L., Collins, A.: The Cognitive Structure of Emotions. Cambridge University Press (1988)
11. RobocupRescue. http://www.robocup.org/robocup-rescue
12. Folkman, S., Lazarus, R.S.: Stress, Coping, and Hope. In: Carr, B.I., Steel, J. (eds) Psychological Aspects of Cancer, pp. 119–127. Springer, US (2013)
13. Gratch, J., Marsella, S.: A domain-independent framework for modeling emotion. Cogn. Syst. Res. **5**, 269–306 (2004)
14. Lim, M., Dias, J., Aylett, R., Paiva, A.: Creating adaptive affective autonomous NPCs. Auton. Agent. Multi. Agent. Syst. **24**, 287–311 (2012)
15. Dias, J., Mascarenhas, S., Paiva, A.: FAtiMA modular: towards an agent architecture with a generic appraisal framework: In: Bosse, T., Broekens, J., Dias, J., van der Zwaan, J. (eds). Emotion Modeling, pp. 44–56. Springer (2014)

16. Rao, A.S., Georgeff, M.P.: Modelling rational agents within a BDI-architecture. In: International Conference on Principles of Knowledge Representation and Reasoning, pp. 473–484 (1991)
17. Valent, P.: The ash wednesday bushfires in victoria. Med. J. Aust. 291–300 (1984)
18. Vázquez, C., Cervellón, P., Pérez-Sales, P., Vidales, D., Gaborit, M.: Positive emotions in earthquake survivors in El Salvador (2001). J. Anxiety Disord. **19**, 313–328 (2005)

Dynamic System of Rating Alternatives by Agents with Interactions

Radomír Perzina and Jaroslav Ramík

Abstract Business process simulation models usually incorporate several essential components that reflect customer behavior for modeling system inputs and outputs and ranking and/or rating given alternatives. In this paper we deal with a dynamic system of rating given number of alternatives based on agent-based simulation with interactions among agents which is based on a set of parameters and on pairwise comparisons method. Our system is able to replicate various examples of processes, e.g. financial market evaluation, evaluation of products' demand and supply, evaluation of political parties in general elections, evaluation of universities etc. A simple simulation experiment is presented and discussed.

Keywords Dynamic system · Agents · Ranking alternatives · Pairwise comparison matrix · Simulation

1 Introduction

In agent-based modeling (ABM), a system is modeled as a collection of autonomous decision-making entities—agents. Each agent individually assesses its situation and makes decisions on the basis of some rules. Agents may execute various behaviors appropriate for the system they represent, for example, producing, consuming, or selling, or a subjective evaluation. Repetitive competitive interactions between agents are a feature of agent-based modeling, which relies on the power of computers to explore dynamics out of the reach of pure mathematical methods, [1]. At the simplest level, an agent-based model consists of a system of agents and the relationships between them. Even a simple agent-based model can exhibit complex behavior

R. Perzina (✉) · J. Ramík
Faculty of Business Administration in Karviná, Silesian University in Opava, Univerzitní nám. 76, 733 40 Karviná, Czech Republic
e-mail: perzina@opf.slu.cz

J. Ramík
e-mail: ramik@opf.slu.cz

G. Jezic et al. (eds.), *Agent and Multi-Agent Systems: Technology and Applications*, Smart Innovation, Systems and Technologies 58,
DOI 10.1007/978-3-319-39883-9_14

patterns, see [2, 3], and provide valuable information about the dynamics of the real-world system that it emulates, [4]. In addition, agents may be capable of evolving, allowing unanticipated behaviors to emerge, [5].

A process simulation models usually incorporate several essential components that reflect agent behavior for modeling system inputs and outputs and ranking and/or rating given alternatives, see e.g. [6–8]. In this paper we deal with a dynamic system of rating given number of alternatives based on agent-based simulation with interactions among agents which is dependent on a set of parameters. Our system is able to replicate various examples of processes, e.g. financial market modeling, see [7, 8], auction models, [2], demand and supply models, [6], evaluation of political parties in general elections, evaluation of universities or other public institutions etc.

The structure of the paper can briefly be described as follows. In Sect. 2 the basic problem is formulated. In each time moment, a finite set of alternatives is ranked by each agent from the finite set of agents. Then, the total ranking is calculated. In the course of time, the individual agents interact with each other according to the given system of rules based on some parameters. A detailed agent-based model is described in Sect. 3. Finally, a simple simulation experiment is presented and discussed in Sect. 4. The conclusion section finalizes the paper.

2 The Problem

We shall deal with the following problem: Let $X = \{x_1, x_2, ..., x_n\}$ be a finite set of alternatives ($n > 1$). We consider a set of agents $K = \{1, 2, ..., |K|\}$, e.g. brokers, customers, electors, students etc. Here, by symbol $|K|$ we denote the (finite) number of elements of the set K.

Each agent $k \in K$ makes successively his/her decisions in every discrete time moment $t = 1, 2, ..., T$ by a pairwise comparisons matrix (PC matrix) $A(k, t)$. The aim is to get *global rating*, or, *ranking* (which is a subproblem), of the alternatives in each time moment t, using the information given by each agent k in the form of an $n \times n$ *individual pairwise comparison matrix*

$$A(k, t) = \{a_{ij}(k, t)\}. \tag{1}$$

In a given time t each agent k evaluates the pair of alternatives x_i, x_j by a positive real number $a_{ij}(k, t)$, for all i and j.

If $a_{ij}(k, t) > 1$, then x_i "is better then" x_j. The higher is $a_{ij}(k, t)$, the stronger is his/her evaluation that x_i "is better then" x_j.

On the other hand, if $0 < a_{ij}(k, t) < 1$, then x_j "is better then" x_i. The lower is the value of $a_{ij}(k, t)$, the stronger is his/her evaluation that x_j "is better then" x_i.

If $a_{ij}(k, t) = 1$, then both alternatives x_j, x_i are evaluated equally.

PC matrix (1) is assumed to be *reciprocal*, which is a natural requirement, see e.g. [9, 10]. Hence, we have

$$a_{ji}(k,t) = \frac{1}{a_{ij}(k,t)}, \text{ for all } i,j \in \{1,2,...,n\}, k \in K, t = 1,2,...,T. \quad (2)$$

The global rating of the alternatives $x_1, x_2, ..., x_n$ in time t is associated with the *global priority vector* $w(t) = (w_1(t), ..., w_n(t))$ which is calculated from the *global PC matrix* $A(t) = \{a_{ij}(t)\}$, by aggregation (i.e. by the geometric average over all agents) of individual PC matrices as follows, see [9]:

$$a_{ij}(t) = \left(\prod_{k=1}^{|K|} a_{ij}(k,t) \right)^{\frac{1}{|K|}}, \text{ for all } i,j \in \{1,2,...,n\}, t = 1,2,...,T. \quad (3)$$

The weights w_j of the global priority vector $w(t) = (w_1(t), ..., w_n(t))$ are calculated as row geometric averages of the global PC matrix as follows, [9]:

$$w_i(t) = \kappa(t) \left(\prod_{j=1}^{n} a_{ij}(t) \right)^{\frac{1}{n}}, \text{ for all } i \in \{1,2,...,n\}, t = 1,2,...,T. \quad (4)$$

Here, $\kappa(t)$ is a normalizing factor as the global priority vector should be normalized in every time moment $t = 1,2,...,T$, i.e.

$$\sum_{j=1}^{n} w_i(t) = 1, \text{ for } t = 1,2,...,T. \quad (5)$$

From (3) and (6) we obtain easily

$$\kappa(t) = \left(\sum_{j=1}^{n} \left(\prod_{j=1}^{n} a_{ij}(t) \right)^{\frac{1}{n}} \right)^{-1}, \text{ for all } t = 1,2,...,T. \quad (6)$$

The global priority vector is associated with the ranking of alternatives as follows:

$$\text{If } w_i(t) > w_j(t) \text{ then } x_i \succ x_j,$$

where $\succ$ stands for "is better then", [9, 10].

The rating of alternatives is given directly by the value of elements $w_i(t)$ of global priority vector $w(t)$. Our final task is to analyze the time series of global ratings of individual alternatives and look for some regularities and/or irregularities in their behavior.

3 Agent-Based Model

In the course of time, the individual agents interact with each other, e.g. personally, by social networks or otherwise. Particularly, in time $t = 1, 2, ..., T$, each agent $k \in K$ belongs to exactly one of s agent-types, $s \in S = \{1, 2, ..., |S|\}$, where $|S| > 1$, i.e. it belongs to exactly one of disjoint sets $K_1(t), K_2(t), ..., K_{|S|}(t)$ satisfying $K_1(t) \cup K_2(t) \cup ... \cup K_{|S|}(t) = K$, $K_1(t) \cap K_2(t) \cap ... \cap K_{|S|}(t) = \emptyset$. The set of agents $K = \{1, 2, ..., |K|\}$ is supposed to be constant in time $t = 1, 2, ..., T$.

Each group of agents $K_s(t)$ of type $s \in S$ is characterized by a positive number—parameter $\alpha_s > 0$, such that the dynamics of the changes of evaluations of the alternatives by agent $k \in K_s(t)$ is described by the following formula:

$$a_{ij}(k, t + 1)) = a_{ij}(k, t) + \alpha_s[a_{ij}(k, t) - a_{ij}(k, t - 1)] + e(k, t), \tag{7}$$

for all $i, j \in \{1, 2, ..., n\}, k \in K_s(t), t = 1, 2, ..., T$. Here, $e(k, t)$ is an *error* member with normal distribution $N(0, \sigma(k, t))$.

From (7) it follows that the pairwise comparisons of alternatives given by individual agents in time $t + 1$ depend on the previous evaluations in time t, and, on an increment of evaluations between time t and $t - 1$ modified by a constant α_s which is a characteristic parameter of the type s. Moreover, a small random value is added to the right hand side of Eq. (7) in order to model small non-specific effects.

In each time $t = 1, 2, ..., T$, the agents may meet, or interact, at random, and there is a probability that one agent may convince the other agent to follow his opinion. In addition, there is also a small probability that an agent changes his opinion independently. A key property of this model is that direct interactions between heterogeneous agents may lead to substantial opinion swings from the set of type $K_r(t)$, into a set of type $K_s(t)$, or, vice versa. Here, $r, s \in S\}, r \neq s$. As an example, we consider $S = 2$, $K_1(t)$ is the set of optimistic traders, $K_2(t)$ is the set of pessimistic traders.

The above mentioned swings from the set of type $K_r(t)$, into a set of type $K_s(t)$, are modeled as follows.

For each group of agents $K_s(t)$ of type $s \in S, t = 1, 2, ..., T$, we calculate the *group PC matrix* $A^s(t) = \{a^s_{ij}(t)\}$ as follows:

$$a^s_{ij}(t) = \left(\prod_{k \in K_s(t)} a_{ij}(k, t) \right)^{\frac{1}{|K_s(t)|}}, \text{ for all } i, j \in \{1, 2, ..., n\}. \tag{8}$$

Again, by $|K_s(t)|$ we denote the number of elements (i.e. agents) in the set $K_s(t)$.

Given $t \in \{1, 2, ..., T\}$, let $k \in K$, $s \in S$. We compute the "distance" $d(k, s, t)$ of PC matrix $A(k, t)$ from $A^s(t)$, i.e.

$$d(k, s, t) = \|\{a_{ij}(k, t)/a^s_{ij}(t)\}\|. \tag{9}$$

Here, $\|...\|$ is a matrix norm, e.g. for an $n \times n$ matrix $A = \{a_{ij}\}$, we use the norm, see e.g. [9]:

$$\|A\| = \max\{\max\{a_{ij}, \frac{1}{a_{ij}}\}|i,j = 1,...,n\}. \tag{10}$$

For $k \in K_s(t)$, denote

$$d(k, s^*, t) = \min\{ d(k, s, t)|s \in \{1, 2, ..., S\}\}. \tag{11}$$

If $d(k, s, t) > d(k, s^*, t)$, then in time $t + 1$ the agent k switches from type s to type s^*, i.e.

$$K_s(t+1) = K_s(t) - \{k\}, K_{s^*}(t+1) = K_{s^*}(t) \cup \{k\}. \tag{12}$$

Therefore, the number of agents will change as follows:

$$|K_s(t+1)| = |K_s(t)| - 1, \text{ and } |K_{s^*}(t+1)| = |K_{s^*}(t)| + 1. \tag{13}$$

If $d(k, s, t) \leq d(k, s^*, t)$, then, in time $t + 1$, the agent k does not switch. In this way we perform all switches of agents in K, i.e. we obtain new sets $K_s(t + 1), s \in \{1, 2, ..., S\}$ in time moment $t + 1$.

Then we continue by computing new group PC matrices, new global PC matrix, global rating of alternatives, etc.

4 Simulation Experiment

Our problem formulated in Sects. 2 and 3 will be illustrated on a small simulation experiment.

Let $X = \{x_1, x_2, x_3, x_4\}$ be a set of four alternatives. We consider a set of agents $K = \{1, 2, ..., 15\}$. Each agent $k \in K$ makes successively his/her decisions in every discrete time moment $t = 1, 2, ..., 100$ by a PC matrix $A(k, t)$. The aim is to investigate global rating of the alternatives in each time moment t, particularly in the final time $t = 100$, which can be interpreted as a time of prediction of ranking of alternatives. To this aim we use information given by each agent k in the form of a 4×4 individual PC matrix.

The simulation starts with 4×4 PC matrix:

$$A(1,1) = \{a_{ij}(1,1)\} = \begin{pmatrix} 1.000 & 2.000 & 4.000 & 6.000 \\ 0.500 & 1.000 & 2.000 & 4.000 \\ 0.250 & 0.500 & 1.000 & 2.000 \\ 0.167 & 0.250 & 0.500 & 1.000 \end{pmatrix}. \tag{14}$$

Here, we use the well known 9 point scale by T. Saaty, see [10] with the following interpretation of comparisons of pairs x versus y:
1 ... x is equally important to y,
3 ... x is slightly more important to y,
5 ... x is strongly more important to y,
7 ... x is very strongly more important to y,
9 ... x is absolutely more important to y.
Values 2, 4, 6 and 8 are intermediate values, e.g. 4 means that x is between slightly more important and strongly more important to y.

The corresponding priority vector $w(t) = (0.513, 0.275, 0.138, 0.074)$, with the corresponding ranking of alternatives:

$$x_1 > x_2 > x_3 > x_4.$$

We shall consider two cases:

(1) Consider $|S| = 1$, i.e. in time t, each agent belongs only to one type of agents, $K = \{1, 2, ..., 15\}$ and $K = K_1(t)$ for all t. The values of parameters α_1 and the standard deviation σ of the error e belong to the interval $[0, 1]$. Then the elements of the pairwise comparison matrix $A(1, t)$ can be understood as stochastic time series and for a more detailed analysis well known Box-Jenkins' methodology can be applied, see e.g. [10].
(2) In time t, each agent belongs to exactly one of $|S| = 2$ agent-types, i.e. two disjoint sets $K_1(t), K_2(t)$ satisfying $K_1(t) \cup K_2(t) = K$, $K_1(t) \cap K_2(t) = \emptyset$. The set of agents $K = \{1, 2, ..., 15\}$, is constant in time $t = 1, 2, ..., 100$. Other initial matrices $A(k, 1)$, $k = 2, 3, ..., 15$, are modifications of $A(1, 1)$ by a small random error e with the normal distribution $N(0, \sigma)$. Moreover, the agents belong to 2 initial types as follows: $K_1(1) = \{1, 2, \ldots, 8\}, K_2(1) = \{9, 10, ..., 15\}$. The values of parameters are set as follows: $\alpha_1 = 0.2, \alpha_2 = 0.3$.

Each type $s \in S = \{1, 2\}$ of agents $K_s(t)$ is characterized by an appropriate positive parameters α_1, α_2 and the dynamics of the changes of evaluations of the alternatives by agent $k \in K_s(t)$ is given by formula (7). For computing distances of an agent's PC matrix to the group PC matrix, we use norm (10). Changes of number of individual agents of a given type are given by formulas (12) and (13). For an illustration, the results of simulation computations for various values of parameters in the form of time series of weights of the alternatives are depicted.

In Fig. 1, we can see that the weights $w_i(t), i = 1, 2, 3, 4$, of alternatives x_i remain approximately constant in the course of time t, $t = 1, 2, ..., 100$. Here, we consider a small error e, i.e. small value of the parameter $\sigma = 0.1$. Consequently, in the final time moment $t = 100$ the vector of weights will not change much:

$$w(100) = (0.556; 0.236; 0.150; 0.059),$$

with the corresponding ranking of alternatives:

$$x_1 > x_2 > x_3 > x_4.$$

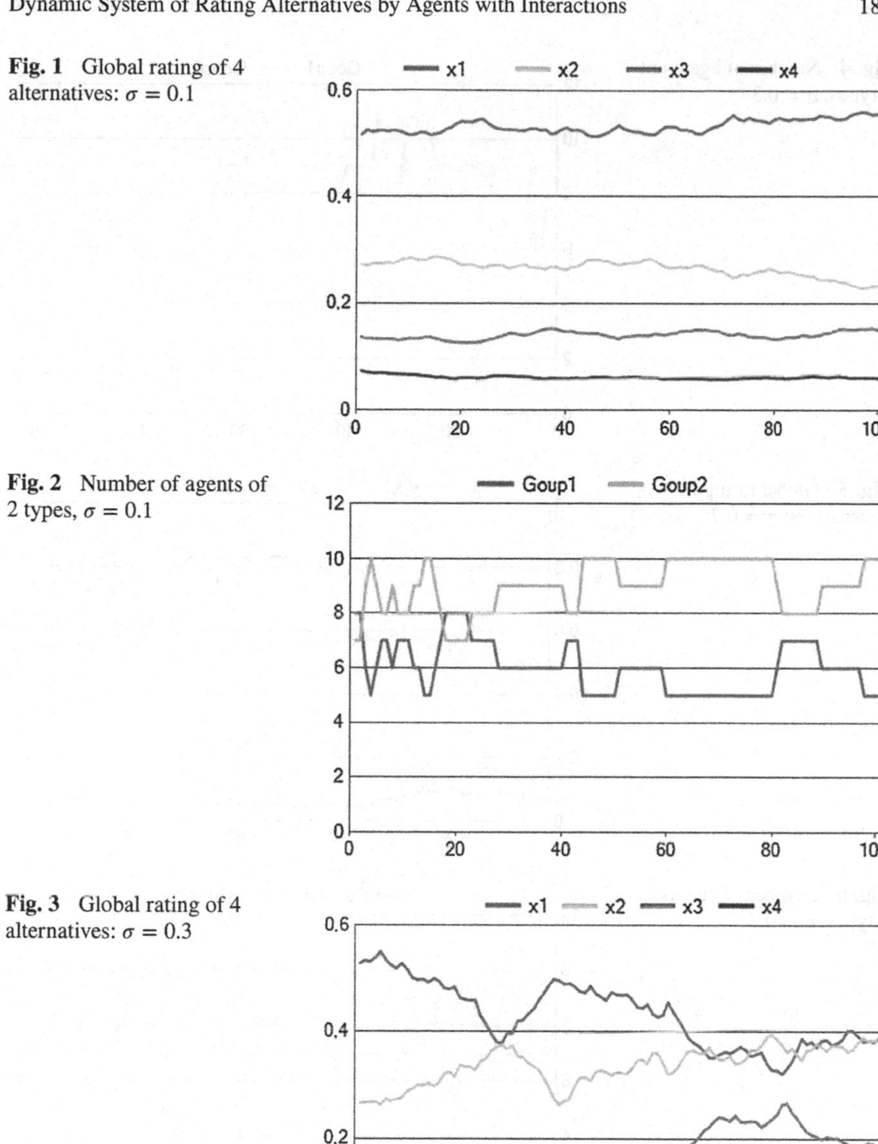

Fig. 1 Global rating of 4 alternatives: $\sigma = 0.1$

Fig. 2 Number of agents of 2 types, $\sigma = 0.1$

Fig. 3 Global rating of 4 alternatives: $\sigma = 0.3$

In Fig. 2, the number of agents of both types are depicted depending on time t.

In Figs. 3 and 4, a similar simulation has been performed, now, with $\sigma = 0.3$. The other parameters remain the same as in the previous case.

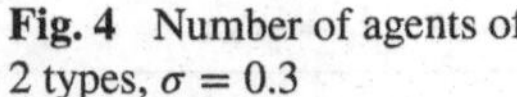

Fig. 4 Number of agents of 2 types, $\sigma = 0.3$

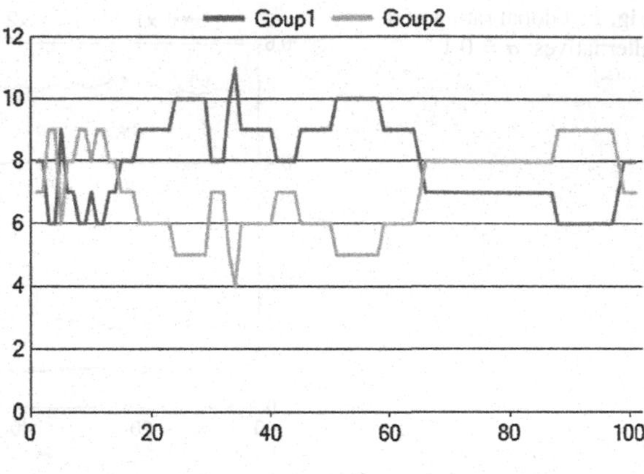

Fig. 5 Global rating of 4 alternatives: $\sigma = 0.7$

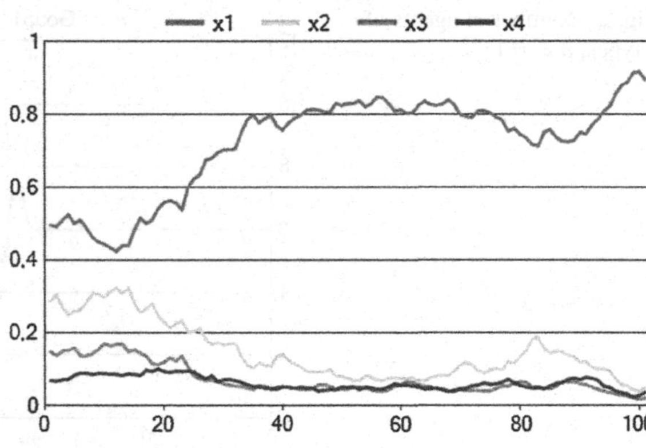

Fig. 6 Number of agents of 2 types, $\sigma = 0.7$

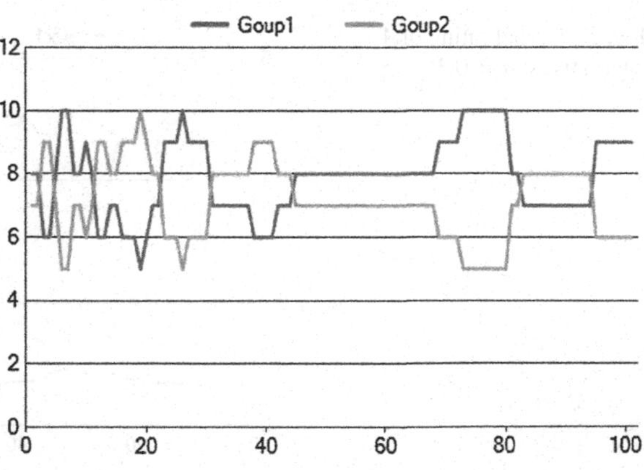

Moreover, in Figs. 5 and 6, we have increased the value of the error to the value $\sigma = 0.7$. As we can see, in the final time moment $t = 100$ the vector of weights is as follows:

$$w(100) = (0.895; 0.049; 0.020; 0.036),$$

with the corresponding ranking of alternatives:

$$x_1 > x_2 > x_4 > x_3.$$

The rating of the alternatives has changed a lot in 100 time moments.

5 Conclusion

This paper deals with the problem of ranking a finite number of alternatives by pairwise comparison matrices. A dynamic system of rating given number of alternatives based on agent-based simulation with interactions among agents which is based on a set of parameters has been presented. This system is able to replicate various examples of business and social processes, e.g. financial market, products' demand and supply, electors' preferences of political parties, etc. A simple simulation example has been presented and discussed.

Acknowledgments This paper was supported by the Ministry of Education, Youth and Sports Czech Republic within the Institutional Support for Long-term Development of a Research Organization in 2016.

References

1. Axelrod, R.: The Complexity of Cooperation: Agent-Based Models of Competition and Collaboration. Princeton University Press, Princeton, NJ (1997)
2. Anthony, P., Jennings, N.R.: Evolving bidding strategies for multiple auctions. In: Proceedings of 15th European Conference on Artificial Intelligence, pp. 182–187. Netherlands (2002)
3. Walsh, W.E., et al.: Analyzing complex strategic interactions in multi-agent games. In: Proceedings of AAAI Workshop on Game-Theoretic and Decision-Theoretic Agents (2002)
4. Dumas, M., La Rosa, M., Mendling, J., Reijers, H.: Fundamentals of Business Process Management. Springer, Berlin (2013)
5. Bonabeau, E.: Agent-based modeling: methods and techniques for simulating human systems. PNAS **99**, 7280–7287 (2002)
6. Barnett, M.: Modeling and simulation in business process management (2013). http://news.bptrends.com/publicationfiles
7. Sperka, R., Spisak, M.: Transaction costs influence on the stability of financial market: agent-based simulation. J. Bus. Econ. Manage. **14**(Supplement 1), S1–S12 (2013)
8. Wooldridge, M.: MultiAgent Systems: An Introduction, 2nd edn. Wiley, Chichester (2009)
9. Ramik, J.: Pairwise comparison matrix with fuzzy elements on alo-group. Inf. Sci. **297**, 236–253 (2015)
10. Saaty, T.L.: Multicriteria Decision Making—the Analytical Hierarchy Process, vol. I. RWS Publications, Pittsburgh (1991)

Traffic Speed Prediction Using Hidden Markov Models for Czech Republic Highways

Lukáš Rapant, Kateřina Slaninová, Jan Martinovič and Tomáš Martinovič

Abstract One of the main tasks of Intelligent Transportation Systems is to predict state of the traffic from short to medium horizon. This prediction can be used to manage the traffic both to prevent the traffic congestions and to minimize their impact. This information is also useful for route planning. This prediction is not an easy task given that the traffic flow is very difficult to describe by numerical equations. Other possible approach to traffic state prediction is to use historical data about the traffic and relate them to the current state by application of some form of statistical approach. This task is, however, complicated by complex nature of the traffic data, which can, due to various reasons, be quite inaccurate. The paper is focused on finding the algorithms that can exploit valuable information contained in traffic data from Czech Republic highways to make a short term traffic speed predictions. Our proposed algorithm is based on hidden Markov models (HMM), which can naturally utilize data sources from Czech Republic highways.

Keywords Markov chains · Hidden markov models · Traffic speed prediction · ASIM · FCD

L. Rapant (✉) · K. Slaninová · J. Martinovič · T. Martinovič
IT4Innovations National Supercomputing Center, VŠB - Technical University of Ostrava, 17. listopadu 15/2172, 708 33 Ostrava - Poruba, Czech Republic
e-mail: lukas.rapant@vsb.cz

K. Slaninová
e-mail: katerina.slaninova@vsb.cz

J. Martinovič
e-mail: jan.martinovic@vsb.cz

T. Martinovič
e-mail: tomas.martinovic@vsb.cz

G. Jezic et al. (eds.), *Agent and Multi-Agent Systems: Technology and Applications*, Smart Innovation, Systems and Technologies 58,
DOI 10.1007/978-3-319-39883-9_15

1 Introduction

One of the main tasks of Intelligent Transportation Systems is to predict state of the traffic in short to medium horizon. This prediction can be then applied in various areas of traffic modeling, analysis, and management, for example travel time prediction and traffic congestion prevention. This prediction can be done by two general approaches. The first approach requires us to have a mathematical description of the traffic flow and its starting and boundary conditions. Examples of this approach are articles written by Tang et al. [13, 14] and Costesqua et. al. [5]. This approach is feasible in simple traffic situations but quickly becomes intractable in case of inaccurate or incomplete data regarding both conditions. Other possible approach to traffic state prediction is to use historical data about traffic and relate them to the current state of the traffic by application of some statistical or machine learning approach. Examples of this approach are described by Calvert et al. [3], Huang [9] and some other authors, whose works will be discussed more thoroughly later. This task is, however, also not a simple one. It is complicated by complex nature of the traffic data. They can, depending on type of their source, be inaccurate or not available in required quantity. Despite this fact, this approach is preferable in case of traffic state prediction on Czech Republic highways due to the unavailability of boundary condition data (i.e. there are not measurements on all ramps leading to and out of highway).

A lot of work has already been done in this area. Georgescu et al. [7] propose to use a multi-level explanatory model (MTM) with linear regression to make a short term traffic speed prediction. They use stationary sensors as their data source. On the other hand, Gopi et al. [8] propose to use Bayesian Support Vector Regression (BSVR). BSVR has the advantage in measuring uncertainty of the prediction. They also work with stationary senor data. On the other hand, De Fabritiis et al. [6] work with floating car data. They propose pattern matching approach in combination with artificial neural network and they proved this approach feasible, but requiring higher penetration of the traffic by floating cars. Paper written by Jiang et al. [11] presents a comparison of several approaches to the problem of traffic speed prediction. They compare models coming from machine learning like artificial neural networks to the statistical models like ARIMA and VAR. They conclude that both kinds of models have several advantages and disadvantages, with machine learning methods coming slightly ahead. Data used for their predictions are coming from the stationary sensors. Asif et al. [2] present yet another approach to the traffic speed prediction using ν-Support Vector Regression using clustering approach to historical data and prove it to be preferable to artificial neural network. The closest to our approach is work presented in paper written by Jiang and Fey [10]. They also use hidden Markov model (HMM) and use it in combination with neural network prediction. However, they use it on artificial data coming from the traffic simulation and are working with different kind of HMM.

Important part of any traffic state prediction is traffic data availability. Generally, data sources describing actual traffic situation on Czech highway can be divided into two groups—stationary data sources (ASIM sensors) and floating car data sources (Floating Car Data).

ASIM sensors are placed on certain toll gates (all these toll gates are placed on the highways). They comprise of various sensors like passive infrared detectors and radars. They are able to distinguish individual vehicle types, and measure their speed and intensity. Their measurements are aggregated every five minutes and mean speed and intensity are calculated. They have many advantages. Perhaps most important advantage is that there is no need for equipping vehicles with additional electronic devices. Consequently, a sensor measures speeds of all vehicles going through it. The second important advantage is a detail of the data. ASIM sensors provide separate information about every lane of the monitored road and can distinguish individual vehicle types. However, these sensors have some serious disadvantages. The ASIM sensor network is very sparse. There are only about 120 toll gates equipped with ASIM sensors; all of these are placed on the motorways. There are also other limitations. Electronic toll gates divide roads into fragments of various length, some of them may extend to many kilometers. Thus, data obtained from ASIM sensors exactly describe only traffic situation around the tollgate.

The other available source is Floating Car Data (FCD). The floating car data (FCD) technique is based on the exchange of information between floating cars traveling on a road network and a central data system. The floating cars periodically send the recent accumulated data on their positions, whereas the central data system tracks the received data along the traveled routes. The frequency of sending/reporting is usually determined by the resolution of data required and the method of communication. FCD have specific discretization. For example, D1 highway is divided into the sections (TMC segments) with length from several hundred meters to few kilometers. The traffic speed is calculated each minute as a mean of speed of all floating cars that passed through the section in the last minute.

This approach has also several advantages and drawbacks. The number of cars equipped with a GPS unit has doubled over the past five years. It can be expected that the trend will continue. It implies that the number of potential data sources will increase. Moreover, data from GPS receivers is not limited to the predefined places so the coverage is much larger than in the case of the stationary data. Drawbacks come mostly from the use of GNSS technology. GPS device can fail to provide precise outputs or the outputs can be intentionally distorted. This imprecision can have an impact on positioning, ranging from meters to tens of meters. Because GNSS is based on satellite technology, GPS receiver has to be able to receive signals from several satellites. This can prove to be difficult in some cases (cities, forests).

Both of these data sources are available to us thanks to the viaRODOS system.[1] Nature of these data sources limits us in several ways. Given the positioning of the ASIM sensors and properties of FCD, it is impossible for us to use any kind of exact model. Moreover, because FCD contain only information about speed, we are limited

[1] http://www.rodos-it4i.cz/defaultEN.aspx.

only to traffic speed prediction. Therefore, focus of this paper is to find the algorithms that can exploit all the possible information contained in the historical and current traffic data available for Czech Republic highways to make a short term traffic speed predictions on highways. We propose a statistical approach based on hidden Markov model, because it provides a framework, which allows us to naturally utilize both of the data sources. Theoretical foundations of these methods can be found in Sect. 2. Description of the algorithms and experimental results are presented in Sects. 3 and 4 respectively.

2 Hidden Markov Models

A hidden Markov model (HMM) is a kind of a statistical Markov model in which the system being modeled is assumed to be a Markov process with unobserved (hidden) states (for more thorough description see [12]). A HMM can be presented as a form of dynamic Bayesian network. It can be modeled as a directed graph, where vertices represent random variables and edges represent dependencies. In opposite to simpler Markov models, the state is not directly visible to the observer. But the output, dependent on the state, is visible. Each state has a probability distribution over the possible outputs. Therefore, the sequence of outputs generated by HMM gives some information about the sequence of states. The name 'idden' refers to the state sequence through which the model passes, not to the parameters of the model. The model is still referred to as a 'hidden' Markov model even in case if the parameters are known. General architecture of an HMM is shown in Fig. 1.

HMM structure comprises of two layers of vertices. Each vertex represents a random variable that can adopt any of a finite number of values. The random variables $x(t)$ represent the hidden states at time t. The random variables y(t) represent the emitted observation at time t. The oriented edges in the graph represent conditional dependencies. Figure 1 also shows, that the model has Markov property, i.e. the conditional probability distribution of the hidden variable $x(t)$ at time t, depends only on the value of the hidden variable $x(t-1)$, earlier states have no influence. In case of random variable $y(t)$, it again depends only on variable $x(t)$. In our case, we consider both values of x and y to be discrete, but emissions y can be modeled as continuous (using Gaussian distribution).

The HMM has three types of parameters: transition probabilities and emission probabilities and initial state probability. The transition probabilities determine the probability, that the hidden state $x(t)$ is chosen given the hidden state $x(t-1)$.

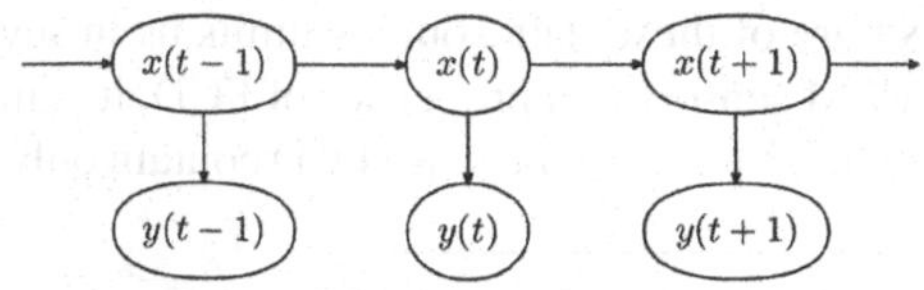

Fig. 1 Snapshot of hidden Markov model

The hidden state space is assumed to consist of N values which are modeled as a categorical distribution. Therefore, it is possible for each of the N possible states that are represented by a hidden variable at time t to determine a transition probability from this state to each of the N possible states at $t+1$. These transition probabilities are usually represented by a $N \times N$ matrix A called transition or Markov matrix. Rows of this matrix represent transition probabilities from the state denoted by row number to all other states (including itself) and naturally sum to 1. Emission probabilities are defined for each of the N possible states. They represent the distribution of the observed variable at a particular time given the state $x(t)$. The size and representation of these probabilities depends on the form of the observed variable. In our case the observed variable is discrete with M possible values, governed by a categorical distribution. Therefore, these probabilities can be represented by a $N \times M$ matrix B called emission matrix.

Initial state probability P is vector of length N and contains probabilities, that HMM begins in certain state.

HMM can perform many inference tasks. We are especially interested in two tasks: estimating the transition and emission probabilities for a hidden Markov model and finding probability of observing a certain emission sequence. The first task is to estimate matrices A and B given some emission sequence and their approximations A' and B'. This task can be performed by Baum-Welsh algorithm (description can be found in [4]). The second task is one of finding probability of observing a certain emission sequence $O = (o_1, o_2, \dots, o_t)$ i.e. what is $P(O|\pi)$, where $\pi = (A, B, P)$ is the hidden Markov model. This task can be efficiently solved by the Forward-backward algorithm (description can be found in [4]).

3 Algorithm

Now let us present you how our algorithm works. As it has been mentioned in Introduction, our algorithm is based on use of both historical and current traffic data. These data are represented by speed time series coming from the two available sources: ASIM sensor network and FCD. Due to the scarcity of ASIM sensors, it is more interesting to predict the traffic speed based on the FCD, which are available for all sections of the Czech Republic highways. However, these data can be quite inaccurate due to the lower penetration of the traffic by floating cars or other technical issues. Therefore, ideal prediction algorithm can utilize both the accurate but strongly localized ASIM sensor data and widely available FCD, which suffer from the fact that they only measure speed of part of the traffic flow.

Solution for this problem was found in the form of hidden Markov model. Both data sources can be utilized in the framework of HMM. ASIM sensor data can be utilized for training of the model and roughly represent the hidden state sequence. The prediction is then done on the FCD, which are regarded as emission. General progress of our algorithm can be summarized into following steps:

1. Create approximation of transition and emission matrices.
2. Train HMM on historical data from the FCD and approximation matrices.
3. Generate n predictions (emissions) and compute their likelihood.
4. Choose m most probable sequences and compute their weighted mean.

Let us go through individual steps of the algorithm more thoroughly. In the first step, ASIM data are utilized to create approximated transition matrix A' from the nearest ASIM sensor to the selected FCD section. This matrix is calculated like transition matrix for the standard Markov chain from the historical data (in our case we have rounded speed to the km/h to create categorical distribution, for example 100 km/h represent one possible state of the model). Our approximated emission matrix B' has same size as transition matrix (i.e. relate speeds from ASIM to speeds from FCD) and is based on measurement error of FCD, i.e. in physics, measurements tend to have Gaussian distribution of errors. Therefore, approximated emission matrix relates hidden states to emissions using Gaussian distribution of probability in each row, centered on the true value, so it is most probable that hidden state emit itself, less probable that it emits value slightly smaller or greater and so on. During our experiments, it has been found that Gaussian with slightly negative kurtosis performs the best in this task.

The second step, training of the model, is performed by standard Baum-Welsh algorithm. Both approximated matrices A' and B' and historical FCD speed time series (i.e. emissions) from the predicted segment are used as an input. As the result, we receive true transition and emission matrices (A and B) of the model. Note that trained model is only suitable for predicting the speed on FCD section on which it was trained.

In the next step, n emission sequences of certain length are generated and connected to the last values of the predicted FCD time series creating extended emission sequence e_i. This step is done to maintain the continuity between generated sequence and time series. Then each extended emission sequence probability $P(e_i|\pi)$ is calculated using Forward-backward algorithm. The m most probable extended emission sequences are chosen and their weighted mean is calculated as:

$$p = \frac{1}{W}\sum_{i=1}^{m} w_i e_i, \tag{1}$$

where p is predicted sequence, w_i are individual weights and W is sum of all weights. This step is included to diminish probability of getting the false classification of predicted emission sequence. In our case, we were using linearly decreasing weights w. The result of this weighted mean p is our predicted emission sequence corresponding to FCD speed values.

4 Experimental Results

Our algorithm was implemented in Matlab environment and utilized parallelization to accelerate the computation. From the performance perspective, only problematic part is training of HMM which can easily take tens of minutes. However, given enough historical data, it is not necessary to frequently update the model. The parallelized computation of prediction with trained model takes only tens of seconds.

The algorithm was tested on data coming from Czech D1 highway from the period of 3.1.2014 to 30.4.2014. We have chosen section from 43rd to 47th km in the direction of Prague. This section contains one ASIM sensor, which was used for the calculation of the approximated transition matrix A'. We have used data from January to March for training and data from April for evaluation. Root Mean Square Error (mathematical description can be found in [1]) was chosen as a measure for quality of the prediction. Five minutes were chosen as a prediction step, due to the aggregation period of ASIM sensors, which is 5 min long. Each prediction produces six such steps, so the traffic speed during next 30 min is predicted.

The first experiment was performed on one such 30 min prediction at randomly chosen time in April. We have done this prediction 100 times and compared the results in terms of accuracy. Graph of RMSE of individual predictions can be seen in Fig. 2.

With mean RMSE equaling to 5.32 and none higher than 8.6, the algorithm performs comparably to other contemporary approaches. The best and the worst case prediction determined by their RMSE are shown in the Fig. 3. Prediction starts at a time window 20; previous time windows are the last values from the predicted FCD time series.

In the best case, traffic speed is predicted almost perfectly both in terms of magnitude and trend. The worst case scenario is still capable of capturing the trend but it is

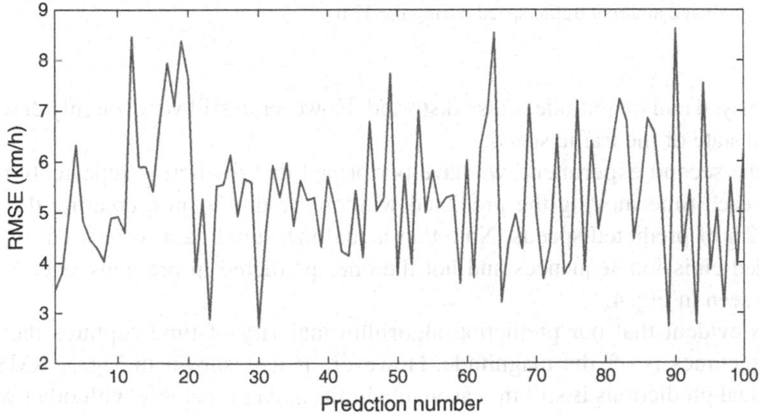

Fig. 2 RMSE of individual predictions

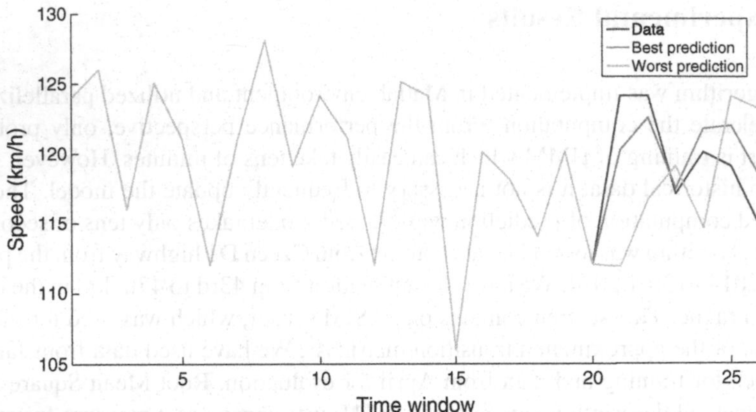

Fig. 3 The best and the worst case speed prediction

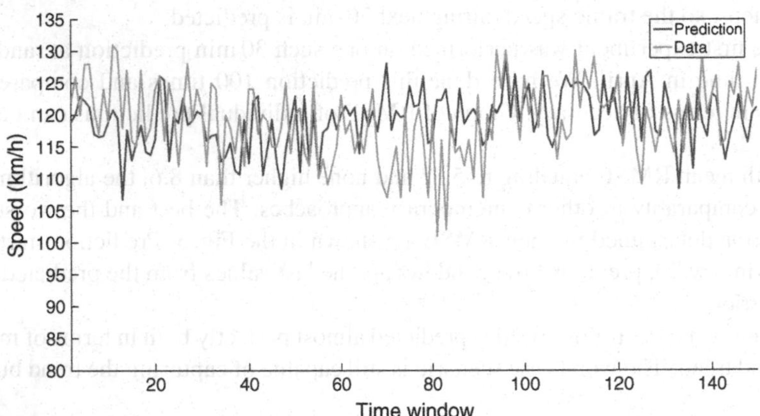

Fig. 4 Predicted and real traffic speed during the 12 h

a bit delayed and magnitude is also distorted. However, it still very roughly describes the real state of the traffic speed.

In the second experiment, we have performed this prediction repeatedly for 25 times, each time moving the predicted window by the 30 min, creating the more than 12 h of predicted speeds. Note that actual measured data were used to create extended emission sequences and not the ones predicted in previous step. Results can be seen in Fig. 4.

It is evident that our prediction algorithm majority of time captures the trend but sometimes is off the magnitude. However, as it is shown in Fig. 5, RMSE of individual predictions is still in reasonable limits and comparable with other works. However, this problem with magnitude of speed will be dealt with in near future. It may be a result of too small training set used for training or consequence of some generalization in our model design.

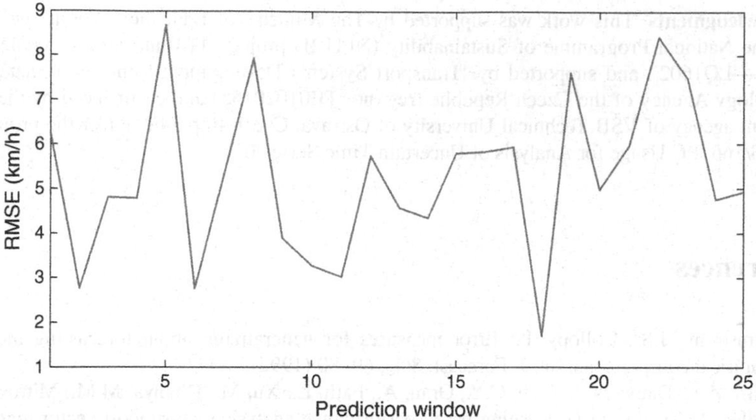

Fig. 5 RMSE of individual predicted windows

5 Conclusion

In this article, we have presented an approach to traffic speed prediction on Czech Republic highways. This approach is based on hidden Markov model, because it is not only useful tool for prediction, but can be used to naturally utilize all data sources from Czech Republic highways. In our algorithm both ASIM sensors and FCD were used as valuable data sources. We have utilized both accuracy of the ASIM data and availability of the floating car data, while overcoming some of their issues.

In comparison with other approaches like that of Jiang [10], our algorithm fared comparably, although with some issues with magnitude of speed. However, even with this issue, RMSE was at worst the same as in case of other approaches. Main advantage of our algorithm is, that in contrast with many other algorithms, it works with real world traffic data from FCD, which are much more difficult to handle than the data from stationary sensors or the data from traffic simulations. Results have proven, that our model describes the behavior of FCD in Czech Republic well.

This algorithm is still at the beginning of development and will utilize many improvements in the future. The first improvement will be to resolve the issue of speed magnitude. This will be done by enlarging the training set and trying to find some calibration method to better relate the magnitudes between model and extended emissions. Related to the enlarging of the dataset is the task to implement more efficient training of HMM. Then we would be able to train the HMM on much larger data sets. We also plan to find other possible approaches to utilization of both data sources in the structure of HMM.

Acknowledgments This work was supported by The Ministry of Education, Youth and Sports from the National Programme of Sustainability (NPU II) project 'IT4Innovations excellence in science—LQ1602', and supported by 'Transport Systems Development Centre' co-financed by Technology Agency of the Czech Republic (reg. no. TE01020155) and co-financed by the internal grant agency of VŠB Technical University of Ostrava, Czech Republic, under the project no. SP2016/166 'PC Usage for Analysis of Uncertain Time Series II'.

References

1. Armstrong, J.S., Collopy, F.: Error measures for generalizing about forecasting methods: empirical comparisons. Int. J. Forecast. **8**(1), 69–80 (1992)
2. Asif, M.T., Dauwels, J., Goh, C.Y., Oran, A., Fath, E., Xu, M., Dhanya, M.M., Mitrovic, N., Jaillet, P.: Unsupervised learning based performance analysis of v-support vector regression for speed prediction of a large road network. In: Proceedings of 15th International IEEE Conference on Intelligent Transportation Systems 2012, pp. 983–988. Anchorage (2012)
3. Calvert, S.C., Taale, H., Snelder, M., Hoogendoorn, S.P.: Application of advanced sampling for efficient probabilistic traffic modelling. Trans. Res. Part C Emerg. Technol. **49**, 87–102 (2014)
4. Cappe, O., Moulines, E., Ryden, T.: Inference in Hidden Markov Models. Springer, New York (2007)
5. Costesequea, G., Lebacque, J.P.: A variational formulation for higher order macroscopic traffic flow models: numerical investigation. Trans. Res.Part B Methodol. **70**, 112–133 (2014)
6. de Fabritiis, C., Ragona, R., Valenti, G.: Traffic estimation and prediction based on real time floating car data. In: Proceedings of 11th International IEEE Conference on Intelligent Transportation Systems 2008, pp. 197–203. Beijing (2008)
7. Georgescu, L., Zeitler, D., Standridge, C.R.: Intelligent transportation system real time traffic speed prediction with minimal data. J. Ind. Eng. Manag. **5**(2), 431–441 (2012)
8. Gopi, G., Dauwels, J., Asif, M.T., Ashwin, S., Mitrovic, N., Rasheed, U., Jaillet, P.: Bayesian support vector regression for traffic speed prediction with error bars. In: Proceedings of the 16th International IEEE Annual Conference on Intelligent Transportation Systems, pp. 137–141. Hague (2013)
9. Huang, M.L.: Intersection traffic flow forecasting based on v-gsvr with a new hybrid evolutionary algorithm. Neurocomputing **147**, 343–349 (2015)
10. Jiang, B., Fei, Y.: Traffic and vehicle speed prediction with neural network and hidden markov model in vehicular networks. In: Proceedings of IEEE Intelligent Vehicles Symposium 2015, pp. 1082–1087. Seoul (2015)
11. Jiang, H., Zou, Y., Zhang, S., Tang, J., Wang, Y.: Short-term speed prediction using remote microwave sensor data: machine learning versus statistical model. Math. Probl. Eng. **2016**, 1–13 (2016)
12. Rabiner, L.R.: A tutorial on hidden markov models and selected applications in speech recognition. Proc. IEEE **77**(2), 257–286 (1989)
13. Tang, T.Q., Li, J.G., Huang, H.J., Yang, X.B.: A car-following model with real-time road conditions and numerical tests. Measurement **48**, 63–76 (2014)
14. Tang, T.Q., Caccetta, L., Wu, Y.H., Huan, H.J., Yang, X.B.: A macro model for traffic flow on road networks with varying road conditions. J. Adv. Trans. **48**, 304–317 (2014)

Part III
Business Process Management

Business Process Modeling of Logistic Production Systems

Petr Suchánek and Robert Bucki

Abstract The paper highlights the problem of the logistic system modeling which consists of similar structures representing a synthetic manufacturing plant. Routes for products as well as matrixes of life and state are given. Structures differ from each other as they have various transport times from one production stand to another. Additionally, tool replacement times and machine maintenance times are given. There are also dedicated buffer stores which make the manufacturing process possible even if the subsequent machine is busy. Manufacturing strategies and heuristic algorithms are responsible for the control of the system. The main criterion is to minimize the total manufacturing time by means of searching for such a sequence of operations which often leads to avoiding bottlenecks. The main goal of the paper is to build a mathematical model of the above characterized system including proposals for algorithms for optimizing production in relation to orders.

Keywords Logistic system · Distributed system · Modeling · Simulation · Information approach · Mathematical modeling · Heuristic algorithms · Manufacturing strategies · Optimization · Business process

1 Introduction

Efficiency and productivity have always been decisive success factors in the field of manufacturing industries. More and more complex machinery and plants are the main engineering challenge relating to manufacturing matters. A high level of efficiency is demanded at each engineering stage as it leads to developing more

P. Suchánek (✉)
School of Business Administration in Karviná, Silesian University in Opava, Karviná
Czech Republic
e-mail: suchanek@opf.slu.cz

R. Bucki
Institute of Management and Information Technology, Bielsko-Biala, Poland
e-mail: rbucki@wsi.net.pl

G. Jezic et al. (eds.), *Agent and Multi-Agent Systems: Technology and Applications*, Smart Innovation, Systems and Technologies 58,
DOI 10.1007/978-3-319-39883-9_16

sophisticated methods. Innovative manufacturing technologies increase production efficiency and product quality. Dynamically developing production and business environment is forcing companies to adapt to this environment and optimize processes taking place in it. Processes must be optimized comprehensively because production must be directly tied to the purchase and distribution processes. Nowadays, universally known patterns show that the model of achieving high efficiency is the one in which production processes are directly linked to the specific order. In other words, there must be a direct link between the internal and external environment of the company [1]. The general aim of all businesses is to achieve the maximum efficiency with the minimal amount of resources [2]. Development of information technology has allowed widespread use of simulation methods which are important support systems for management, decision making and optimization within complex enterprise systems [3]. Simulation methods are based on mathematical models which are a necessary starting point for simulations using computer technology. Simulations can be used predictively i.e. to make it possible to model and analyze even non-existent, planned processes, which helps to manage the initial investment or support real-time control when they are used directly in connection with ERP (Enterprise Resource Planning), CRM (Customer Relationship Management), etc. [4]. It is also possible to use a variety of methods for comprehensive optimization using simulation, however we should be aware of the fact that there is a need to create an adequate simulation model. Moreover, then it is necessary to verify whether subsequent simulations and their outcomes are applicable [5, 6]. If production, purchasing, distribution, and other logistic approaches are discussed, then simulation models based on the principles of Petri nets can be used for the purposes of the simulation [7], multi-agent approach [8], heuristic approach [9, 10], hybrid simulation approaches [11], stochastic methods [12], etc. Information technology can process simulation models described by a large number of parameters and, for this purpose, commonly used tools and methods for data processing are marked as the so-called "big data" [13]. The most common key parameter within the simulation is its time when the timeout is always a lossy component both during production and distribution and during preparing and implementing new logistic processes [14]. Rapidly changing market needs often require a rapid adjustment of production technologies and, therefore, it is important to investigate the influence of the diversity of products on the complexity of assembly processes which is also a typical example for a simulation application [15]. In this context, flexible manufacturing is discussed, which means that flexibility in case manufacturing: is understood as the ability to deal with slightly or greatly mixed parts;

- allows variations in assembling components and variations in the sequence of processes;
- changes the production volume;
- changes the design of a certain product.

In many cases, the objective is to optimize enterprise processes to make them more effective, not to find a completely effective solution. The heuristic approach can be used for these purposes. A heuristic is understood as a technique designed for solving a problem more quickly when classic methods are too slow, or for finding an approximate solution when classic methods fail to find any exact solution. This is achieved by trading optimality, completeness, accuracy or precision for speed. In a way, it can be considered a shortcut. The general procedure always concerns creating a map of the system described by the adequate mathematical model which is subsequently used as a basis for the simulation software. In this paper the authors present a particular mathematical model based on a heuristic approach that can be converted into a simulation program and applied for a particular purpose in practice. Similar approaches can be found in e.g. [16] or [17]. Due to the maximum extent of the article only the mathematical model is given. It is subject to subsequent extension and is to be programmed similarly as in [16] a [17] in order to its possible use for practical goals and further research in the discussed field.

2 Mathematical Model

Let us assume that there is a matrix of N orders for M customers:

$$Z^k = \left[z^k_{m,n}\right], \quad m = 1, \ldots, M, \quad n = 1, \ldots, N, \quad k = 1, \ldots, K \tag{1}$$

where: $z^k_{m,n}$—the nth order for the mth customer at the kth stage. There are Π available manufacturing plants where orders can be made.

Let us assume the structure of the πth manufacturing system can be defined by means of the following matrix:

$$E = \left[e^{\pi}_{i,j}\right], \quad i = 1, \ldots, I, \quad j = 1, \ldots, J, \quad \pi = 1, \ldots, \Pi \tag{2}$$

where: $e^{\pi}_{i,j}$—the machine placed in the ith row of the jth column in the πth manufacturing plant.

The elements of the above matrix take the following values:

$$e^{\pi}_{i,j} = \begin{cases} 1 & \text{if the machine in the } i\text{th row of the } j\text{th column in the } \pi\text{th manufacturing} \\ & \text{plant exists and is equipped with a tool able to perform a predefined operation,} \\ 0 & \text{otherwise.} \end{cases}$$

It is assumed that each manufacturing machine marked as $e^{\pi}_{i,j}$ equipped with its dedicated tool can perform the same operation on each nth product. Moreover, it is further assumed that the time of this operation does not differ in case of various products.

Let us introduce the matrix of buffer stores in the manufacturing system:

$$B^{\pi}=\left[b^{\pi}_{i,j}\right],\quad i=1,\ldots,I,\quad j=1,\ldots,J,\quad \pi=1,\ldots,\Pi \tag{3}$$

where: $b^{\pi}_{i,j}$—the buffer store behind the machine placed in the ith row of the j column in the πth manufacturing plant.

The elements of the matrix of buffer stores take the values given below:

$$b^{\pi}_{i,j}=\begin{cases}1 & \text{if the buffer store behind the machine in the } i\text{th row of the } j\text{th} \\ & \text{column in the } \pi\text{th manufacturing plant exists,} \\ -1 & \text{otherwise.}\end{cases}$$

At the same time:

$b^{\pi}_{i,0}$—the entrance buffer store for the machine placed before the first column of the ith row in the πth manufacturing plant;

$b^{\pi}_{i,J}$—the final buffer store placed behind the machine in the last column of the ith row in the πth manufacturing plant.

The capacity of the buffer store is shown in the following matrix:

$$\Upsilon=\left[\upsilon\left(b^{\pi}_{i,j}\right)_{n}\right],\quad i=1,\ldots,I,\quad j=1,\ldots,J,\quad \pi=1,\ldots,\Pi,\quad n=1,\ldots,N \tag{4}$$

where: $\upsilon\left(b^{\pi}_{i,j}\right)_{n}$ —the capacity of the buffer store placed behind the jth column in the ith row in the πth manufacturing plant for the nth product (expressed in pieces).

It is assumed that the route of machines and buffer stores for manufacturing the nth product is the same in each πth manufacturing plant, however, the transport time of the product from $e^{\pi}_{i,j}$ to $e^{\pi}_{i',j'}$ through the $b^{\pi}_{i,j}$ (if it exists), $j<j'$ may differ. Moreover, these times represent transport times of products to a buffer store, manipulation times in the buffer stores, storing times in buffer stores and transporting products to the subsequent machine.

Times of storing semi-products in buffer stores depend on the course of manufacturing and are added to the total manufacturing time.

The matrix of routes takes the following form:

$$D=\left[d^{\eta}_{j,n}\right],\quad j=1,\ldots,J,\quad n=1,\ldots,N,\quad \eta=1,\ldots,\mathrm{H} \tag{5}$$

where: $d^{\eta}_{j,n}$—the number of the ith row the jth column with the machine dedicated for making the nth product in the ηth step of the manufacturing process.

Actually, $d^{\eta}_{j,n}=i$ if the operation is carried out in it and $d^{\eta}_{j,n}=-i$ otherwise.

The matrix of routes of buffer stores takes the following form:

$$B=\left[b^{\eta}_{j,n}\right],\quad j=1,\ldots,J,\quad n=1,\ldots,N,\quad \eta=1,\ldots,\mathrm{H} \tag{6}$$

where: $b^{\eta}_{j,n}$—the number of the *i*th row the *j*th column of the buffer store dedicated for storing the *n*th product after the *η*th step of the manufacturing process.

Actually, $b^{\eta}_{j,n}=i$ if the buffer store is used for storing the *n*th product and $b^{\eta}_{j,n}=-i$ if it either does not exist, is not used for storing the *n*th product or is not necessary in the manufacturing system any more.

It is assumed that the FIFO method is implemented for storing semi-products in buffer stores.

Let us define the life matrix of tools in machines in the *π*th manufacturing plant:

$$G=\left[g(\pi)_{i,j}\right],\quad i=1,\ldots,I,\quad j=1,\ldots,J,\quad \pi=1,\ldots,\Pi \tag{7}$$

where: $g(\pi)_{i,j}$—the life of the tool placed in the machine in the *i*th row of its *j*th column in the *π*th manufacturing plant (expressed in conventional time units).

Let us define the state matrix of tools in machines in the *π*th manufacturing plant:

$$S^k=\left[s(\pi)^k_{i,j}\right],\quad i=1,\ldots,I,\quad j=1,\ldots,J,\quad \pi=1,\ldots,\Pi,\quad k=1,\ldots,K \tag{8}$$

where: $s^{\pi}_{i,j}$—the state of the tool placed in the machine in the *i*th row of its *j*th column in the *π*th manufacturing plant at the *k*th stage (expressed in conventional time units). Let us define the flow capacity matrix of tools in machines in the *π*th manufacturing plant:

$$P^k=\left[p(\pi)^k_{i,j}\right],\quad i=1,\ldots,I,\quad j=1,\ldots,J,\quad \pi=1,\ldots,\Pi,\quad k=1,\ldots,K \tag{9}$$

where: $p^{\pi}_{i,j}$—the flow capacity of the tool placed in the machine in the *i*th row of its *j*th column in the *π*th manufacturing plant at the *k*th stage (expressed in conventional time units).

Consequently, the flow capacity of the tool placed in the machine in the *i*th row of its *j*th column in the *π*th manufacturing plant at the *k*th stage can be calculated as follows:

$$p(\pi)^k_{i,j}=g^{\pi}_{i,j}-s(\pi)^k_{i,j} \tag{10}$$

Let us introduce the matrix of times representing the times of transporting the nth element within the πth manufacturing system during the production process:

$$\vec{T}(\pi) = \left[\begin{array}{c}\tau(\pi)^n{}_{W\to b_{i,0}} \\ b_{i,0}\to e_{i,1} \\ W\to e_{i,1}\end{array}\;\vdots\;\begin{array}{c}\tau(\pi)^n{}_{e_{i,j}\to b_{i,j}} \\ b_{i,j}\to e_{i',j'} \\ e_{i,j}\to e_{i'',j'}\end{array}\;\vdots\;\begin{array}{c}\tau(\pi)^n{}_{e_{I,J}\to b_{I,J}} \\ b_{I,J}\to Z \\ e_{I,J}\to Z\end{array}\right] \quad (11)$$

$i=0, 1, \ldots, I, j=0, 1, \ldots, J,, j<j', n=1, \ldots, N, \pi=1, \ldots, \Pi,$

where:

$\tau(\pi)^n{}_{e_{i,j}\to b_{i,j}}$ $b_{i,j}\to e_{i',j'}$ $e_{i,j}\to e_{i',j'}$	the time of transporting the nth product to its buffer store ($e_{i,j}\to b_{i,j}$), from the buffer store to the subsequent machine ($b_{i,j}\to e_{i',j'}$) or directly to the subsequent machine if there is no buffer store for this operation ($e_{i,j}\to e_{i',j'}$),
$\tau(\pi)^n{}_{W\to b_{i,0}}$ $b_{i,0}\to e_{i,1}$ $W\to e_{i,1}$	the time of transporting the nth product from the store of charges to the first buffer store ($W\to b_{i,0}$), from the first buffer store to the first machine ($b_{i,0}\to e_{i,1}$) or directly to the first machine ($W\to e_{i,1}$),
$\tau(\pi)^n{}_{e_{I,J}\to b_{I,J}}$ $b_{I,J}\to Z$ $e_{I,J}\to Z$	the time of transporting the nth product to the last buffer store ($e_{I,J}\to b_{I,J}$), from the buffer store to the matrix of orders ($b_{I,J}\to Z$) or directly to the matrix of orders ($e_{I,J}\to Z$)

Let introduce the matrix of manufacturing times in the πth manufacturing plant:

$$T^n_{pr} = \left[\tau(n)^{pr}_{i,j}\right] \quad (12)$$

where: $\tau(n)^{pr}_{i,j}$—the production time of the nth product in the machine placed in the ith row of the jth column.

Let us introduce the matrix of replacement times of tools times in the πth manufacturing plant:

$$T^n_{pr} = \left[\tau(n)^{repl}_{i,j}\right] \quad (13)$$

where: $\tau(n)^{repl}_{i,j}$—the replacement time of the tool in the machine placed in the ith row of the jth column in case of making the nth product.

It is assumed that there is a sufficient number of tools which can be replaced.

Let us introduce the matrix of maintenance times of machines in the πth manufacturing plant:

$$T_{\text{main.}} = \left[\tau^{main.}_{i,j}\right] \quad (14)$$

where: $\tau_{i,j}^{main.}$—the maintenance time of the machine placed in the ith row of the jth column in πth manufacturing plant.

It is assumed that the maintenance time of the machine placed in the ith row of the jth column in discussed manufacturing plants differ.

3 Manufacturing Strategies

Orders can be made in any of the Π manufacturing plants. The adjustment of the nth order for the mth customer to the πth manufacturing plant is expressed by the following matrix:

$$\Xi = \left[\xi_{(m,n)\rightarrow\pi}\right], \quad m=1,\ldots,M, \quad n=1,\ldots,N, \quad \pi=1,\ldots,\Pi \qquad (15)$$

where: $\xi_{(m,n)\rightarrow\pi}$—adjustment of the nth order of the mth customer to the πth manufacturing plant.

At the same time elements of this matrix take the following values:

$$\xi_{(m,n)\rightarrow\pi} = \begin{cases} 1 & \text{if the } n\text{th order of the } m\text{th customer is adjusted to the } \pi\text{th manufacturing plant,} \\ 0 & \text{otherwise.} \end{cases}$$

On some occasions it is reasonable to make orders in a manufacturing plant situated closer to them in order to minimize the final product cost by means of minimizing transport costs.

To solve this problem from the point of mathematical modeling it is necessary to introduce the matrix responsible for choosing the adequate manufacturing plant for the given order n of the mth customer on the basis of the chosen criterion. As an example, the distance criterion can be the basis for the matrix which is responsible for choosing the right manufacturing plant:

$$Y = [y_{m\rightarrow\pi}], \quad m=1,\ldots,M, \quad \pi=1,\ldots,\Pi \qquad (16)$$

where: $y_{m\rightarrow\pi}$—the distance from the mth customer to the πth manufacturing plant (expressed in km).

At the same time if $y_{m\rightarrow\pi} = -1$, then it is not possible to match the mth customer's order with the πth manufacturing plant for some reasons (e.g. the lack of transport means, natural barriers, business conflicts, etc.).

4 Heuristic Algorithms

Control of the production system requires implementing adequate methods which can be grouped as follows:

1. Algorithm α_1 chooses the nearest πth manufacturing plant for the mth customer's order.
2. Algorithm α_2 chooses the biggest nth order placed the nearest the πth manufacturing plant.
3. Algorithm α_3 chooses the πth manufacturing plant characterized by the lowest total state of tools in all machines for the nearest mth customer.
4. Algorithms α_1 and α_2 at random.
5. Algorithms α_1 and α_3 at random.
6. Algorithms α_2 and α_3 at random.
7. Algorithms α_1,α_2 and α_3 at random.
8. The mth customer and the πth manufacturing plant at random.
9. The nth order and the πth manufacturing plant at random.

5 Conclusions

Simulations are now crucial and, in many cases, an indispensable element of management systems. The basis of the simulation is to create a mathematical model which allows the exact expression of monitored parameters. Moreover, it lets us compare them with different states of the system. One possible approach for creating simulation models is a heuristic approach which is effectively usable for the purpose of optimizing logistics. The paper presents an example of a mathematical model of the specific logistic system which consists of similar structures representing the synthetic manufacturing plant. In the general form, this model is useful for a variety of practical cases and it can be customized for other specific ones. The mathematical model presented in the paper is the basis for creating the simulation model.

Acknowledgement This paper was supported by the Ministry of Education, Youth and Sports in the Czech Republic within the Institutional Support for Long-term Development of a Research Organization in 2016.

References

1. Shu, L.B., Liu, S.F., Li, L.P.: Study on business process knowledge creation and optimization in modern manufacturing enterprises. In: First International Conference on Information Technology and Quantitative Management, pp. 1202–1208. Elsevier Science Bv, Amsterdam (2013)
2. Dlouhy, M.: Efficiency and resource allocation within a hierarchical organization. In: Proceedings of 30th International Conference Mathematical Methods in Economics, Pts I And Ii, pp. 112–116. Silesian Univ Opava, School Business Administration in Karvina, Karvina (2012)

3. Fanti, M.P., Iacobellis, G., Ukoyich, W., Boschian, V., Georgoulas, G., Stylios, C.: A simulation based decision support system for logistics management. J. Comput. Sci. **10**, 86–96 (2015)
4. Stumvoll, U., Nehls, U., Claus, T.: A simulation-based decision support system to update material planning parameters of an ERP-System in real-time. In: Simulation in Produktion und Logistk 2013, pp. 569–578. Heinz Nixdorf Inst, Paderborn (2013)
5. Werke, M., Bagge, M., Nicolescu, M., Lindberg, B.: Process modelling using upstream analysis of manufacturing sequences. Int. J. Adv. Manuf. Technol. **81**(9–12), 1999–2016 (2015)
6. Wang, H.B., Wang, Z.W.: Study on the method and procedure of logistics system modeling and simulation. In: Proceedings of the 2nd International Conference on Science and Social Research (Icssr 2013), pp. 776–780. Atlantis Press, Paris (2013)
7. Qian, Z.W., Sun, H.T.: A simulation study on production logistics balance based on petri net plus flexsim. In: Proceedings of the 2013 The International Conference on Education Technology and Information System (Icetis 2013), pp. 1029–1034 (2013)
8. Sperka, R., Spisak, M.: Trading agents' negotiation in business management using demand functions: simulation experiments with binomial distribution. Knowledge-Based and Intelligent Information & Engineering Systems 18th Annual Conference. KES-2014, pp. 1436–1444. Elsevier Science Bv, Amsterdam (2014)
9. Melo, M.T., Nickel, S., Saldanha-da-Gama, F.: An efficient heuristic approach for a multi-period logistics network redesign problem. TOP **22**(1), 80–108 (2014)
10. Josvai, J.: Optimization methods of sequence planning using simulation. In: Simulation in Produktion und Logistk 2013, pp. 71–76. Heinz Nixdorf Inst, Paderborn (2013)
11. de Keizer, M., Haijema, R., Bloemhof, J.M., van der Vorst, J.G.A.J.: Hybrid optimization and simulation to design a logistics network for distributing perishable products. Comput. Ind. Eng. **88**, 26–38 (2015)
12. Chramcov, B.: The optimization of production system using simulation optimization tools in witness. Int. J. Math. Comput. Simul. **7**(2), 95–105 (2013)
13. Golzer, P., Simon, L., Cato, P., Amberg, M.: Designing global manufacturing networks using Big Data. In: 9th Cirp Conference on Intelligent Computation in Manufacturing Engineering —Cirp Icme '14, pp. 191–196 (2015)
14. Seebacher, G., Winkler, H., Oberegger, B.: In Plant logistics efficiency valuation using discrete event simulation. Int. J. Simul. Model. **14**(1), 60–70 (2015)
15. Modrak, V., Marton, D., Bednar, S.: Modeling and determining product variety for mass-customized manufacturing. In: 5TH CATS 2014—CIRP Conference on Assembly Technologies and Systems, pp. 258–263. Elsevier Science Bv, Amsterdam (2014)
16. Bucki, R., Chramcov, B., Suchanek, P.: Heuristic algorithms for manufacturing and replacement strategies of the production system. J. Univers. Comput. Sci. **21**(4), 503–525 (2015)
17. Bucki, R., Suchanek, P.: The method of logistic optimization in E-commerce. J. Univers. Comput. Sci. **18**(10), 1238–1258 (2012)

Application of a Business Economics Decision-Making Function in an Agent Simulation Framework

Roman Šperka

Abstract The aim of the paper is to introduce a novel multi-agent simulation approach and its application allowing to deal with the trading processes of a virtual company. The motivation is to use implemented simulation framework as a basic part of an information system, operating as an integrated component of an actual ERP system implemented within a company so as to investigate and to predict the chosen business metrics of a company. Such a system serves to enable the management of a company to support their decision-making processes. The paper firstly presents the contemporary importance of simulation method. Secondly, the paper characterizes concrete multi-agent model of a virtual company, agents, and the decision-making function. Thirdly, the simulation framework application MAREA will be introduced. Lastly, the simulation results and their comparison with actual data, and the verification possibilities of the simulation model are described. It will be demonstrated that the proposed approach to customer behaviour in an agent-based simulation model could properly contribute to a better decision-making process.

Keywords Trading processes · Virtual company · Simulation · Agent-based · Model

1 Introduction

Due to the ongoing impact of globalization, the importance of business system modelling has recently undergone a period of rapid growth. The management teams of business companies have to increase the flexibility and tempo of decision-making in order to maintain pace with market developments. The complexity of business operations often does not allow measures to be taken without knowing the

---

R. Šperka (✉)
Department of Informatics and Mathematics, School of Business Administration in Karviná, Silesian University in Opava, Univerzitní nám. 1934/3, 733 40 Karviná, Czech Republic
e-mail: sperka@opf.slu.cz

G. Jezic et al. (eds.), *Agent and Multi-Agent Systems: Technology and Applications*, Smart Innovation, Systems and Technologies 58,
DOI 10.1007/978-3-319-39883-9_17

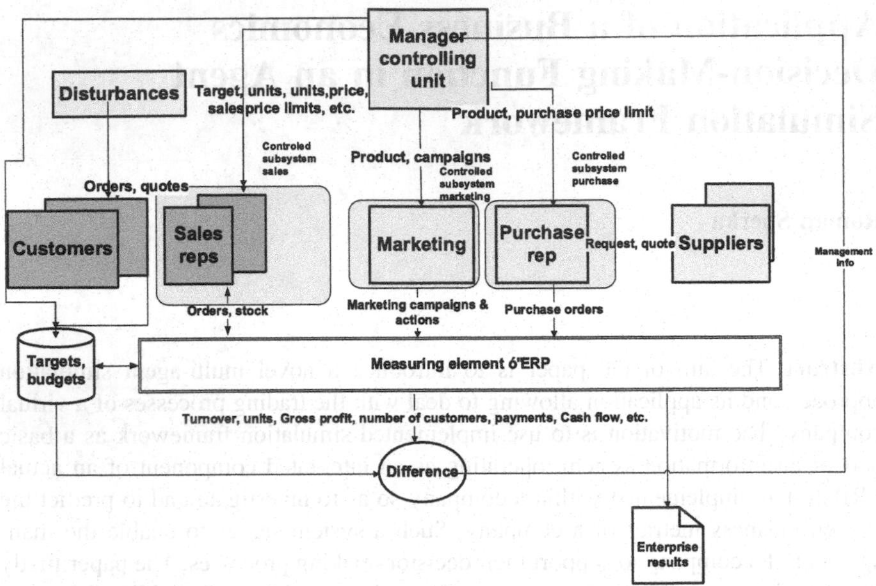

Fig. 1 Generic model of a business company. *Source* own

impacts of such decisions. This is precisely where modelling and simulations demonstrate their importance (e.g., [1]). While analytical modelling approaches are based mostly on mathematical theories [2], our approach concentrates on experimental simulations. The generic business company used for these simulations, using the control loop paradigm [3–6] is presented in Fig. 1. Generic model of a business company.

Agent-based modelling and simulation (ABMS) provides some opportunities and benefits resulting from the use of multi-agent systems as a platform for simulations with the aim of investigating consumer behaviour. They are characterized by a distributed control and data organization, which enables the representation of complex decision-making processes with only a few specifications. Many scientific works devoted to the area of ABMS have been published in the recent past. They concern the analysis of company positioning and the impact on consumer behaviour (e.g., [7–9]). The reception of the product by the market has been discussed in [10, 11]. More general deliberations on ABMS in the investigation of consumer behaviour are shown in e.g., [12–14]. The main advantage of ABMS models is that no complete analytical solution is needed to model sales processes; this is possible by virtue of basic agent properties such as pro-activeness, autonomy and social behaviour.

The motivation behind this paper is to present trading processes simulation in the form of a multi-agent-based model of a virtual trading company. The broader purpose of the research presented is to define a model which could be used as a simulation platform aimed at supporting management decisions. The general idea comes from the research by [3]. He proposed the integration of actual system models with management models working together in real-time. The actual system (e.g., ERP system) outputs proceed to the management system (e.g., simulation framework) from where they are used to investigate and predict important company results (metrics, KPIs). Actual and simulated metrics are compared and evaluated in a management model that identifies the steps required to respond in a manner that drives the system metrics towards their desired values. Many other researchers, e.g., [15, 16] use a similar approach to support management decisions. In our case, we used a generic control loop model of a company and implemented a multi-agent simulation framework, which represents the sales and management parts of the system, namely the trading processes and the negotiations between sales and customers. The simulation runs are based on the actual data of an active trading company.

The paper is structured as follows. In Sect. 2, the general model structure of a virtual company is described. Section 3 concentrates on the introduction of a simulation framework application—MAREA. The simulation results and their comparison with actual data are presented in Sect. 4. The last section of the paper is conclusion.

2 General Model Structure

To ensure the outputs of business processes simulations a simulation framework was implemented and used to trigger the simulation experiments. The framework covers business processes supporting the selling of goods by company sales representatives to the customers—seller-to-customer negotiation (Fig. 1). It consists of the following types of agents: sales representative agents (representing sellers, seller agents), customer agents, an informative agent (provides information about the company market share, and company volume), and manager agent (manages the seller agents, calculates KPIs). Disturbance agent is responsible for the historical trend analysis of sold amount (using his influence on customer agent). All the agent types are developed according to the multi-agent approach. The interaction between agents is based on the FIPA contract-net protocol [17].

The number of customer agents is significantly higher than the number of seller agents in the model because the reality of the market is the same. The behavior of agents is influenced by two randomly generated parameters using the normal distribution (an amount of requested goods and a sellers' ability to sell the goods). In the lack of real information about the business company, there is a possibility to randomly generate different parameters (e.g., company market share for the product, market volume for the product in local currency, or a quality parameter of the seller). The influence of randomly generated parameters on the simulation outputs while using different types of distributions was previously described in [6].

In the text to follow, the seller-to-customer negotiation workflow is described and the formal definition of a decision function is proposed. Decision function is used during the contracting phase of agents' interaction. It serves to set up the limit price of the customer agent as an internal private parameter. One stock item simplification is used in the implementation. Participants of the contracting business process in our multi-agent system are represented by the software agents—the seller and customer agents interacting in the course of the quotation, negotiation and contracting. There is an interaction between them. The behavior of the customer agent is characterized in our case by proposed customer decision function (Eq. 1).

At the beginning disturbance agent analyses historical data—calculates average of sold amounts for whole historical year as the base for percentage calculation. Each period turn (here we assume a week), the customer agent decides whether to buy something. His decision is defined randomly. If the customer agent decides not to buy anything, his turn is over; otherwise he creates a sales request and sends it to his seller agent. Requested amount (generation is based on a normal distribution) is multiplied by disturbance percentage. Each turn disturbance agent calculates the percentage based on historical data and sends the average amount values to the customer agent. The seller agent answers with a proposal message (a certain quote starting with his maximal price: *limit price*1.25*). This quote can be accepted by the customer agent or not. The customer agents evaluate the quotes according to the decision function. The decision function was proposed to reflect the enterprise market share for the product quoted (a market share parameter), seller's ability to negotiate, total market volume for the product quoted etc. (in, e.g., [6]). If the price quoted is lower than the customer's price obtained as a result of the decision function, the quote is accepted. In the opposite case, the customer rejects the quote and a negotiation is started. The seller agent decreases the price to the average of the minimal limit price and the current price (in every iteration is getting effectively closer and closer to the minimal limit price), and resends the quote back to the customer. The message exchange repeats until there is an agreement or a reserved time passes.

The customer decision function for the *m*th seller pertaining to the *i*th customer determines the price that *i*th customer accepts (adjusted according to [6]).

$$c_n^m = \frac{\tau_n T_n \gamma \rho_m}{O v_n} \tag{1}$$

c_n^m price of *n*th product offered by *m*th seller,
τ_n market share of the company for *n*th product $0 < \tau_n < 1$,
T_n market volume for *n*th product in local currency,
γ competition coefficient, lowering the success of the sale $0 < \gamma < 1$,
ρ_m *m*th sales representative ability to sell $0,5 \leq \rho_m \leq 2$
O number of sales orders for the simulated time,
v_n average quantity of the *n*th product, ordered by *i*th customer from *m*th seller

$$\mathrm{O} = ZIp \tag{2}$$

Z number of customers,
I number of iterations,
p mean sales request probability in one iteration

Customer agents are organized in groups and each group is being served by concrete seller agent. Their relationship is given; none of them can change the counterpart. Seller agent is responsible to the manager agent. Each turn, the manager agent gathers data from all seller agents and stores KPIs of the company. The data is the result of the simulation and serves to understand the company behavior in a time—depending on the agents' decisions and behavior. The customer agents need to know some information about the market. This information is given by the informative agent. This agent is also responsible for the turn management and represents outside or controllable phenomena from the agents' perspective.

3 Simulation Framework Application

In this section, an enhanced software prototype of a framework based on agent-based trading company control loop is introduced. The prototype is based on the research results presented herein above. The Enterprise Resource Planning system (ERP) using the REA ontology approach is used as a measuring element in the framework. The system has been developed in cooperation between Silesian University in Opava, School of Business Administration in Karvina, Czech Republic and REA technology Copenhagen, Denmark. After the prototype tests at the end of the year 2011, prototype was presented at the beginning of 2012 for the first time [18].

Application enables users to set up trading company parameters (see the example in Table 1) and run trading simulation for a specific time to interpret the development of KPIs of this company. The application consists of two main components, the Simulation of a multi-agent system (MAS) and the ERP system—see Fig. 2. A simulation designer can either use the ERP system directly, or can program intelligent agents to perform the same activities that a human user can perform. For example, a simulation designer can use the ERP system directly to create initial data for a simulation, then start the agent platform to run a simulation, then using the ERP system inspect the simulation results, and even adjust the data (within the rules implemented in the ERP system) and then start the agent platform to continue running the simulation. Both agents and a human user can read data from the ERP system, write data to the ERP system, and perform actions, such as sending a purchase order.

The main screen called Simulation monitor (Fig. 2) consists of five panes:

- Logo pane with sponsors of the project. You can hide the logo by clicking it. Clicking the upper area of the window shows the sponsor's logo again.

Table 1 Parameterization of agent-based simulation

Agent type	Agent count	Parameter name	Parameter value
Customer agent	500	Maximum disc. turns	10
		Mean quantity	40 m
		Quantity s. deviation	32
Seller agent	25	Mean ability	1
		Ability st. deviation	0.03
		Minimal price	0.36 EUR
Market info	1	Item market share	0.15 EUR
		Item market volume	1.033.535 EUR
		Competition coefficient	0.42
		No items sold in one iteration	1.330
		Iterations count	52
Manager agent	1	Purchase price	0.17 EUR
Disturbance agent	1		

Source own

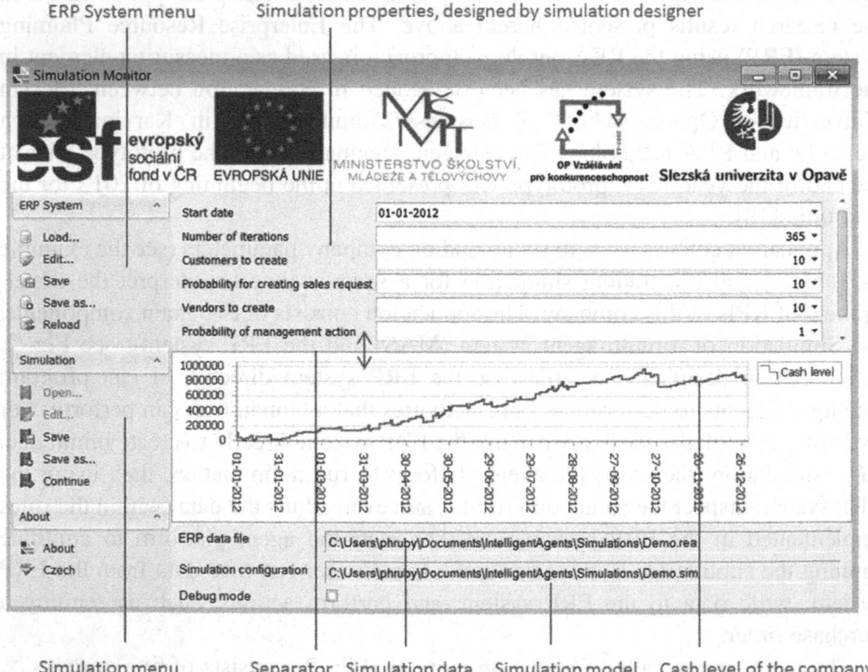

Fig. 2 Main components of the framework. *Source* adapted from [16]

- Simulation properties. These properties are determined by the agent simulation model, and different simulation models might contain different simulation properties.
- ERP system menu
- Simulation menu
- A graph of the cash level of the company, indicating the progress of the simulation. Please note that this graph is updated by ERP transactions that influence cash level, i.e. purchase, sale, marketing, education, initial cash and initial stock (in case it had nonzero monetary value). That is, if the simulation runs but none of these transactions occur, the cash level graph in the Simulation monitor will not be updated.

User can save trading company ERP data to.REA file for future use. This can be done also for the agent simulation model, containing the agent script and simulation properties to the.SIM file. Simulation monitor allows to start, pause, resume, and continue the simulations, as well opens the ERP system and Simulation Editor. The ERP system (Fig. 2) can be opened by clicking the Edit button in the ERP System menu in the Simulation Monitor. It enables to manipulate with trading company parameters and entities traded.

Simulation of negotiation between agents about sales and purchases is one of the key functions of the multi-agent simulation system. The messages the agents send to each other during negotiation are recorded in the ERP system. All messages about sales (from the initial request to closing the deal) are part of the Sales request entity; likewise all messages about purchase (from the initial request to closing the deal) are part of the Purchase request entity.

The ERP system has been configured to calculate KPIs by summing up other values. For example, Cash level is calculated as a total of all transactions that change Cash level—payments for purchases, income from sales, payment of bonuses, initial cash, etc. Turnover and Gross profit is calculated as a total of gross profits and turnovers of specific product types. The values of the most important KPIs in all simulation steps can be exported to an Excel file and analysed later by typical Excel tools like a contingency table or by a data analysis like histograms etc. The negotiation steps can also be exported to Excel in order to analyse the customer and sales representative behavior.

4 Experimental Results

At the start of simulation experiments phase some parameters were set. Agent count and their parameterization are listed in Table 1.

Agents were simulating one year—52 weeks of interactions. As mentioned above —manager agent was calculating the KPIs. Simulation results were compared with the real data from an anonymous computer selling company with 30 employees (from Slovakia). The real data were taken from the company's accounting

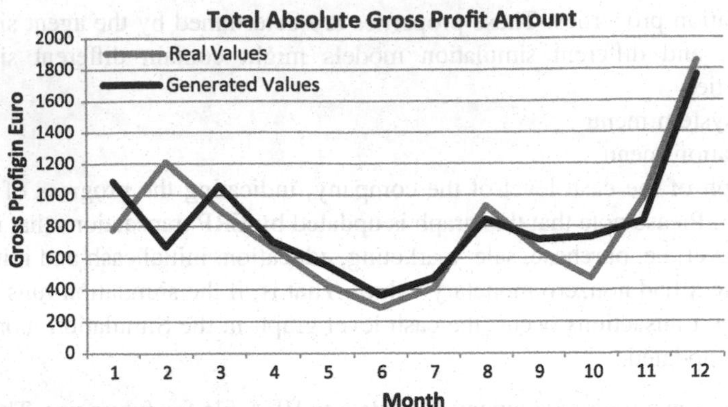

Fig. 3 The generation values graph—monthly. *Source* adapted from [15]

information system. For the comparison of simulated and real data, monthly averages were used.

Total gross profit was chosen as another representative KPI. Figure 3 contains the month sums of total gross profit for real and generated data. As can be seen from this figure, the result of simulation was quite similar to the real data.

To prove the relationship between the real and generated data—two instruments were chosen—Correlation Analysis to show the correlative relation between them and Chi-Square Test for Independence to show the similarity of distribution for both data series.

Correlation coefficient for total gross profit amount was 0.857, which represents very strong correlation between real and generated data. Also the Chi-Square Test for Independence has proven that the distribution of real and generated values is very similar. There is a frequency histogram of gross profit for real and generated values in Fig. 4.

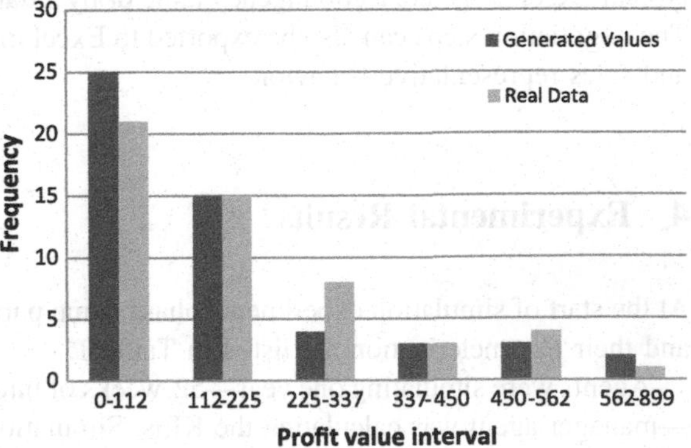

Fig. 4 Gross profit frequency histogram. *Source* adapted from [15]

5 Conclusion

An innovative approach for virtual company simulation was proposed in the paper. Agent-based modeling and simulation techniques were used to implement a multi-agent company model in the form of simulation framework. We proceed many simulation experiments with the aim to validate the results obtained with the real data from an existing trading company. We set the simulation parameters according to the real state of this company. The statistics prove that using the simulation framework is it possible to predict several KPIs of a company.

The presented MAREA application serves for the decision support of company's management as well as for educational purposes. It enables users to get familiar with the principles of trading using business company virtual model. The setup of the application provides possibilities to edit the company parameters and to run trading simulations. This allows users to analyse trading behavior back-to-back according to the parameters setup. The validation shows reasonable results with the necessity to integrate some kind of mechanism dealing with seasonal differences in KPIs. Future research will concentrate on the log files analysis to give us feedback about processes in the running simulation experiment.

Acknowledgments This paper was supported by the Ministry of Education, Youth and Sports Czech Republic within the Institutional Support for Long-term Development of a Research Organization in 2016.

References

1. Suchanek, P., Vymetal, D.: Security and disturbances in e-commerce systems. In: Proceedings of the 10th International Conference Liberec Economic Forum 2011 (2011). ISBN: 978-80-7372-755-0
2. Liu, Y., Trivedi, K.S.: Survivability Quantification: The Analytical Modeling Approach. Department of Electrical and Computer Engineering. USA, Durham: Duke University. http://people.ee.duke.edu/~kst/surv/IoJP.pdf (2016). Accessed 21 Jan 2016
3. Barnett, M.: Modeling & Simulation in Business Process Management. Gensym Corporation, pp. 6–7. http://w.businessprocesstrends.com/publicationfiles/11-03%20WP%20Mod%20Simulation%20of%20BPM%20-%20Barnett-1.pdf (2003). Accessed 21 Jan 2016
4. Šperka, R., Spišák, M., Slaninová, K., Martinovič, J., Dráždilová, P.: Control loop model of virtual company in BPM simulation. In: Proceedings of the Advances in Intelligent Systems and Computing. Soft Computing Models in Industrial and Environmental Applications, vol 188, p. 515–524. Springer, Germany, Berlin (2012). doi:10.1007/978-3-642-32922-7_53, ISSN: 2194-5357, ISBN: 978-3-642-32921-0
5. Slaninová, K., Martinovič, J., Dráždilová, P., Vymětal, D., Šperka, R.: Analysis of agents` behavior in multiagent system. In: Proceedings of the 24th European Modeling and Simulation Symposium. EMSS 2012, pp. 169–175. Vienna, Austria. 19–21.9.2012. Universita di Genova, Italy, Rende (2012). ISBN: 978-88-97999-09-6
6. Vymětal, D., Spišák, M., Šperka, R.: An influence of random number generation function to multiagent systems. In: Proceedings of the Agent and Multi-Agent Systems. Technologies and

Applications. LNAI 7327, pp. 340–349. Springer, Germany, Berlin (2012). doi:10.1007/978-3-642-30946-5, ISSN: 0302-9743, ISBN: 978-3-642-30946-5

7. Tay, N., Lusch, R.: Agent-based modeling of ambidextrous organizations: virtualizing competitive strategy. IEEE Trans. Intell. Syst. **22**(5), 50–57 (2002)
8. Wilkinson, I., Young, L.: On cooperating: Firms relations networks. J. Bus. Res. Issue **55**, 123–132 (2002)
9. Casti, J.: Would-be Worlds. How Simulation is Changing the World of Science. Wiley, New York (1997)
10. Goldenberg, J., Libai, B., Muller, E.: The Chilling effect of network externalities. Int. J. Res. Mark. **27**(1), 4–15 (2010)
11. Heath, B., Hill, R., Ciarallo, F.: A survey of agent-based modeling practices (January 1998 to July 2008). J. Artif. Soc. Soc. Simul. **12**(4), 5–32 (2009)
12. Adjali, I., Dias, B., Hurling R.: Agent based modeling of consumer behavior. In: Proceedings of the North American Association for Computational Social and Organizational Science Annual Conference. University of Notre Dame. Notre Dame (2005)
13. Ben, L., Bouron, T., Drogoul, A.: Agent-based interaction analysis of consumer behavior. In: Proceedings of the first international joint conference on Autonomous agents and multiagent systems: part 1, pp. 184–190. ACM. New York (2002)
14. Collings, D., Reeder A., Adjali, I., Crocker, P., Lyons, M.: Agent based customer modelling. Computing in Economics and Finance. No. 1352. http://econpapers.repec.org/paper/scescecf9/1352.htm(1999). Accessed 21 Jan 2016
15. Šperka, R., Spišák, M.: Transaction costs influence on the stability of financial market: agent-based simulation. J. Bus. Econ. Manag. vol. 14, Supplement 1, S1-S12. Taylor & Francis, United Kingdom, London (2013). doi:10.3846/16111699.2012.701227, ISSN: 1611-1699
16. Šperka, R., Vymětal, D.: MAREA—an education application for trading company simulation based on REA Principles. In: Proceedings of Advances in Education Research, vol. 30. Information, Communication and Education Application. ICEA 2013, Hong Kong, China. 1–2.11.2013. Delaware: Information Engineering Research Institute (IERI), USA, pp. 140–147 (2013). ISBN: 978-1-61275-056-9
17. Foundation for Intelligent Physical Agents, FIPA: FIPA Contract Net Interaction Protocol. In Specification, http://www.fipa.org/specs/fipa00029/SC00029H.html (2002). Accessed 21 Jan 2016
18. Vymětal, D., Scheller, C.: MAREA: multi-agent rea-based business process simulation framework. In: Conference Proceedings, International Scientific Conference ICT for Competitiveness, pp. 301–310. OPF SU, Karviná (2012). ISBN: 978-80-7248-731-8

Reduction of User Profiles for Behavioral Graphs

Kateřina Slaninová, Jan Martinovič and Martin Golasowski

Abstract Visualisation of relations between the users is an important part of business process analysis. The authors focused on behavioral graphs to represent relations between the users based on their behavior in the system. The behavior is determined by sequences of activities the users have performed. The proposed method deals with the problem of the behavioral graph complexity. This problem is solved by reduction of user profiles. Several methods were tested to determine what method is more suitable for the analysis of this type of event logs. The approach was tested on an event log recorded by a virtual company model developed as a multi-agent system MAREA.

Keywords Business process modeling · Business process analysis · Complex networks · User behavior · Behavioral patterns · Behavioral graphs · MAREA

1 Introduction

Analysis of users' behavior plays an important role in business process analysis. It is the analysis focused on resource perspective which is very valuable for obtaining information about main roles in the company. It helps to answer such questions like who is doing which activities, who is cooperating with whom, how many times the resource is executing the activities per case, etc. Modern applications like informa-

K. Slaninová (✉) · J. Martinovič · M. Golasowski
IT4Innovations Supercomputing Center, VŠB - Technical University of Ostrava,
17. listopadu 15/2172, 708 33 Ostrava-poruba, Czech Republic
e-mail: katerina.slaninova@vsb.cz

J. Martinovič
e-mail: jan.martinovic@vsb.cz

M. Golasowski
e-mail: martin.golasowski@vsb.cz

G. Jezic et al. (eds.), *Agent and Multi-Agent Systems: Technology and Applications*, Smart Innovation, Systems and Technologies 58,
DOI 10.1007/978-3-319-39883-9_18

tion systems, enterprise systems, or e-commerce systems record transactions and executed activities in a systematic way, which can be harnessed for this type of analysis.

Process mining techniques typically focus on performance and control-flow issues. A standard log file usually consists of records with information about executed events that have occurred in the system. These records may contain various attributes such as timestamp (date and time when the event happened), a resource (originator), the type of event, and additional information. If a resource of such event is a person, we can extract the relevant records and obtain information about his/her behavior. Social network analysis then takes part of process mining techniques, focused on activity performers, on users.

Social network analysis is one of the techniques used in organizational mining, focused on the organizational perspective of process mining [8, 12]. Using information about the originator (device, resource, or person) who initiated the event (activity), these techniques are able to show more about people, machines, organizational structures (roles and departments), work distribution, and work patterns.

Process mining techniques focused on performers and their activities are not new, for example knowledge based semantification of business communications [4]. Besides, fundamental approaches related to social network analysis exist. Attributes like person as an originator enable us to discover real process work flow models of the organization or to construct models that explain some aspects of person's behavior. Several techniques can be used to find valuable information from social networks, like centralities, clique identification, etc.

Relationship between the users derived from event logs can be set from various points of view, and therefore, the weight of the relations can be defined by several ways. According to van der Aalst [13], the weight is usually determined on the basis of (possible) causality (handover of work, subcontracting), joint cases, joint activities, or special event types [12]. Such networks are based on user profiles [10]. Therefore, a resulted graph of social network with each type of weight has a different meaning.

Weight determined by (possible) causality represents how work flows among users within particular cases. As an example, we can mention handover of work, which represents handover between the users. Another one is subcontracting, which is described by number of times a user executed an activity in-between two activities executed by another user. Both examples have also their modifications. The visual representation of such relation type can be for example by Subcontracting graph [12].

Relation weight determined by joint cases ignores causal dependencies in a case (transaction). It is only based on a frequency of activities that users performed in a common case (transaction). The more frequent is common work of both users on cases, the stronger the relationship between them is. We can mention Working together graph as an example of such type of relation [12].

In our previous work, we have proposed a new type of social network based on user profiles, behavioral graph (network) [7]. Behavioral graph is a social network based on user profiles defined on the basis of user behavior in the system. The user profile is a vector of all cases (or behavioral patterns), in which originators participated, not only the transactions (sequences of events executed by the originator in the system), typical for working together graph. The paper is focused on this new type of social network/graph, with the focus on reduction of the user profiles to obtain behavioral patterns from the transactions recorded in the log file.

The content of the paper starts with Sect. 2 focused on brief introduction of the proposed methodology for the extraction of user behavioral network (graph). Determination of sequence similarity, so important for detection of user behavioral patterns, follows in Sect. 3. Also algorithms for finding representative sequences and reduced user's profiles construction are discussed in this section. Then, analysis of agent's behavior during the simulation of business processes in a virtual company model developed as multi-agent system by MAREA environment is presented in Sect. 4. Finally, Sect. 5 concludes the paper.

2 Proposed Methodology

The main goal of the methodology is to find latent ties between the users in the user network on the basis of their behavior in the system, and a transparent visualization of the user groups with similar behavior. The proposed approach consists of several steps and requires the solution of the following sub problems:

1. Definition of user behavior in the system,
2. Creation of user profiles based on their behavior within the system,
3. Finding of user behavioral patterns,
4. Construction of user profiles based on user behavioral patterns,
5. Finding of user groups with similar behaviors by using the behavioral patterns and the transparent visualization of latent ties between them.

The first phase, including the data pre-processing, depends on the characteristics of the used data collection and on the system. The output of this phase is a set of originators (users) U and set of sequences performed by the originators in the analyzed system S.

Let $U = \{u_1, u_2, \ldots, u_n\}$, be a set of users (originators), where n is number of users u_i. Then, sequences of events (traces) $\sigma_{ij} = \left\langle e_{ij1}, e_{ij2}, \cdots e_{ijm_j} \right\rangle$, are sequences of events executed by the user u_i in the system, where $j = 1, 2, \ldots, p_i$ represents type of the sequence, and m_j is length of j-th sequence. Thus, set $S_i = \{\sigma_{i1}, \sigma_{i2}, \cdots \sigma_{ip_i}\}$ is a set of all sequences executed by the user u_i in the system, and p_i is number of that sequences.

Sequences σ_{ij} extracted with relation to certain user u_i are mapped to set of sequences $\sigma_l \in S$ without this relation to users: $\sigma_{ij} = \left\langle e_{ij1}, e_{ij2}, \ldots, e_{ijm_j} \right\rangle \rightarrow \sigma_l = \langle e_1, e_2, \ldots, e_{ml} \rangle$, where $e_{ij1} = e_1, e_{ij2} = e_2, \ldots, e_{ijm_j} = e_{ml}$.

The second phase is focused on the construction of base user profiles. A base user profile of the user (originator) $u_i \in U$ is a vector $b_i \in N^{|S|}$ represented by a row i from matrix $B \in N^{|U|\times|S|}$. A detailed description of all the steps of the methodology was presented in [5].

This paper is focused on the third and the fourth step of the methodology, finding user behavioral patterns and the construction of reduced user profiles. A detailed description of these phases is presented in Sect. 3 of this paper. In a case study presented in Sect. 4, we have tested two methods for finding behavioral patterns, and have compared two approaches for setting the similarity between the sequences. Finally, the user groups with similar behavior were identified in user network, the transformation of the user networks, and new latent ties uncovered between the users were examined.

3 User Behavior Patterns and Reduced User Profiles

One of the aims of the approach is to discover user behavioral patterns. Behavioral patterns are representatives of clusters of similar sequences (traces) performed by users. Let C be a set of clusters consisted of similar event sequences. Let $R = \{\rho_1, \rho_2, \dots, \rho_r\}$ be a set of representatives of clusters ρ_k, where k=1, 2, ..., r. Then, a representative $\rho_k \in R$ of a cluster C_k is a sequence, which describes similar event sequences $\sigma_l = \langle e_1, e_2, \dots e_{ml} \rangle$ in the cluster C_k (it has the maximal average similarity to the all sequences in the cluster C_k). Representatives $\rho_k \in R$ are called behavioral patterns.

Behavioral patterns were used to create reduced user profiles. Let U be a set of users u_i. Thus, a set $S_i \left\{ s_{i_1}, s_{i_2}, \dots, s_{i_j} \right\}$ is a set of all sequences in which the user $u_i \in U$ participates. The S_i set is mapped to a set $\Pi_i = \{\pi_1, \pi_2, \dots \pi_{s_i}\}$ which contains all behavioral patterns $\Pi_i \subseteq R$ for a particular user u_i. A sequence σ_{ij} is mapped to a representative pattern π_k of its own cluster. A reduced user profile of the user $u_i \in U$ is a vector $p_i \in N^{|R|}$ represented by a row i from matrix $P \in N^{|U|\times|R|}$.

3.1 *Determination of Sequence Similarity*

Two well known methods of sequence comparison, the *Longest Common Substring* (LCS) [2] and the *Longest Common Subsequence* (LCSS) [3] were used for the determination of the sequence similarity.

Moreover, two different approaches were used for determination of the similarity between sequences (traces) x and y. The output of both methods LCS and LCSS is a common subsequence z of compared sequences x and y, while $(z \subseteq x) \wedge (z \subseteq y)$. The first approach for determination of the sequence similarity $Sim_{seq}(x, y)$ between the sequences x and y was counted by Eq. 1:

$$Sim_{seq}(x, y) = \frac{2l(z)}{l(x) + l(y)} \tag{1}$$

where $l(x)$ and $l(y)$ are lengths of the compared sequences x and y, and $l(z)$ is a length of a subsequence z.

The second approach for the determination of the sequence similarity $SimExt_{seq}(x, y)$ between the sequences x and y was counted by Eq. 2:

$$SimExt_{seq}(x, y) = \frac{l(z)^2}{l(x)l(y)} \frac{Min(l(x), l(y))^2}{Max(l(x), l(y))^2}, \tag{2}$$

where $l(x)$ and $l(y)$ are lengths of the compared sequences x and y, and $l(z)$ is a length of a subsequence z.

Both Eqs. 1 and 2 take into account possible differences between $l(x)$ and $l(y)$. Due to this reason, z is adapted so that $Sim_{seq}(x, y)$ and $SimExt_{seq}(x, y)$ is strengthened in the case of similar lengths of sequences x and y, and analogically weakened in the case of higher difference of $l(x)$ and $l(y)$. However, the difference between the two approaches should be in the distribution of the weights, which is important while analyzing different types of logs. This difference was also examined in Sect. 4.

The extraction of the clusters with similar sequences was realized by hierarchical agglomerative clustering (bottom-up clustering approach) [1]. The clustering method, which is hierarchically applied on weighted and undirected graphs, often gives clusters near to a complete graph. Due to this fact it is possible, that the sequences with the high value of relation weight w_{seq} (in this case $Sim_{seq}(x, y)$ or $SimExt_{seq}(x, y)$) may appear in one cluster. However, even the sequences, which are not similar at all, are always similar to their neighbor sequences.

This problem was solved by Algorithm 1 for finding the most suitable representative sequence of the cluster. The algorithm allows an additional, and a more appropriate division of the clusters, for a selected similarity threshold θ.

The next step was the more precise division of the obtained sequence clusters C_ρ and finding the representatives for these obtained clusters using our Algorithm 1.

The extraction of clusters with similar sequences, by hierarchical agglomerative clustering (HAC) used a dendrogram, constructed during the previous phase of weighting the relations between sequences, as its input. However, the weights in the dendrogram do not correspond to the weight values of the relations between the sequences w_{seq}. Due to this reason, we have developed the algorithm for a further, and more appropriate, division of the clusters obtained by HAC for a selected similarity threshold θ, see Algorithm 2.

Algorithm 1 Determination of representative in cluster of sequences

Goal: The determination of sequence representatives ρ (behavioural patterns) which will be used for a description of the sequence clusters C, for selected similarity threshold θ.

Input:

- A cluster C, where a similarity between the objects inside the cluster is given by values $Sim(c_l, c_k)$.
- A selected threshold θ for a sequence similarity w_{seq} in clusters.

Output:

- A set of representatives $R = \{\rho_1, \rho_2, \ldots, \rho_r\}$ and a set of clusters C_ρ which correspond to the appropriate representatives.

1. For each object, belonging into a cluster C, determine its average similarity to other objects inside that cluster. Similarity between the objects c_l and c_k is given by the condition $Sim(c_l, c_k) > \theta; \forall c_l, c_k \in C$, where θ is the selected threshold for the object similarity.
2. Select the object with the maximum average similarity counted in Step 1. The selected object is labelled ρ.
3. Create a new cluster C_ρ, which contains the objects from cluster C, which meets the condition $Sim(c_l, c_k) > \theta$ and contains the representative ρ.
4. Add a representative ρ into the set R and record the appropriate cluster C_ρ, which is created in Step 5.
5. $C = C - C_\rho$.
6. If $C \neq \emptyset$, then repeat to Step 1 with a new cluster C.

4 Case Study

The output from a company model transformed into a multi-agent system by means of MAREA workbench [9] was used in the use case. MAREA (Multi-Agent, REA based system) is a complex modeling environment consisting of REA Based ERP system used for registration of the all economic activities simulating both controlling and controlled subsystems of the modeled company.

The intelligent agents were programmed to be able to simulate the human activities in the modeled company environment. For example, the sales representatives and the customers can negotiate on the price the manager can decide on management actions such as marketing campaigns etc. Each activity of the agents is represented by accompanying message which is simultaneously registered in the event log. The messages are presented by means of MAREA message viewer and can be compared with other ERP outputs. A detailed description of the simulation model framework was presented in [11].

A real data from one high tech company selling IT accessories was used in this use case. UTP cable sales statistics were adapted as an put into the ERP base indicators such as product market share, selling price limit, average purchase price, estimated market volume in the region where the company is active, etc.

The experiments followed all the steps of the proposed methodology presented in Sect. 2. Set of users (in this use case software agents) U and set of sequences (traces)

Algorithm 2 Finding clusters with similar sequences after hierarchical clustering

Goal: A more appropriate division of clusters C_{HAC}, obtained by hierarchical agglomerative clustering (HAC), using a selected threshold θ for the object similarity.
Input:

- A set of objects $O = \{o_1, o_2, \ldots, o_n\}$ of size n, a set of clusters $C_{HAC} = \{C_1, C_2, \ldots, C_{n-1}\}$ of size $n - 1$ and a binary tree which represents a linkage between clusters and objects. It holds true that leave nodes of the binary tree $o_l \in O$ are individual objects. Non-leave nodes $c_k \in C_{HAC}$ represent clusters (or sets of clusters) of similar objects $C_i = \{o_{i1}, o_{i2}, \ldots, o_{im}\}$ from the set O.
- A selected threshold θ for object similarity in clusters.

Output:

- A set of clusters C_{seq}.

1. The definition of two empty sets:
   - $S_{used} = \emptyset$—a set of used objects $o_i \in O$,
   - $C_{seq} = \emptyset$—a set of found clusters C_j consisting of objects $o_{j_i} \in O$ which meet the condition $Sim(C_i, C_j) > \theta$.
2. Select a cluster with the least amount of objects from the set C_{HAC} and label it C_{actual}, $C_{actual} \in C_{HAC}$ If more than one cluster with the same amount of objects exist then select the first cluster.
3. $C_{HAC} = C_{HAC} - \{C_{actual}\}$—Revoke the cluster C_{actual} from the set C_{HAC}.
4. $C_{test} = C_{actual} - S_{used}$—A cluster C_{test} consists of all the objects within the cluster C_{actual} which have not been matched with any cluster from the set C_{seq}.
5. Find a set C_{result}, which is an output from Algorithm 1 'Determination of Representative in Cluster of Sequences'. The algorithm has the cluster C_{test} and threshold θ at its input. If θ leads to the division of the cluster C_{test} into more clusters, then $|C_{result}| > 1$.
6. If $|C_{result}| >$ then $C_{seq} = C_{seq} \bigcup C_{result}$ and $C_{used} = S_{used} \bigcup C_{test}$.
7. If $C_{HAC} \neq \emptyset$ then return into Step 2.
8. If $C_{result} = 1$ then $C_{seq} = C_{seq} \bigcup C_{result}$.

S_i performed by the users were extracted from the log file of the simulation (step 1), and base user profiles were created (step 2).

The main focus was put on the third and the fourth step of the methodology. The similarity of the extracted sequences S_i was compared by LCSS and LCS method and with the two approaches for the similarity determination Sim_{seq} (see Eq. 1) and $SimExt_{seq}$ (see Eq. 2). The distribution of the similarity weight is presented in Fig. 1.

The difference between these two methods is in their distinctive approach to the comparison of the sequences. The LCSS method is more immune to slight distortions within the compared sequences, while the LCS method is a bit stricter. It can be seen from the different weight distribution in Fig. 1. Moreover, two different approaches for the determination of the relation weight were tested. Both Figs. 1 and 2 show that relation weight computed by Sim_{seq} allows better distribution through the whole interval than $SimExt_{seq}$ in this case. We must emphasize here that this is true for this type of log. The second approach may fit better for other domains (we have already tested it successfully for example in [6]).

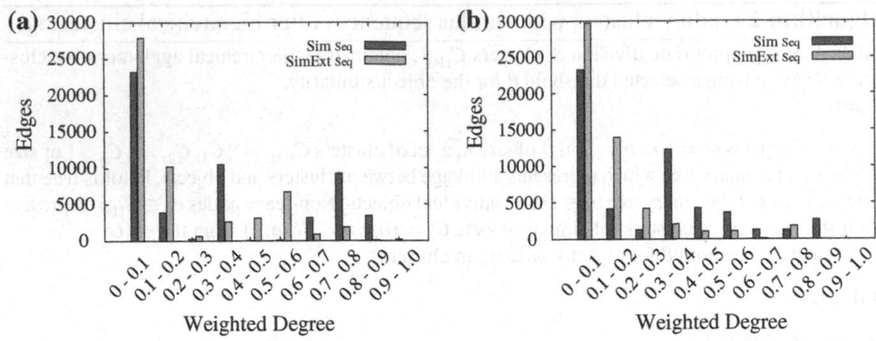

Fig. 1 Weight distribution—sequence similarity. **a** LCSS, **b** LCS

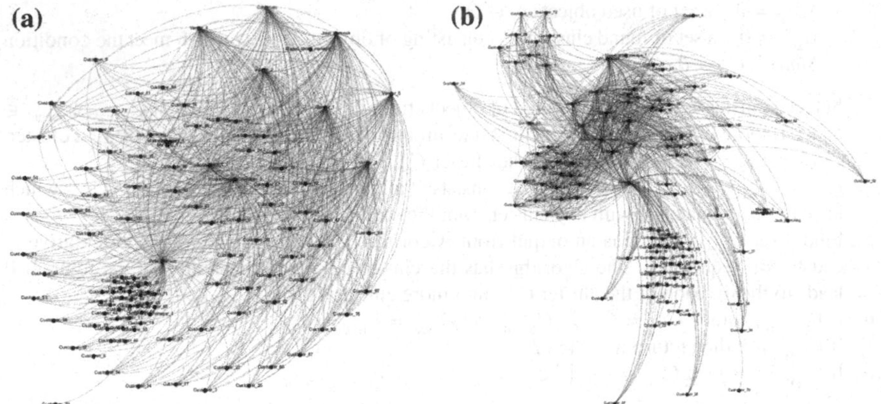

Fig. 2 User network—LCSS, $\theta = 0.7$. **a** Base user network, **b** Reduced user network

The following Table 1 for Sim_{seq} and Table 2 for $SimExt_{\text{seq}}$ show the clusters of sequences extracted after hierarchical agglomerative clustering.

Both tables contain information about found clusters (components) and their quantity for the selected threshold θ. The Algorithm 1 ensures that the similarity between the sequences inside the clusters will be higher than θ.

As we can see from the statistics presented in Tables 1 and 2, the distribution of relation weights counted by Sim_{seq} or $SimExt_{\text{seq}}$ influences the structure and the amount of the found clusters. Also the selection of the method LCSS or LCS is very important for the output we finally obtain.

The following Table 3 shows characteristics of final user networks we can obtain using reduced user profiles constructed by representative sequences from the found behavioral patterns (clusters discussed above). The user networks were constructed by representatives obtained by LCSS method and Sim_{seq}. Besides common graph characteristics presented here, we can see how the user network changes by using

Table 1 Components of similar sequences—Sim_{seq}

θ	Component size (count)	Isolated nodes	Sum of clusters	Avg. nodes per cluster	Avg. size of clusters
LCS					
0.9	2 (6)	403	6	2.000	1.500
0.8	48(1), 11(1), 14(1), 13(1), 11(2), 9(1), 8(2), 7(1), 4(1), 3(1), 2(27)	266	36	4.139	8.778
0.7	48(1), 14(1), 13(1), 11(1), 10(1), 9(2), 5(2),4 (1), 3(5), 2(37),	198	52	4.173	10.909
0.6	54(1), 27(1), 12(1), 10(1), 9(2), 5(1), 4(3), 3(8), 2(35)	183	53	4.377	12.700
LCSS					
0.9	4 (1), 3(5), 2(55)	286	61	2.115	2.500
0.8	7(1), 5(1), 4(7), 3(13), 2(62)	212	84	2.417	3.667
0.7	21(1), 13(1), 12(1), 11(2), 9(1), 6(2), 4(3), 3(13), 2(46)	183	70	3.314	8.200
0.6	115(1), 8(1), 6(1), 4(6), 3(6), 2(42)	160	57	4.474	19.857

Table 2 Components of similar sequences—$SimExt_{seq}$

θ	Component size (count)	Isolated nodes	Sum of clusters	Avg. nodes per cluster	Avg. size of clusters
LCS					
0.9		415	0		1.000
0.8		415	0		1.000
0.7	38(1), 10(2), 9(1), 8(1), 7(1), 6(1), 4(1), 2(2)	319	10	9.600	9.445
0.6	38(1), 10(2), 9(1), 8(1), 7(2), 6(1), 4(1), 3(1), 2(8)	297	18	1.317	6.556
LCSS					
0.9	2(1)	413	1	2.000	1.500
0.8	4(1), 3(4), 2(30)	339	35	2.171	2.500
0.7	12(1), 6(1), 4(2), 3(5), 2(36)	302	45	2.511	2.492
0.6	12(1), 6(1), 4(3), 3(11), 2(47)	258	63	2.492	4.667

reduced user profiles. It is shown mainly by updated and new edges presented in this table.

Figure 2 then confirms that using reduced user profile not only several relations between the users in the user network strengthen, but also some new latent ties reveal. This figure is presented for selected threshold $\theta = 0.7$.

Table 3 Description of user network—LCSS, Sim_{seq}

	Base network	Reduced network			
		0.6	0.7	0.8	0.9
Nodes	117	117	117	117	117
Edges	766	2733	1062	811	766
Avg. degree	13.094	46.718	18.154	13.863	13.094
Avg. weighted degree	0.570	13.806	2.367	0.862	0.641
Graph density	0.113	0.403	0.156	0.120	0.113
Modularity	0.331	0.011	0.299	0.321	0.310
Avg. Clust. Coef.	0.902	0.978	0.910	0.904	0.902
Updated edges		2731	1060	805	760
New edges		1967	296	45	0

5 Conclusion

The reduction of behavioral graphs by reduced user profiles was presented in this paper. This issue is a part of the proposed methodology for finding latent ties between the users in the user network based on their behavior within the system. The methodology includes among the other steps finding behavioral patterns, and construction of reduced user profiles based on behavioral patterns, on which was this paper focused. One of the important factors in these steps is the selection of the suitable method for the determination of the relation weight between the sequences. The influence of the two selected methods LCSS and LCS on the resulting behavioral graph was tested, including two types of approaches for the determination of relation weight.

The experiments done on the event log recorded by a virtual company model developed as the multi-agent system MAREA showed that LCSS method and Sim_{seq} are the most suitable for this type of log. The new (reduced) user/agent network was created with reduced user profiles. In comparison with base network (the network constructed with base user profiles) is the reduced user network more clear with visible relations (new or stronger ties) between the users. These experiments confirmed our assumption that it is very important to select the right method for determination of the relation between the sequences to set the reduced user profiles.

Since the presented approach was successfully tested and verified on the simulation model of the virtual company, it is planned to be used for the business analysis of the event log collection from a real ERP system.

Acknowledgments This work was supported by The Ministry of Education, Youth and Sports from the National Programme of Sustainability (NPU II) project ‘IT4Innovations excellence in science—LQ1602’.

References

1. Aggarwal, C.C., Reddy, C.K.: Data Clustering: Algorithms and Applications. Chapman & Hall/CRC data mining and knowledge discovery series. CRC Press, Boca Raton, FL (2014)
2. Gusfield, D.: Algorithms on Strings, Trees and Sequences: Computer Science and Computational Biology. Cambridge University Press (2008)
3. Hirschberg, D.S.: Algorithms for the longest common subsequence problem. J. ACM **24**, pp. 664–675 (1977). http://doi.acm.org/10.1145/322033.322044
4. Meimaris, M., Vafopoulos, M.: Knowledge-based semantification of business communications in erp environments. Lect. Notes Comput. Sci. **7652**, 159–172 (2013)
5. Slaninová, K.: User behavioural patterns and reduced user profiles extracted from log files. In: Proceedings of 13th International Conference on Intellient Systems Design and Applications, ISDA 2013, pp. 289–294. IEEE Computer Society (2013)
6. Slaninová, K., Kocyan, T., Martinovič, J., Dráždilová, P., Snášel, V.: Dynamic time warping in analysis of student behavioral patterns. In: Proceedings of the Dateso 2012 Annual International Workshop on Databases, Texts, Specifications and Objects. CEUR Workshop Proceedings, pp. 49–59 (2012)
7. Slaninová, K., Vymětal, D., Martinovič, J.: Analysis of event logs: Behavioral graphs. In: Web Information Systems Engineering—WISE 2014 Workshops. Lecture Notes in Computer Science, vol. 9051, pp. 42–56. Springer-Verlag (2015)
8. Song, M., van der Aalst, W.M.P.: Towards comprehensive support for organizational mining. Decis. Support Syst. **46**(1), 300–317 (2008)
9. Šperka, R., Vymětal, D.: Marea—an education application for trading company simulation based on rea prin-ciples. In: Proceedings of the Advances in Education Research. Information, Communication and Education Application, vol. 30, pp. 140–147 (2013)
10. van der Aalst, W.M.P.: Process Mining: Discovery, Conformance and Enhancement of Business Processes, 1st edn. Springer, Heidelberg (2011)
11. Vymětal, D., Scheller, C.: Marea: Multi-agent rea-based business process simulation framework. In: Conference proceedings of International Scientific Conference ICT for Competitiveness, pp. 301–310 (2012)
12. van der Aalst, W.M.P., Reijers, H.A., Song, M.: Discovering social networks from event logs. Comput. Support. Coop. Work **14**(6), 549–593 (2005)
13. van der Aalst, W.M.P., van Dongen, B.F., Herbst, J., Maruster, L., Schimm, G., Weijters, A.J.M.M.: Workflow mining: a survey of issues and approaches. Data Knowl. Eng. **47**(2), 237–267 (2003)

Part IV
Learning Paradigms and Applications: Agent-Based Approach

Intelligent Agents and Game-Based Learning Modules in a Learning Management System

Kristijan Kuk, Dejan Rančić, Olivera Pronić-Rančić and Dragan Ranđelović

Abstract Many researchers have taken a great deal of effort to promote high quality game-based learning applications, such as educational games, animations, simulations, animated or interactive simulation mechanisms in learning management system (LMS), and so on. The Bloom's taxonomy strategy was successfully implemented as effective gaming model in two different game-based learning applications (puzzle and platformer games) that motivate and actively engage college students in order to make learning process more enjoyable. During using the game-based modules in LMS Moodle, special attention was paid both to the integration of game-play aspects and the relationship between learning styles and game genres. In this paper we shall describe the proposed approach and introduce an adaptation and personalization of player as student model based on game genres. We have analyzed learning styles and teaching strategies that match the game features which resulted in embedding the analysis personalization and teaching strategies into the game. This article presents the effectiveness of agent-based approach in teaching strategies of Moodle gaming education resources.

Keywords Learning management system · Intelligent agent · Game-based learning · Learning styles · Game genres

K. Kuk · D. Ranđelović
Academy of Criminalistic and Police Studies, Belgrade, Serbia
e-mail: kristijan.kuk@kpa.edu.rs

D. Ranđelović
e-mail: dragan.randjelovic@kpa.edu.rs

D. Rančić (✉) · O. Pronić-Rančić
Faculty of Electronic Engineering, Niš, Serbia
e-mail: dejan.rancic@elfak.ni.ac.rs

O. Pronić-Rančić
e-mail: olivera.pronic@elfak.ni.ac.rs

G. Jezic et al. (eds.), *Agent and Multi-Agent Systems: Technology and Applications*, Smart Innovation, Systems and Technologies 58,
DOI 10.1007/978-3-319-39883-9_19

1 Introduction

The advantages of *LMS* (Learning Management Systems) such as Moodle, to support the presential lectures in higher education. The convergent use of the *ITS* (Intelligent Tutoring Systems) in *LMS* approaches can potentiate the learning process, making the LMS an intelligent learning environment [1]. The features provided by the *LMS* can be enhanced by using *AI* (Artificial Intelligence) techniques, and using cooperative intelligent agents (working in the background) or animated pedagogical agents (interacting with the user) [2].

Pedagogical agents are autonomous software entities that provide learning process support through interaction with the students, lecturers and other participants in the learning process, as well as cooperation with other similar agents. Personalized approaches based on intelligent agent technology imply that each student has its own personal (pedagogical) agent, tutor respectively that directs student towards learning. The role of such agent is to gather, maintain and analyze information on assigned students and on the basis of such information perform adaptation of the contents provided to the student. The whole process is based on the model of students that is in the case of game represented as the player model, both in the other approaches to personalization, and in case of agent. Pedagogical agent can be realized as a window of *Help option* in special class of game-based learning modules as stated in the paper [3].

The aforementioned results show that educational games can be an effective tool to complement the educational instruments available to teachers, in particular for spurring user motivation and for achieving learning goals at the lower levels in the Bloom's taxonomy [4]. In Bloom's taxonomy, there's a presumption that one must master the low-tier elements before reaching true learning, in the form of being able to evaluate and create. However, in most of the digital game environments, players are often expected to create, improvise and learn from their mistakes first, and later on to learn about the details. This contradiction shows where the interesting parts are, in the details. In order to learn and understand the game, inclusive of all the dimensions outlined in Bloom's taxonomy players have to perform competently [5]. But rather than the linear order that makes Bloom's taxonomy neat and useful for curriculum planning and evaluation purposes, the game-based approach looks to take a different route in how players gather and use knowledge. Games are often classified into genres, which purport to define games in terms of having a common style or set of characteristics, e.g. as defined in terms of perspective, gameplay, interaction, objective, etc.

An adaptive *AI* can drastically increase the replayability of our game, and make the game experience much more intense and personalized for the players. A technique we call player modeling is borrowed from the similar notion of student modeling in intelligent tutoring system research. Following section briefly address game genres based on Bloom's educational objectives and applied for education in *LMS*.

2 Related Work

Moodle supports plugin based structure where a new idea can be implemented as plugin. A game is a great plugin in Moodle. Game plugins help in making assessment test creative and bring game-based learning in Moodle which seeks student's interest. It makes use of questions, quizzes and glossaries to create a variety of interactive games. These games take questions from either quiz or glossary to form a board where student can play the game. The first component is the activity module named Game [6] by Vasilis Daloukas created in 2008. There are 8 games in this module available as a plugin for Moodle. The games are: *Hangman, Crosssword, Cryptex, Millionaire, Sudoku, Snakes and Ladders, Hidden Picture, Book with questions.*

Kumar was to create a generic game, as a plugin for Moodle [7] that can be used by any available course. During the project he thought of some games that can be very handy in a particular course like physics and mathematics. He is proposing four new games: *Tic-Tac-Toe*, *The weakest link*, *Guess-in-time* and *Anagram*.

There are many types of game designers. Game-based learning (*GBL*) allows game designer to clearly evaluate student performance because they are able to leverage specific achievements that are connected to Learning Objectives (*LO*) embedded within the game. GBL is a type of game play that has defined learning outcomes. Generally, game based learning is designed to balance subject matter with gameplay. Gamification is using game design principles to change non game-like classrooms into fun and engaging gamelike environments, for the purpose of motivating and changing learner behaviors. Stefan Göbel et al. [8] describes the concept and use of Narrative Game-based Learning Objects (*NGLOB*) for the personalization and adaptation of Story-based Digital Educational Games (*DEG*). They characterize the potential of personalized and adaptive *DEG* with *NGLOBs* combining learner modeling, player modeling and storytelling of the Bat Cave application.

Educational Game Learning Objects could basically be reused in different games that use different surroundings and settings. The work of Minović et al. [9] in game-based learning area is moving from traditional web-based *LMS* towards game-based *LMS*, with the intention of integrating the upsides of using games in university education.

Both students and teachers can be provided with the results of the students' activity by the agent-supported Moodle *LMS*, collecting data on the Moodle database. Authors Scutelnicu et al. have presented a novel methodology for incorporating software agents into an *LMS*—Moodle. Their research [10] is focused on the integration of software agents with Moodle to improve its capabilities by creating a monitoring agent based tool for its forum. This approach is demonstrated by a prototype of *JADE* agents for automated forum monitoring.

3 Theoretical Background

Online educational games are usually e-learning systems that use *Flash* or *HTML* technology to create educational game-based applications for enhancing learning outcomes. Aside from the obvious fact that flash games allow hours of free and varied game play, they do a lot to advance various genres of games and actually help students to learn new things and challenge themselves by advancing to higher levels. *Flash* or *HTML* gaming also introduces games to people who don't typically play games, and it could act as a kind of a gateway into console/PC platform. Although there are plenty of exceptions, the majority of flash games on Newgrounds social media website clearly fall into one of several genres and subgenres [11]. Flash authors can pick one from dozens of genres and subgenres when start designing their game. Several larger genres are used in the Newgrounds "games" web page to help users find the style of the game they want to play. To the interest of our teaching curriculum, and on the basis on the game style we have selected following list of game genres and their basic features:

- **Action games**: This genre, initially defined by games such as *Pong* and *Space In-vaders*, is characterized by fast-paced events and movements which usually have to be performed reflexively. Since action games are dynamic, complex and multi-stage games, every level of Bloom's hierarchy is reached repeatedly throughout a number of levels of play. The ultimate goal and the rules of the game must be familiar to the player.
- **Platform games**: A platform game (or platformer) is a video game characterized by requiring the protagonist to run and jump to and from surfaces (platforms, floors, stairs etc.) while avoiding different obstacles represented on a single or scrolling game screen. Traditionally, these were 2D games that achieved a great popularity on earlier gaming platforms. *Klonoa*: *Empire of Dreams*, the first handheld title in its series, was also a puzzle-platformer.
- **Puzzle**: These games fall into such computer game genres as puzzle, arcade, and card/dice. Puzzle games are video games that emphasize puzzle or problem solving. Since they can involve the exercise of logic, memory, pattern matching, reaction time, etc., different problem solving skills can be tested. With their easy objectives puzzle and arcade games are commonly played by those who do not normally play video games. Namely, players are required to solve puzzles based on repetitive manipulation of words, numbers or shapes to complete tasks. Players need to understand the concepts in order to predict what will happen, based on the moves they make. These are typically 2D games with few backgrounds. The main format of the game usually remains the same although sometimes the games have several levels on different screens. Puzzle-platformer hybrids such as *Achievement Unblocked*, which use action elements also belong to this genre.

Bloom's educational objectives are realized to a variety of extents in different broad genres of video games. Authors John Sherry and Angela Pacheco take each

Table 1 Game genres based on Bloom's educational objectives

	Action/Platformer	Puzzle
Knowledge	x	x
Analysis	x	x
Comprehension	x	x
Application	x	x
Synthesis	x	
Evaluation	x	

objective in turn, describing the genre and indicating which levels of Bloom's hierarchy are best related to each genre [12]. Based on their Bloom's hierarchy of learning outcomes, our approach of game genres based on Bloom's educational objectives for previously given genre games is shown in Table 1.

4 Student Model Based on Game Genres

The main component of adaptive e-learning systems is a student model. Student modelling is the process of taking into account student cognition in the design of an *ITS*. The basic idea of our model is to integrate methods and concepts of player and student models: the game maintains a profile of each player that captures the skills, weaknesses, preferences, and other characteristics of that player. Existing types of student models are classified with regard to their diagnostic capabilities and the domain-dependency of their updating mechanisms. Domain-specific information and domain-independent information are two major groups of information collected in the student models [13]:

- The domain-specific information model is referred to as the student knowledge model. It describes the students' knowledge level, their understanding of domain knowledge or curriculum elements, the errors that the students made, the students' knowledge development process, records of learning behaviors, records of evaluation or assessment, and so forth.
- The domain-independent information is information about the student' skills, so it is based on their behavior. Learning goals (evaluating the learners' achievements), cognitive capabilities (inductive reasoning skill and associative learning skill), motivational states the learners are driven by, background and experience, and preferences might be incorporated in it.

Some genres, such as puzzle, combine aspects of platformer games with other genres. Platformers put emphasis on running and jumping through certain environment. They can either focus on solving environmental puzzles. Puzzle platformers are characterized by their use of a platform game structure to drive a game whose challenge is derived primarily from puzzles. After performing an analysis of existing platformer and puzzle games used in the teaching process topic of course

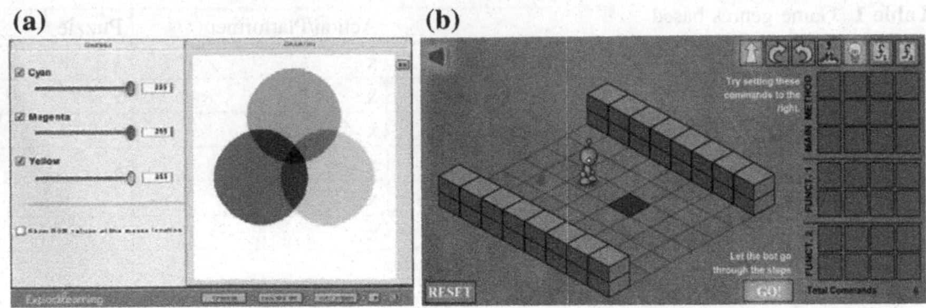

Fig. 1 A screenshot of puzzle-platformer games. **a** Subtractive Colors Gizmo simulation. **b** Light bot game (Color figure online)

Fundamentals of IT. We came to the conclusion that it was necessary to introduce some of those con-temporary teaching resources for units: *Color Fundamentals* and *Programmer-style logic*. We considered that the main objective of computer techniques was awaking interest through two next puzzle platformer games (Fig. 1):

- **Puzzle**: *Subtractive Colors*—online interactive simulation tool (called *Gizmo*) to explore what color light they observe when they subtract each color one at a time from the white light they created.
- **Action/Platformer**: *Light bot*—online game in which you have to control a robot and help it light up all the blue tiles by giving it correct set of commands.

The goal of this experiment is to discuss the effect of different learning style characteristics on game genres in a game-based learning application. There are totally 92 students participating in this study, with the average age of 20–21 years. There are five main phases for implementing this experiment: The goal of this experiment is to discuss the effect of different learning style characteristics on game genres in a game-based learning application. There are totally 92 students participating in this study, with the average age of 20–21 years. There are five main phases for implementing this experiment:

1. First, all 92 students of the 1st year of Academy of Criminology and Police Studies attending *Fundamentals of IT* Course on Department of Informatics and Computer Sciences have to finish pre-test and Index of Learning Styles Questionnaire (*ILSQ*);
2. Second, according to the analyses of the obtained results, the students are divided into two groups, equal by pre-test results. The aim is to generate two groups based on the previous knowledge. One group is called Basic Group (B) with 50 % or less right answers and the other is Advanced Group (A) with more than 50 % of right answers during the pre-test.
3. Third, on the grounds of the experience the authors have created Teaching strategies of gaming for the student towards his learning style and knowledge

Table 2 Results *FS ILSQ*

Learning style	*Vi*	*Ve*	*Sen*	*Int*	*Seq*	*Glo*	*Act*	*Ref*
Percentage (%)	87	13	83	17	74	26	65	35
Number of students	20	3	19	4	17	6	15	8

level so we have formed four groups of students: *Act*-A, *Act*-B, *Sen*-A, *Sen*-B, *Vi*-A, *Vi*-B, *Seq*-A and *Seq*-B;

4. Forth, on the basis of given strategy formed groups used various types of game genres (*Subtractive Colors* and *Light bot*) for playing/learning without time limit. After playing acquired knowledge was tested by post-test;
5. Fifth—analyses of the obtained results, discussion and suggestions.

In the first phase the Felder and Silverman index of learning styles questionnaire (*FS ILSQ*) [14] was given to 23 students and the results are shown in Table 2. We marked with *Vi* for Visual, *Ve* for Verbal, *Sen* for Sensitive, *Int* for Intuitive, *Seq* for Sequential, *Glo* for Global, *Act* for Active, *Ref* for Reflexive.

As shown in the Table 2, four major groups were selected amongst surveyed students: Visual, Sensitive, Sequential and Active. It resulted that most of the surveyed students (71 out of 92 students) preferred learning style with features of these groups. That was the reason why only 71 students were participating in the further experiment. In order to assess current knowledge level for the given field, students were assessed via test from the red teaching units maker in the experiment as pre-test. Current knowledge level of a teaching unit that a student has is marked via two values: Basic for the achieved result 30–50 %, and Advanced for the achieved result >50 %.

After exploring student's categories of learning style according to the results of *FS ILSQ* and assessment of students' general knowledge prior to playing, process of students profile development is continued by mapping the acquired results to the value of the personalization vector game genre. The vector presents metadata which provide teaching strategy for making personalized game-based learning experience, according to the requirements described in Table 3 for understanding visual design, including physical techniques such as balance and symmetry.

Values of the recommended genre shown in the table above as teaching strategies are based on the assumption of the authors and their experience in utilization of such genre of teaching material in teaching strategy. In order to test the validity of proposed teaching strategy, after playing of given genre we have tested students for the same teaching unit with the same set of questions as in the pre-test. This test in the experiment is labelled as post-test. Comparison of the knowledge on the pre-test and acquired one on the post-test, is presented as the value of the accomplished learning progress after the game. Average progress accomplished for each learning style group is shown in the Fig. 2. Students with average knowledge are marked with A, and students with basic knowledge with B.

Analyzing the results, we could come to the conclusion that the strategy of certain game genre application, as given in the table, has significant influence to two

Table 3 Teaching strategies of gaming for the student towards his learning style and knowledge level

Learning style	Description of learning style	Previous knowledge level	Teaching strategy
Act	• Interactive curriculum type • Curriculum processing—individually routed steps for the realization of the task • Level of system feedback—particular or feedback on demand • Presentation of lessons with lots of interactivity and collaboration	Basic	Subtractive colors
		Advanced	Light bot
Sen	• Lot of practical work (simulations and games, learning based on problems and role playing, role playing, discussion panel, brainstorming)	Basic	Light bot
		Advanced	Light bot
Vi	• Presentation of lessons by multimedia, including videos and animation, simulations and games, charts, algorithms	Basic	Subtractive colors
		Advanced	Light bot
Seq	• Course organization in linear way. The content is presented orderly, step by step, consist of small orderly steps that are logically associated to the problems being solved	Basic	Subtractive colors
		Advanced	Subtractive colors

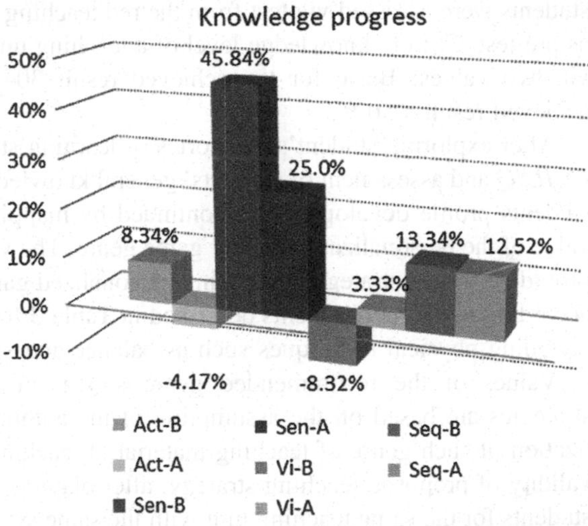

Fig. 2 The comparative knowledge progress for various learning styles after gaming

groups of students with learning style marked as *Sen* and *Seq*. High level of perception with Sensitive students and a high level of understanding of the information with the sequential students played important roles in achieving these results. Such statement leads to the conclusion that selecting game genre for this type of students does not affect the current knowledge level of the teaching unit, but that implementation of any game will certainly help in better understanding of the teaching material. On the grounds of the example of given teaching strategy and on the basis of the results from the Fig. 2, we can notice that there was no progress achieved at the Advanced level with style marked as *Act* and the Basic level with *Vi* learning style, so these two cases were expelled from the final conclusion. Other set values in the teaching strategy gave positive results so by simple counting in most cases we can issue a recommendation.

Recommendations for utilization of the following game genres on the basis of performed experiments are the following:

- For the knowledge level Basic—students with the learning style group *Act/Seq* are useful to use puzzle type of the game,
- For the knowledge level Advanced—students with the learning style group *Vi/Sen* are useful to use action/platformer type of the game.

5 Intelligent Agents in LMS Moodle via Activities Game Block

Moodle platform is used by many reputed educational institutes for managing their curriculum. All information in a course is organized in separate blocks. Activities block is used by the teacher to add an activity (a forum, an assignment, quizzes and so on). Activity modules are activities that teachers add to their courses to supply resources for students. Assignment activity has two items and a point of information in the common module setting. One of them is the visible—assignment activity which chooses whether to "show" or "hide" the assignment. Usually, teachers or trainers can sort submissions by group or view all course participants, but the task of the intelligent agent itself is to create the two necessary groups. If a grouping is selected, students assigned to groups within the grouping will be able to see together one of the presented games.

While there are many factors to consider when designing a gamification experience in *LMS* Moodle, a very important first step is to target learning styles group membership. If there are differences in learning styles between groups of students, then intelligent agent must use learning style information to aid their planning and preparation for organized activity module activities. Knowledge of game genre preferences can provide a bridge to course success in a game education Moodle mode. Based on the present recommendations in previous section, the task of implemented intelligent agent is to determine group membership for each student/participant for the course. Enabling user group setting allows the intelligent agent to automatically add students to groups in Moodle courses based on learning style membership.

The Index of Learning Styles (*ILS*) questionnaire is a 44-question multiple-choice survey, and each of the paired learning dimensions have eleven questions designated to them. Every question provides two answer options; these answers represent both aspects of the paired dimension. The output from the *ILS*

Table 4 Relevant patterns for each learning style dimension developed by Sabine Graf

Active/Reflective	Sensing/Intuitive	Visual/Verbal	Sequential/Global
content_visit (−)	content_visit (−)	content_visit (−)	outline_visit (−)
content_stay (−)	content_stay (−)	ques_graphics (+)	outline_stay (−)
outline_stay (−)	example_visit (+)	ques_text (−)	ques_detail (+)
example_stay (−)	example_stay (+)	forum_visit (−)	ques_overview (−)
selfass_visit (+)	selfass_visit (+)	forum_stay (−)	ques_interpret (−)
selfass_stay (−)	selfass_stay (+)	forum_post (−)	ques_develop (−)
selfass_twice_wrong (+)	exercise_visit (+)		navigation_skip (−)
exercise_visit (+)	ques_detail (+)		navigation_overview_visit (−)
exercise_stay (+)	ques_facts (+)		navigation_overview_stay (−)
quiz_stay_results (−)	ques_concepts (−)		
	ques_develop (−)		
forum_visit (−)	quiz_revisions (+)		
forum_post (+)	quiz_stay_results (+)		

questionnaire results in a user's score for particular learning style preferences, ranging from −11 to 11 for each axis. Learning Styles Tests module (*LSTest*) [15] from 2009 is very useful in evaluating learning styles as Moodle plugin. The problem with this module is that it was developed for older versions of Moodle.

However, while the questionnaire asks students about how they think they prefer to behave and learn, the proposed approach gathers data about how students really behave and learn by observing them during their learning process. Sabine Graf et al. [16] proposes an automatic approach for identifying students' learning styles in *LMSs* as well as a tool that supports teachers in applying this approach. Their system *DeLeS* is based on Moodle futures which include content objects, outlines, examples, self-assessment tests, exercises, and discussion forums.

The author is using "+" and "−" for indicating a high and low occurrence of the respective pattern from the viewpoint of an active, sensing, visual, and sequential learning style which present on Table 4. Based on this, information and formulas for the respective learning style in DeLeS can be used to get hints for detecting students' learning styles by intelligent agent. After that the agent itself generates two groups of students: Group A—*ct/Seq* and Group B—*Vi/Sen*, adding each student to the group according to the learning style preferences. Then, some informal feedback testing from a sample of knowledge level is done.

As given in the previous section, a crucial factor for the agent's decision is also a student's current level of knowledge for the lesson unit in which game module is included to supplement the lesson's regular topics [17]. The standard quiz reports are located under the Results. The *Grade/x* column of quiz attempts can give us the information about current knowledge level of a lesson unit. Two values *Grade* and *X* determine the knowledge level, which helps the agent to group the students in Basic or Advanced group.

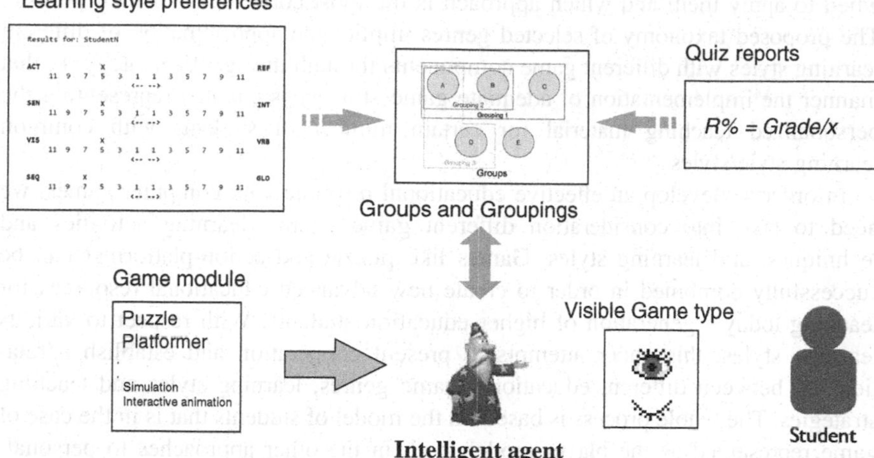

Fig. 3 A concept of intelligent agent in LMS with game module

The parameters can be taken from the test results database—self assignment tests for lesson unit. Afterwards, the agent decides about displaying a certain game from the game set based on the teaching strategies given in Table 1. Implement agent in our approach can be realized by groups (see Fig. 3). In Moodle, groups can be separate (the work of the group is visible only to group members) or visible (group work is visible to the entire class). A grouping consists of one or more groups. With groupings we can restrict access to content or activities to a specific set of groups. Agent will definitely let you know what students like, so even informal focus groups can be very helpful in deciding on player association's gaming strategy. The agent's decision is based on the following two *IF THEN* rule:

$$IF\ group = Act\ or\ Seq\ and\ knowledge\_level = Basic\ THEN\ type\_game = Puzzle \quad (1)$$

$$IF\ group = Vi\ or\ Sen\ and\ knowledge\_level = Advanced\ THEN\ type\_game = Platformer \quad (2)$$

6 Conclusion

Educational researches have shown Moodle as a good *LMS* platform of multimedia. It can be used effectively for the 21st century students. The most used contribution method is activity modules, but there is no unique method of using good educational resources via Moodle activity modules. The most effective approach depends upon the task, context and learner's personality. The learning will be more effective if learners can choose from a wide range of possible learning methods, if they know when to apply them and which approach is the most convenient for their perusal. The proposed taxonomy of selected genres implies the approximation of different learning styles with different game components through the agent's strategy. In that manner the implementation of adequate game strategy is enabled representing the personalized teaching material for certain number of students with common learning style/styles.

In order to develop an effective educational resource with computer games, we need to take into consideration different game genres, learning activities and techniques, and learning styles. Games like puzzle and action-platformer can be successfully combined in order to create new advanced educational resources for teaching today's generation of higher education students. With respect to various learning styles, this paper attempts to present cooperation and establish a relationship between different educational game genres, learning styles and teaching strategies. The whole process is based on the model of students that is in the case of game represented as the player model, both in the other approaches to personalization, and in case of intelligent agent. Also, this paper presents implementation agent in *LMS* Moodle as personal guide to classification games in library of *Explore Learning Gizmos* math and science simulations.

In the future our research will be focused on making an approach for automatic identifying teaching strategies of gaming. The approach will be based on inferring students' gaming styles from different teaching methods behavior in an online course for college grade level and IT discipline.

References

1. Palomino, C.E.G., Silveira, R.A., Nakayama, M.K.: An intelligent LMS model based on intelligent tutoring systems. In: Intelligent Tutoring Systems, pp. 567–574. Springer International Publishing (2014)
2. Jaques, P.A., Vicari, R.M.: Estado da arte em ambientes inteligentes de aprendizagem que consideram a afetividade do aluno. Revista Informática na Educação, pp. 15–38 (2005)
3. Kuk, K., Milentijević, I., Rančić, D., Spalević, P.: Pedagogical agent in multimedia interactive modules for learning—MIMLE. Expert Syst. Appl. **39**, 8051–8058 (2012)
4. Bellotti, F., Berta, R., De Gloria, A.: Designing effective serious games: opportunities and challenges for research. Int. J. Emerg. Technol. Learn. (iJET) **5** (2010)
5. Leo, L.C.: Bloom's taxonomy vs. game-based learning: toward a preliminary. Doctoral Student and Teaching Fellow (2003)
6. Moodle. http://docs.moodle.org/20/en/Game_module. Accessed 5 Jan 2015
7. Kumar, P.: Moodle plugin for game based learning. Doctoral dissertation, Indian Institute of Technology Bombay (2013)
8. Göbel, S., Wendel, V., Ritter, C., Steinmetz, R.: Personalized, adaptive digital educational games using narrative game-based learning objects. In: Entertainment for Education. Digital Techniques and Systems, pp. 438–445. Springer, Berlin, Heidelberg (2010)
9. Minović, M., Milovanović, M., Starčević, D.: Modelling knowledge and game based learning: model driven approach. J. Univers. Comput. Sci. **17**(9), 1241–1260 (2011)
10. Scutelnicu, A., Lin, F., Kinshuk, T.C.L., Graf, S., McGreal, R.: Integrating JADE agents into Moodle (2007)

11. Newgrounds. http://newgrounds.wikia.com/wiki/List_of_Flash_Game_Genres. Accessed 28 Dec 2015
12. Sherry, J.L., Pacheco, A.: Matching computer game genres to educational outcomes (2004)
13. Esichaikul, V., Lamnoi, S., Bechter, C.: Student modelling in adaptive e-learning systems. Knowl. Manage. E-Learn. Int. J. (KM&EL) **3**(3), 342–355 (2011)
14. Felder, R.M., Silverman, L.K.: Learning and teaching styles in engineering education. Eng. Educ. **78**(7), 674–681 (1988)
15. Rubio Reyes, B.: Activity Module: Learning Styles Tests (LSTest). http://moodle.org/mod/data/view.php?d=13&rid=3098. Accessed 19 Dec 2009
16. Graf, S., Kinshuk, Liu, T.-C.: Supporting teachers in identifying students' learning styles in learning management systems: an automatic student modelling approach. Educ. Technol. Soc. **12**(4), 3–14 (2009)
17. Kuk, K., Milentijević, I., Rančić, D. Spalević, P.: Designing intelligent agent in multilevel game-based modules for e-Learning computer science course. In: e-Learning Paradigms and Applications. Subtitle: Agent-based Approach. Series: Studies in Computational Intelligence, vol. 528. Springer, Germany (2014)

Robot-Oriented Generative Learning Objects: An Agent-Based Vision

Vytautas Štuikys, Renata Burbaitė, Vida Drąsutė and Kristina Bespalova

Abstract The paper presents a juxtaposing of the robot-oriented generative objects (GLOs) with software (SW) agents and identifies the capabilities to introduce more intelligence to the educational environment based on those GLOs. The main contribution of the paper is the agent-based architecture of the system and its partial implementation, enabling to solve the prescribed tasks more efficiently (with a less user's intervention and a higher robot's accuracy). Also the case study and some experimental results are described.

Keywords Generative learning objects · Software agents · Educational robots

1 Introduction

In the technology enhanced learning (TEL), the agent-based approaches aim at enhancing the functionality and capabilities of the educational systems. There might be human beings as agents (HAs), hardware agents (HWAs) and software agents (SWAs). Typical examples of HWAs are educational robots (ERs), the extremely active branch in educational research [1, 2]. They are equipped with the sensor system to react to the working (educational) environment. The robot's functionality is predefined by its control program to perform the physical task (such as obstacle finding and passing, line following, etc.). In fact, the program is the learning object (further

V. Štuikys (✉) · R. Burbaitė · V. Drąsutė · K. Bespalova
Kaunas University of Technology, Kaunas, Lithuania
e-mail: vytautas.stuikys@ktu.lt

R. Burbaitė
e-mail: renata.burbaite@ktu.lt

V. Drąsutė
e-mail: vida.drasute@ktu.lt

K. Bespalova
e-mail: kristina.bespalova@ktu.lt

G. Jezic et al. (eds.), *Agent and Multi-Agent Systems: Technology and Applications*, Smart Innovation, Systems and Technologies 58,
DOI 10.1007/978-3-319-39883-9_20

LO) to be understood and learnt by students (e.g. for Computer Science subjects). It is possible to understand that deeply only through the explorations of the robot's environment which, in this case, is also treated as a part of the educational environment. ER operates autonomously (to the extent of the control program capabilities), though the interaction with the HA is not excluded. Therefore, the environment with the embodied control program, possibly enriched with the additional feedback capabilities due to the adequate use of sensors, can be seen as a generic agent (meaning that it covers both the hardware and software notions).

The capabilities of agent-based technology can be introduced into the TEL systems either integrating those features within the educational software internally at the design time, or introducing the SWAs externally. Of course, there might be some compromise between those approaches. In general, SWAs are defined as computer systems situated in an environment and being able to achieve the prescribed objectives by: (i) acting autonomously, i.e. by deciding themselves (fully or partially: the term semi-autonomy is also used) what to do, and (ii) being sociable, i.e. by interacting with other software agents or users. The attribute 'autonomy' also means that after deciding what to do the action is to be performed automatically.

With respect to the interfacing and communicating capabilities of the so-called generative LOs (GLOs) [3, 4], they can be thought in some cases as a simplest form of SWAs (the more extensive discussion on that will be given later). Note that the GLOs notion has been introduced by Boyle, Morales et al. [3] and nominates the paradigm shift from the component-based reuse model to generative reuse model. Here, by GLOs we mean those that are implemented using meta-programming techniques [4].

In this paper, we consider the robot-oriented meta-programming-based GLOs (in [4], with some additional attributes, they are treated as smart LOs). Typically, they are designed in advance and reside in the personal library accessible through the Internet by the human agent. Structurally and semantically, they represent a family of the related robot control programs to describe the variability of a given teaching task, such as 'line follower', 'pick-carry', etc. The main distinguishing feature of those GLOs, as compared to the component-based traditional LOs (e.g. taken from the digital libraries), is that they specify not only the content itself, but also the context relevant to this content explicitly in the same unified specification. The variability space, to some extent, may contribute to enforcing and enlarging the possibilities for autonomy in terms of possibilities for adaptation and decision making (the main agents attributes).

Therefore, the aim of this paper is first to analyze those GLOs with the agent technology in mind and then to enhance the existing distributed educational environment by introducing SWAs for enforcing the intelligence capabilities of both the environment and GLOs. The main contribution of the paper is the agent-based architecture of the system and its partial implementation, enabling to solve the prescribed tasks more efficiently (with a less user's intervention and a higher robot's accuracy).

The remaining part of the paper includes: the related work, agent-based analysis of GLOs, description of the extended architecture of the educational system, a case study, discussion and evaluation and conclusion.

2 Related Work

We categorize the related work as follows: (A) relevant to agent-based e-learning, (B) SWAs in the LOs domain, (C) educational robots as agent-based LOs.

A. The researchers emphasize the definitions, goals, and functionality of pedagogical [5] or e-learning [6] agents, agent-based approach in e-learning paradigms and applications [7]. The mentioned literature resources raise the main issues to use SWAs in e-learning domain, such as adaptability, personalization, effectiveness, etc. Kim and Baylor [8] highlight the role of pedagogical agents and define three categories: expert, motivator, and mentor agents. The paper [5] emphasizes a software engineering models and their applications in e-learning. The paper [9] presents the intelligent ontology based on the using of SWAs. The papers [10, 11] present a multi-agent-based architecture for secure and effective e-learning and the conceptual design of a smart classroom based on multi-agent systems. The researchers [12] focus on learners' characteristics and groups formation principles purification (refinement) using agent-based approaches. The paper [12] highlights intelligent learning environments (ILE) with implemented SWAs.
B. The researchers propose some denominations of LOs that are in the relations with software agents: agent-based LO [13], personalized LO [14], intelligent LO [15] intelligent learning resources [16]. The general properties of the intelligent LOs are:

   (i) *To respond* to changes in its environment [16]; (ii) *To act* without the direct intervention of humans [13]; (iii) *To control* over their actions [13]; (iv) *To be* self-contained, reusable, adaptable, flexible, customizable, contextualized [14, 15].

   Therefore, the most important issues are related to integration LOs into ILEs and making them more intelligent from a viewpoint of adaptation and personalization aspects.
C. Werfel [17] interprets ERs as embodied teachable agents that open new possibilities as the social and educational tools. The paper [18] discusses the issues of engaging learners into the process by using the robotic educational agent. Lin et al. [19] present the usage of the physical agent (robot) as interactive media integrated into the traditional e-book with the purpose to contextualize the content. The papers [20, 21] focus on the intelligent robotic tutor that can be treated as physical educational agent. Therefore, ERs have a great potential and not well enough exploited capabilities for e-learning such as those to increase learning motivation, to apply modern educational methods, to extend the learning environments facilities, etc.

Though the analysis is by no means comprehensive, we able to conclude that SWA-based and ER-based approaches cover many different aspects of e-learning and gaining ground on it. Nevertheless, so far, little is known on the integrative aspects of LOs (GLOs), ERs and SWAs within educational systems.

3 GLOs and SW Agent Domains Analysis: Problem Statement

In this section we introduce a series of definitions for both the GLOs and SW agent domains (because there is no unified view of them), analyze the definitions aiming at the identification of similarities and differences of the two. Then, on the basis of making the juxtaposition of relevant attributes, we formulate the problem to be considered in this paper.

Definition 1A Generative LO (GLO) is "an articulated and executable learning design that produces a class of learning objects" [3]. In this paper, we consider those GLOs whose are defined by Definitions 1B–1D.

Definition 1B GLO is the executable meta-level specification (meaning parameterized) to enable producing of concrete LOs on demand automatically according to the pre-specified parameter values and the context of use. In this context, the meta-programming techniques [4] are regarded as a relevant technology to develop the meta-specification. Note that LOs may be chunks of text, pictures, computer programs, etc.

Definition 1C Robot-oriented GLO is the executable meta-level specification to enable producing of concrete LOs being the robot control programs (CPs) generated on demand automatically according to the pre-specified parameter values and the context of use.

Definition 1D Context-aware GLO is the executable meta-level specification (as defined by Definition 1B or Definition 1C) whose parameters are enriched by the pedagogical and technological context information to support the automatic content generation and adaptation. Context-aware robot-oriented GLO is also treated as Smart LO [22].

By the context information we mean the technological context (such as robot specific attributes), pedagogical context (e.g. teaching goal, model, student's profile, age, etc.), or some combination thereof.

Property 1 GLO interface is the graphical representation of parameters (P) and their values (V). Parameters represent variants of either the context or content attributes. Parameter values define the GLO's semantics.

Property 2 Variability space for decision making is expressed as P × V.

Property 3 GLO's meta-body implements the variability space through the constructs of the meta-language and target language [23].

Definition 2A Agents are "… programs that engage in dialogs and negotiate and coordinate transfer of information" (Michael Coen [24]).

Definition 2B Intelligent agents are defined as: "…software entities that carry out some set of operations on behalf of a user or another program with some degree of independence or autonomy, and in so doing, employ some knowledge or representation of the user's goals or desires." (IBM [24]).

Definition 2C Software agents are entities that "differ from conventional software in that they are long-lived, semi-autonomous, proactive, and adaptive". (Software Agents Group at MIT [24]).

Definition 2D An autonomous agent is "a system situated within and a part of an environment that senses that environment and acts on it, over time, in pursuit of its own agenda and so as to effect what it senses in the future" [25].

Definition 2E The agent is "a hardware or (more usually) software based computer system that enjoys the properties: *autonomy*, *social ability*, *reactivity,* and *proactiveness*" [26].

Property 4 *Autonomy* is the property when agents operate without the direct intervention of humans or others, and have some kind of control over their actions and internal state.

Property 5 *Social ability* is the property when agents interact with other agents (and possibly humans) via some kind of agent-communication language.

Property 6 *Reactivity* is the property when agents perceive their environment (which may be the physical world, a user via a graphical user interface, a collection of other agents, the Internet, or perhaps all of these combined), and respond in a timely fashion to changes that occur in it.

Property 7 *Proactiveness* is the property when agents do not simply act in response to their environment; they are able to exhibit goal-directed behavior by taking the initiative by [26].

Definition 3A *Environment* is a part of the agent based system in which the agent is operating through communication.

Definition 3B *GLO environment is the space where the interactions occur among the following entities: user and his/her PC, PC's operating system, meta-language processor, and remote server where GLO resides.*

Definition 3C *Educational robot's environment* is the robot itself and its working surrounding that typically holds the following properties. It is: *observable*—new action can be based on the most recent percept; *deterministic*—predicting effects of actions are easy; *episodic*—there is no need to look ahead beyond the end of the previous episode; *static*—there is no significant time restriction to make a decision.

Definition 3D *Combined environment* is the distributed environment that includes the GLO environment, the robot's environment and the Internet (meaning the direct link "Robot-Server").

Based on the analysis of related work and provided definitions, as well on the simplified functionality for both entities introduced in Fig. 1, we have made a juxtaposing of the entities trying to identify their similarity according to the

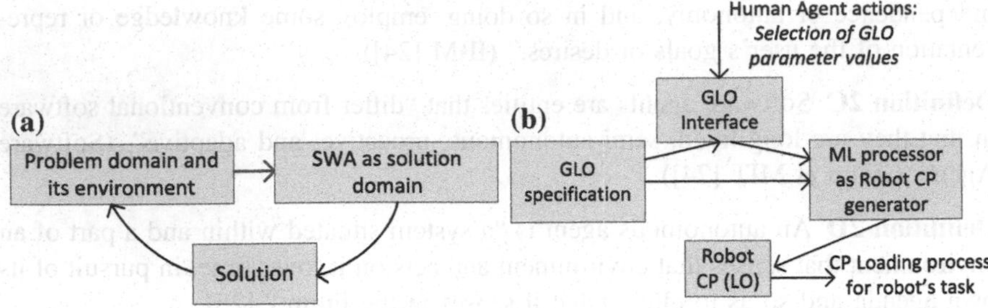

Fig. 1 SWA-based problem solving (**a**) and LO generation from GLO (see Definition 1C) (**b**)

Table 1 Juxtaposing of GLOs and SWA attributes as they are seen in the given definitions

Agent definition	Similarity: matching properties of GLO & SWA	Differences: non-matching properties of GLO & SWA	Explanation and evaluation
Definition 2A	"engage in dialogs and negotiate"	"coordinate transfer of information"	Poor (see Definition 1C)
Definition 2B	"carry out some set of operations on behalf of a user with some degree of independence"	"carry out some set of operations on behalf of another program"	Moderate (see Definitions 1B and 1C)
Definition 2C	"long-lived, semi-autonomous and adaptive"	"proactive", i.e. exhibit goal-directed behavior by taking the initiative	Moderate (see Definition 1D)
Definition 2D	"situated within and a part of an environment and acts on it" through human intervention	"situated within and a part of an environment that senses that environment"	Poor
Definition 2E	"semi-autonomy, social ability, reactivity, and proactiveness"	"autonomy, proactiveness"	Poor

coincidence range: *very poor*, *poor*, *moderate, good* and *excellent* (those are based on our and expert knowledge). Results are outlined in Table 1.

We formulate main results of analysis as follows. Both entities, SWAs and GLOs, are complex software specifications with the different degree of intelligence; though their functionality and structure differ significantly (the first is functioning on the fly dynamically, the second is functioning on the basis of static predefined specification), however, both have also some similarity as it is defined by Property 8.

Property 8 The system that consists of GLO (as a specific small knowledge base for the robot's specific tasks) and meta-language processor (as a semi-autonomous tool for providing the solution automatically) is seen as a week specialized SW agent (meaning a very low-level of intelligence) (see Fig. 1b). It is so, because the following properties hold: *semi-autonomy,* partial *social ability* (both due to the human's intervention), *reactivity.*

The task we consider in this paper is formulated as follows: to propose the concept and its implementation to substitute (partially or fully) the *human actions* in the process of generating robot control programs (CPs) *by the predefined system of SW agents.*

4 Robot-Oriented Agent-Based Educational Environment (ROABEE): Architecture and Processes

In Fig. 2, we outline the system ROABEE as a compound of the user's PC, Server and ER. Within each constituent, there are basic components. The highlighted components have been added to the previously developed system [4], aiming at enhancing its capabilities. The 'thin arrows' indicate the internal interaction among the components, while the 'fat arrows' indicate on the external processes.

The GLOs library represents a set of GLOs, the predesigned and tested entities. From the pedagogy viewpoint, each entity covers a separate topic of the course (Introduction to Programming taught in the gymnasium 10–11th classes). From the functionality and capabilities viewpoint, GLO is the generator (meta-language (ML) processor accomplishes this task) of the content. The latter is the robot's control program (CP), here treated as LO to support the teaching task. In the former system, the user was fully responsible for selecting the parameter values from the GLO interface. In the enhanced system (i.e. ROABEE), this function (fully or partially) is given to the software agents (SWAs). As the GLO interface contains different parameters (pedagogical-social context, technological, content), it is convenient to have a separate SWA for each group of parameters.

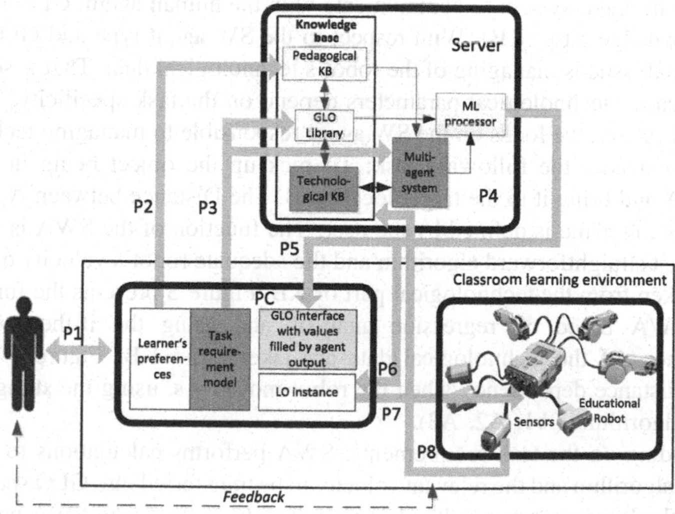

Fig. 2 Architecture of the extended system along with the functioning processes

In the GLO knowledge base (KB) there is the initial knowledge about the tasks to be accomplished by the robot. Again, this knowledge is relevant to the existing types of parameters within GLOs specification. Furthermore, there might be a specific knowledge inherent to a particular task (e.g., the initial technological knowledge of the robot's straightforward moving differs from the knowledge of the curved-line moving task). Some part of the initial knowledge resides within the GLO library (within of the GLO interface for each task). The remaining part of GLO KB is the dynamic knowledge representation the histories of the statistic data accumulated over time by using GLO in real teaching setting.

We present the functionality of the system as a sequence of the cooperating processes indicated by adequate numbers (from 1 to 8, see Fig. 2). The user interacts with PC (P1) to send his/her current profile to the profile DB (P2). Then the task requirement model should be supplied (P3) to the agent-based system and GLOs library to initiate the processes (P4) in the server. The latter activity results in completing the GLO interface (fully and partially depending on the agents' functionality) and generating LO. Both items are sent to PC (P5) for possible checking. Then the generated LO (P6) transfers to the ER for the task accomplishing (P7). Finally, it is possible to transfer the sensor's data to the KB either directly (P8), or through the feedback and user connections.

5 Implementation of Software Agent: A Case Study

As within the GLO interface, there are semantically different parameters (pedagogy-related, technological and content), it is convenient to have separate SW agents for managing selecting values for each kind of parameters. Therefore, we have a multi-agent system to communicate with the human agent, GLO library and GLO knowledge base (KB). With respect to the SW agent type and GLO KB, the most crucial issue is managing of the robot's technological data. That is so because, in many cases, technological parameters depend on the task specificity. Therefore, in our case study, we focus on the SW agent responsible to managing technological data and consider the following task: To pick up the object being in the initial location A and bring it to the target location B. The Distance between A and B and the processing Time is defined by the user. The function of the SWA is to identify the relevant straightforward algorithm and the adequate robot's velocity on the basis of data taken from the technological part of KB. Figure 3 presents the functionality of the SWA based on regression analysis and using the if-then techniques. Dependences of the technological data are taken from KB. Figure 4 shows the velocity-distance dependency when the robot moves 4 s, using the straightforward moving algorithms (A1, A2, A3).

Depending on the task requirements, SWA performs calculations to define the adequate algorithm and the relevant velocity to be transferred into GLO specification.

In Table 2, we present results of experiments to evaluate the SWA performance by measuring the real distances of moving as compared to the given requirements.

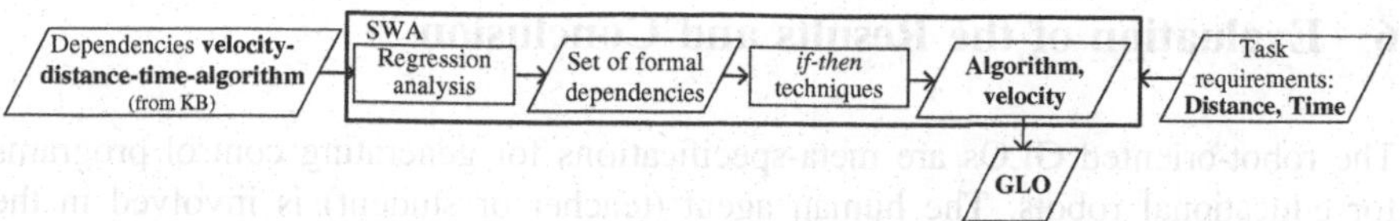

Fig. 3 SWA functioning algorithm

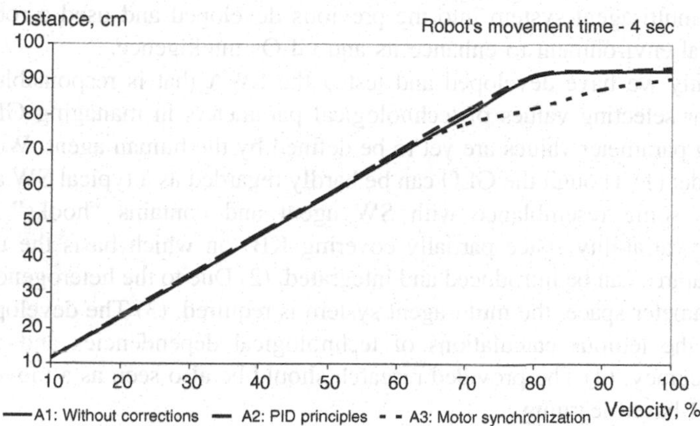

Fig. 4 A fragment of technological data taken from KB

Table 2 Experimental evaluation of SWA performance

Distance[a] (cm)	50	40	30	20	10
Time[b] (s)	Measured distance[c] (cm)				
1	–	–	29.3	19.8	10.0
2	46.5	39.0	29.4	20.3	10.0
3	48.6	39.5	29.5	20.3	10.0
4	48.2	39.5	29.5	20.5	10.0
5	48.6	39.5	29.5	20.5	10.3
6	48.6	39.5	30.0	20.5	10.3
7	49.0	40.0	30.0	20.5	10.0
8	49.4	39.5	30.0	20.5	10.3
9	49.6	39.5	31.0	20.5	10.0
10	49.6	40.5	31.0	20.3	10.2
Average	**48.7**	**39.6**	**29.9**	**20.4**	**10.1**
Standard deviation	**1.0**	**0.4**	**0.6**	**0.2**	**0.1**

Explanation
[a]Distance is taken from task requirements
[b]Time is also taken from task requirements
[c]Measured distance is obtained by experiment

6 Evaluation of the Results and Conclusion

The robot-oriented GLOs are meta-specifications for generating control programs for educational robots. The human agent (teacher or student) is involved in the autonomous generation process through selecting or introducing parameter values via the graphical user interface of GLOs. In this paper, we have identified: (1) some similarity of GLOs to software agents (SWAs), (2) proposed the structure to integrate the multi-agent system into the previous developed and used a robot-based educational environment to enhance its and GLOs intelligence.

Currently we have developed and tested the SWA that is responsible for calculating or selecting values of technological parameters in managing GLOs. The remaining parameter values are yet to be defined by the human agent. We are able to conclude: (1) Though the GLO can be hardly regarded as a typical SW agent, the GLO has some resemblance with SW agent and contains "hooks" (such as parameter variability space partially covering KB) on which basis the true intellectual features can be introduced and integrated. (2) Due to the heterogeneity of the GLO parameter space, the multi-agent system is required. (3) The developed SWA excludes the tedious calculations of technological dependencies and ensures a higher accuracy. (4) The provided research should be also seen as a move towards the STEM-based learning.

References

1. Benitti, F.B.V.: Exploring the educational potential of robotics in schools: a systematic review. Comput. Educ. **58**(3), 978–988 (2012)
2. Alimisis, D.: Educational robotics: open questions and new challenges. Themes Sci. Technol. Educ. **6**(1), 63–71 (2013)
3. Leeder, D., Boyle, T., Morales, R., Wharrad, H., Garrud, P.: To boldly GLO-towards the next generation of learning objects. In: World Conference on E-Learning in Corporate, Government, Healthcare, and Higher Education, pp. 28–33 (2004)
4. Štuikys, V.: Smart Learning Objects for the Smart Education in Computer Science: Theory, Methodology and Robot-Based Implementation. Springer (2015)
5. Veletsianos, G., Russell, G.S.: Pedagogical agents. In: Handbook of Research on Educational Communications and Technology, pp. 759–769. Springer (2014)
6. Gregg, D.G.: E-learning agents. Learn. Organ. **14**(4), 300–312 (2007)
7. Ivanović, M., Jain, L.C.: E-Learning Paradigms and Applications: Agent-based Approach. Springer (2013)
8. Kim, Y., Baylor, A.L.: Research-based design of pedagogical agent roles: a review, progress, and recommendations. Int. J. Artif. Intell. Educ. 1–10 (2015)
9. Davidovsky, M., Ermolayev, V., Tolok, V.: An implementation of agent-based ontology alignment. In: ICTERI, p. 15 (2012)
10. Ahmad, S., Bokhari, M.U.: A new approach to multi agent based architecture for secure and effective e-learning. Int. J. Comput. Appl. **46**(22), 26–29 (2012)
11. Aguilar, J., Valdiviezo, P., Cordero, J., Sánchez, M.: Conceptual design of a smart classroom based on multiagent systems. In: Proceedings on the International Conference on Artificial

Intelligence (ICAI), The Steering Committee of The World Congress in Computer Science, Computer Engineering and Applied Computing (WorldComp), p. 471 (2015)
12. Giuffra, P., Cecilia, E., Ricardo, A.S.: A multi-agent system model to integrate virtual learning environments and intelligent tutoring systems. IJIMAI **2**(1), 51–58 (2013)
13. Mohammed, P., Mohan, P.: Agent based learning objects on the semantic web. In: SW-EL'05: Applications of Semantic Web Technologies for E-Learning, p. 79 (2005)
14. Pukkhem, N., Vatanawood, W.: Personalised learning object based on multi-agent model and learners' learning styles. Maejo Int. J. Sci. Technol. **5**, 3 (2011)
15. Silveira, R.A., Gomes, E.R., Viccari, R.M.: Intelligent learning objects: an agent approach to create reusable intelligent learning environments with learning objects. In: Advances in Artificial Intelligence-IBERAMIA-SBIA 2006, Springer, pp. 17–26 (2006)
16. Mamud, M., Stump, S.: Development of an intelligent learning resource using computer simulation about optical communications. In: Education and Training in Optics and Photonics, Optical Society of America, p. EMB1 (2009)
17. Werfel, J.: Embodied teachable agents: learning by teaching robots. In: The 13th International Conference on Intelligent Autonomous Systems (2013)
18. Brown, L., Kerwin, R., Howard, A.M.: Applying behavioral strategies for student engagement using a robotic educational agent. In: 2013 IEEE International Conference on Systems, Man, and Cybernetics (SMC), pp. 4360–4365. IEEE (2013)
19. Lin, J.-M., Chiou, C.W., Lee, C.-Y., Hsiao, J.-R.: Supporting physical agents in an interactive e-book. In: Genetic and Evolutionary Computing, pp. 243–252. Springer (2015)
20. Corrigan, L.J., Peters, C., Castellano, G.: Identifying task engagement: Towards personalised interactions with educational robots. In: 2013 Humaine Association Conference on Affective Computing and Intelligent Interaction (ACII), pp. 655–658. IEEE (2013)
21. Li, J., Kizilcec, R., Bailenson, J., Ju, W.: Social robots and virtual agents as lecturers for video instruction. Comput. Human Behav. (2015)
22. Burbaite, R., Bespalova, K., Damaševičius, R., Štuikys, V.: Context aware generative learning objects for teaching computer science. Int. J. Eng. Educ. **30**(4), 929–936 (2014)
23. Štuikys, V., Damaševičius, R.: Meta-Programming and Model-Driven Meta-Program Development: Principles, Processes and Techniques. Springer Science & Business Media (2013)
24. Bădică, C., Budimac, Z., Burkhard, H.-D., Ivanovic, M.: Software agents: languages, tools, platforms. Comput. Sci. Inf. Syst. **8**(2), 255–298 (2011)
25. Franklin, S., Graesser, A.: Is it an agent, or just a program?: A taxonomy for autonomous agents. In: Intelligent Agents III Agent Theories, Architectures, and Languages, pp. 21–35. Springer (1997)
26. Wooldridge, M., Jennings, N.R.: Agent theories, architectures, and languages: a survey. In: Intelligent agents, pp. 1–39. Springer (1995)

Part V
Anthropic-Oriented Computing (AOC)

Prediction of the Successful Completion of Requirements in Software Development—An Initial Study

Witold Pedrycz, Joana Iljazi, Alberto Sillitti and Giancarlo Succi

Abstract A lot of requirements are discarded throughout the product development process. However, resources are invested on them regardless of their fate. If it would exist a model that predicts reliably and early enough whether a requirement will be deployed or not, the overall process would be more cost-effective and the software system itself more qualitative, since effort would be channeled efficiently. In this work we try to build such a predictive model through modelling the lifecycle of each requirement based on its history, and capturing the underlying dynamics of its evolution. We employ a simple classification model, using logistic regression algorithm, with features coming from an engineering understanding of the problem and patterns observed on the data. We verify the model on more than 80,000 logs for a development process of over 10 years in an Italian Aeronautical Company. The results are encouraging, so we plan to extend our study on one side collecting more experimental data and, on the other, employing more refined modeling techniques, like those coming from data mining and fuzzy logic.

1 Introduction

During the typically long lifetime of a software system, certain requirements are frequently modified or even completely deleted, due to factors like change of business needs, increase of the complexity of the system itself, changes in mutual dependencies, or disability to deliver the service [1, 2]. However, financial resources, human and machine effort and time is invested on these requirements,

W. Pedrycz · J. Iljazi
Department of Electrical and Computer Engineering,
The University of Alberta, Edmonton, Canada

A. Sillitti
Center for Applied Software Engineering, Genoa, Italy

G. Succi (✉)
Innopolis University, Innopolis, Russia
e-mail: G.Succi@innopolis.ru

G. Jezic et al. (eds.), *Agent and Multi-Agent Systems: Technology and Applications*, Smart Innovation, Systems and Technologies 58,
DOI 10.1007/978-3-319-39883-9_21

regardless their fate and final state. If there would be a model that could predict early enough on the development life cycle that some requirements are more prone to failure than others, the management of the processes would be more cost-effective and deterministic [3]. This paper is focused on trying to predict such failure using a suitable and simple mathematical model.

Empirical software engineering is an established discipline that employs when needed predictive modeling to make sense of current data and to capture future trends; examples of use are predictions of system failures, requests for services [3], estimation of cost and effort to deliver [4, 5]. With respect to requirement modeling, empirical software engineering has so far put most effort in lifecycle modeling [6–8] and volatility prediction [9, 10]. In our view, the prediction of the fate of requirements and of the associated costs has not received an adequate attention; in this work we propose this novel viewpoint that centers its attention on this neglected area of study.

In this work we use logistic regression to learn from historical sequences of data, each sequence with a final label of success or a failure depending on its final evolution state. in an industrial software system development process. The model is experimented on a real data coming from 10 years of requirements properly evaluated in an Italian Aeronautical Company. The experimental data features more than 6,000 requirements and 80,000 logs of their evolution.

As it is shown further, this results in a statistically significant model that is able to detect failures and successes with a satisfactory precision. This analysis opens the path for a future better understanding of the requirement lifecycle dynamics, the identification of the typical patterns that show a future failure, and a broader exploration of classifying techniques for data of software engineering processes. Moreover, keeping in mind the limitations of the model we build, the next step would be also a consideration of the trade-off between specificity and out of sample generalization.

The paper is structured as follows: in Sect. 2 we present the state of the art, in Sect. 3 the methodology followed in this early study is detailed, highlighting the prediction method chosen and the way its performance is assessed. In Sect. 4 it is presented the case study. The paper continues further with Sect. 5 where the limitation of this work are outlined and the future work is presented. The paper is concluded (Sect. 5) by summarizing the results of this work and its importance.

2 State of Art on Empirical Studies on Requirements

Requirements are an important part of the software development process; some of their characteristics are studied broadly and the results of such studies are widely present in the literature. However, as mentioned in the introduction, there is more to explore especially in terms of empirical studies of requirements evolution [11]: in this context the existing research is focused on two main directions. The first is modeling the requirements evolution with the aim of understanding its dynamics early on their specification [6] or during the lifecycle [8, 12], considering important also developing tools to capture this evolution through logs of different stages of development process [13].

The second direction has for long focused on the requirements volatility. There are studies identifying some main reasons that can take a requirement not to succeed or to change frequently, business needs, requirement complexity and dependency, environment of development, etc. [1, 2]. To the volatility issue, a few studies approach it, as a reflection of the socio-technical changes in the world and in industry. For instance, Felici considers changes not an error to deal with, but an unavoidable part of software process development that reinforces the requirement itself and saves the project from failing, by adapting continuously. In this case except the cost accompanying the requirements volatility, the authors try to associate also the benefit driven from the changes [14–16]. There is also a part of studies that give importance to how changes propagate [17], and how they can be related to an increase in defects present and in the overall project cost [18].

With respect to learning from historical data, there are empirical software engineering studies attempting to build predictive models on requirements, some using simple regression models for identifying which are the future volatile trends and which requirements are more prone to change [9, 10], other using different alternatives like building a rule-based model for prediction [7]. Describing changes and stability points through scenario analysis is also another proposed solution [19].

In any case, what is relevant to stress with reference to our work, is that often the deleted requirements are not taken in consideration at all in these analysis, or the deletion process itself is considered simply as one of the changes that may happen [10, 14]. Thus, despite being a fairly easy concept to grasp and challenging to work with, there is almost no literature referring to failing requirements or trying to model the requirement final state. We think that our unicity gives more value to this paper since it shift the attention toward a field explored so little.

3 Methodology

The main goal of this study is to answer to the research question: “Can we predict reliably the final status of a requirement of a software system analysing its historical evolution through stages and the development environment? If so, how early on the life cycle of the requirement can we perform the prediction?” Our aim is to build a model to make this possible. We start by making a relevant assumption, that requirements are independent of each other, meaning that the status of one requirement does not affect the status of other requirements. This assumption is quite stringent, and for certain perspective also unfeasible in the real world, still it is the basis of the followup analysis, and, as we will see, its violation does not affect the validity of our findings.

3.1 Prediction Method

Machine learning is usually the right technique when we do not have an analytical solution and we want to build an empirical one, based on the data and the observable patterns we have [20]. Inside this field, supervised learning is used when this data is labeled, meaning the dependent variable is categorical. This is also our case, since we are dealing with a dichotomy problem, a requirement being successful or failed, thus the main choice is to use supervised learning algorithms in order to build a predictive model. Applying logistic regression algorithm, from the supervised learning pool, would give not only classification of the requirement objects into pass/fail, where pass stands for a success and fail for a never deployed requirement, but also the probability associated with each class.

Logistic regression is an easy, elegant and yet broadly used and known to achieve good results in binary classification problems. Assuming that we have d predictors and n observations, with X being the matrix of observations and Θ the matrix of the weights:

$$\text{for} \quad X = \begin{bmatrix} x_{11} & \dots & x_{1d} \\ \dots & & \\ x_{n1} & & \end{bmatrix} \qquad \theta = \begin{bmatrix} 1 \\ d \end{bmatrix}$$

the LR model is $logit(X\theta) = \frac{1}{1+e^{-X\theta}}$.

In order to learn Θ, the algorithm minimises the uncertainty, by first assuming an independence between the observations. The result of the iterations is the discriminant that divides the input space into pass and fail area. Thus, per any new requirement data point, once learned the model and tuned with the right threshold, we would be able to give a statistically significant probability value of it being a future pass or fail.

3.2 Evaluation of Performance of Predicting Models

To assess the performance of the classifier we use the confusion matrix. The confusion matrix values allow us to calculate the predicting ability and error made in terms of type I and type II error. Clearly, since logistic regression delivers a probability of belonging to a certain class, as we discussed in the section above, in order to determine the label a threshold is used. As the threshold changes, also the ability of classifiers to capture correctly passess and failures is channeled (inclined/biased). We use the ROC curve analysis, through maximising the combination of sensitivity-specificity of our model we obtain the best threshold value and tune the model accordingly. Thus, the metrics we will use to measure performance are: sensitivity, specificity and the Area Under the Curve. Sensitivity shows the ability of the model to capture correctly fail-requirements in the dataset, whereas the Specificity shows the ability to capture passes. The area under the curve is calculated by plotting Sensitivity versus Specificity for threshold value from 0 to 1, whereas the formulas to calculate these two are below:

Confusion matrix	Predict fail	Predict pass
Real fail	True Positive	False Negative
Real pass	False Positive	True Negative

$$Sensitivity = \frac{TruePositive}{TruePositive + FalseNegative}$$

$$Specificity = \frac{TrueNegative}{TrueNegative + FalsePositive}$$

4 Case Study

4.1 Data Description and Processing

The data we use in this research come from an industrial software system, developed in an aeronautical company in Italy, Alenia Aermacchi. The data of the development process are formated as a file of logs, presenting the history of 6608 requirements and 84891 activities performed on them on a time period from March 2002 to September 2011. Every log is characterized by the following attributes summarized in the table:

Timestamp	Date and time in which the requirement was firstly created in the log system
Version	Version of the software product in which the requirement was first created
Developer	Developer that performed this specific action on the requirement object, and entered it in the log system
Activity	The specific activity performed on the requirement
Requirement ID	The identification number, unique per each requirement

There are 13 different activities performed on a requirement throughout the development process, with the most significant numerically and interesting from the process point of view being:

createObject	The requirement is created in the system
createLink	Creates a link between different requirement objects and keeps count of it
modifyObject	The requirement is modified
deleteObject	The requirement object is deleted

The early identification of "deleteObject" is basically what we are after in this study, since it means that the requirement was deleted from the system without ever being deployed, this is what we will call a "fail". Every other activity concluding the lifecycle of the requirement present in database is considered a success, and the requirement itself a "pass". At this point is clear that once processed, we get binomial labeling for the attributes characterizing the requirements lifecycle inside this process. From a first statistical analysis, we notice that there is a dominance of "modifyObject" activities in the requirements labeled as "fail." The early attribute selection is performed based on domain expertise and influenced by any possible pattern that we identified, as mentioned above. The processed final requirement records are the rows of a data frame whose columns are the potential features of the predictive model we aim to build, except the ID that is used to keep track of the specific object. Each requirement record is the characterized by elements presented in the following table.

Requirement ID	The identification number
Opening version	The version where it was first created; it will also ensure a timeline on the requirement objects
Developer	The last developer that worked with the requirement
Number of transitions	The cumulative number of activities performed on the requirement; this is also an approximation of the number of modification a requirement goes through.
Lifetime duration	The longevity of a requirement, calculated as the difference between the closing and the opening dates
Final state	The label either pass or fail

4.2 *Learning of the Model and Evaluation of Performance*

We start with a slightly imbalanced time-ordered dataset of requirements. Each record of it is a requirement object profiled by four features that we will try to use in our model: version, number of Transitions, lifetime, developer. We allow the lack of balance in order to give an initial bias toward the pass-class, considering that, however, a false-pass costs less to the system than a false-fail. To learn the model, stepwise regression is performed on the data. The statistically significant features used by the model are: version, number of transitions and lifetime of the requirement. The steps followed are:

1. Dataset is divided into training and validation sets(respectively 75 %–25 %), preserving the initial class distribution present on the original data, using package "caret"
2. Logistic regression algorithm is run (10-k cross validation, repeated 3 times) on training data to build the model.
3. ROC curve is plotted in order to find the best sensitivity-specificity values by using the criteria "most top-left" on the roc curve. We notice: 79 % predictive ability of the model (AUC = 0.79 with 95 % CI) and the best threshold being 0.27
4. Model is tested on the validation set, where for the best threshold value, the results achieved in terms of predictivity are: sensitivity = 0.74 and specificity = 0.95

Thus, the model built, out of all true failures can capture correctly 74 % of them and out of passes 95 % of them, with the results being under 95 % CI. Below the plot:

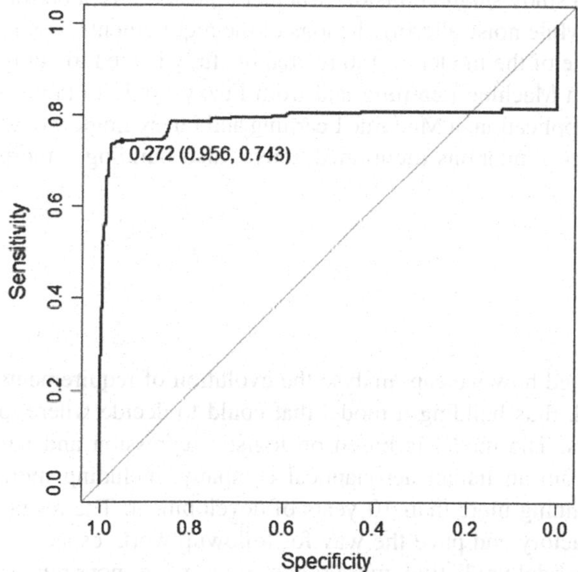

This classification model built and tested on the whole dataset of requirements shows clearly for a possibility of predicting pass and fail requirements.

5 Limitations and Future Work

We think that this work presents a novel idea that can move forward in several directions, especially taking into account its limitations. The first limitation is inherent with the single case study. This work has a very significant value given the size of its dataset—a very rare situation in Software Engineering, and a quite unique situation in the case of requirements evolution, however, it is a single case, and, as usual a replication would be strongly advisable. The second limitation is that we have assumed the requirements independent, while they are not. This situation is similar to the one in which a parametric correlation is used while clearly its requirements are not satisfied—still it has been experimentally verified that the predictive value still hold, as it holds in this specific case. The third limitation is that we build a model having available the whole stories of the requirements, while it would be advisable to be able to predict early whether a requirement is going to fail. Therefore, the future studies should focus on trying to build predictive models using only initial parts of their story. The fourth limitation is that we do not take into account that the information on requirements is collected manually by people, so it is subject to errors, and this study does not take into consideration how robust the requirements are with respect to human errors. So a follow-up study should consider whether a presence of an error, for instance in the form of a white noise altering the logs of the requirements, would alter significantly the outcome of the model. A future area of study is then to apply more refined techniques from Machine Learning and from Fuzzy Logic to build a predictive system. The joint application of Machine Learning and Fuzzy Logic would resolve by itself the last three limitations mentioned above, also building a more robust and extensible system.

6 Conclusions

In this work we have presented how we can analyse the evolution of requirements to predict those that will fail, thus building a model that could to decide where to drop the effort of developers. The model is based on logistic regression and has validated on data coming from an Italian aeronautical company, including more than 6,000 requirements spanning more than 10 years of development. The results of this early study are satisfactory and pave the way for followup work, especially in the direction to make the model predictive, more robust, employing more refined analysis techniques, like those coming from Data Mining and Fuzzy Logic.

Acknowledgments The research presented in this paper has been partially funded by Innopolis University and by the ARTEMIS project EMC2 (621429).

References

1. Harker, S.D.P., Eason, K.D., Dobson, J.E.: The change and evolution of requirements as a challenge to the practice of software engineering. In: Proceedings of 1st ICRE, pp. 266–272 (1993)
2. McGee, S., Greer, D.: Towards an understanding of the causes and effects of software requirements change: two case studies. Requir. Eng. **17**(2), 133–155 (2012)
3. Succi, G., Pedrycz, W., Stefanovic, M., Russo, B.: An investigation on the occurrence of service requests in commercial software applications. Empirical Softw. Eng. **8**(2), 197–215 (2003)
4. Boehm, B., Abts, C., Chulani, S.: Software development cost estimation approaches—a survey. Ann. Softw. Eng. **10**, 177–205 (2000)
5. Pendharkar, P., Subramanian, G., Rodger, J.: A probabilistic model for predicting software development effort. IEEE Trans. Softw. Eng. **31**(7) 2005
6. Yu, E.S.K.: Towards modelling and reasoning support for early-phase requirements engineering. In: Proceedings of 3rd IEEE International Symposium on Requirements Engineering, pp. 226–235, 6–10 Jan 1997
7. Le Minh, S., Massacci, F.: Dealing with known unknowns: towards a game-theoretic foundation for software requirement evolution. In: Proceedings of 23rd International Conference: Advanced Information Systems Engineering (CAiSE 2011). London, UK, 20–24 June 2011
8. Nurmuliani, N., Zowghi, D., Fowell, S.: Analysis of requirements volatility during software development lifecycle. In: Proceedings of ASWEC, pp. 28–37 (2004)
9. Loconsole, A., Borstler, J.: Construction and Validation of Prediction Models for Number of Changes to Requirements, Technical Report, UMINF 07.03, Feb. 2007
10. Shi, L., Wang, Q., Li, M.: Learning from evolution history to predict future requirement changes. In: Proceedings of RE 2013, pp. 135–144
11. Ernst, N., Mylopoulos, J., Wang,Y.: Requirements Evolution and What (Research) to Do about It. Lecture Notes in Business Information Processing, vol. 14, pp. 186–214 (2009)
12. Russo, A., Rodrigues, O., d'Avila Garcez, A.: Reasoning about Requirements Evolution using Clustered Belief Revision. Lecture Notes in Computer Science, vol. 3171, pp. 41–51 (2004)
13. Saito, S.; Iimura, Y., Takahashi, K., Massey, A., Anton, A.: Tracking requirements evolution by using issue tickets: a case study of a document management and approval system. In: Proceedings of 36th International Conference on Software Engineering, pp. 245–254 (2014)
14. Anderson, S.; Felici, M.: Controlling requirements evolution: an avionics case study. In: Proceedings of 19th SAFECOMP, pp. 361–370 (2000)
15. Anderson, S.; Felici, M.: Requirements evolution from process to product oriented management. In: Proceedings of 3rd PROFES, pp. 27–41 (2001)
16. Anderson, S.; Felici, M.: Quantitative aspects of requirements evolution. In: 26th Annual International Computer Software and Application Conference, COMPSAC 2002, pp. 27–32, 26–29 Aug 2002
17. Clarkson, J., Simons, C., Eckert, C.: Predicting change propagation in complex design. In: Proceedings of DETC'01, ASME 2001 Design Engineering Technical Conferences and Computers and Information in Engineering Conference. Pittsburgh, Pennsylvania, 9–12 Sept 2001
18. Javed, T., Maqsood, M., Durrani, Q.S.: A study to investigate the impact of requirements instability on software defects. ACM SIGSOFT Softw. Eng. Notes **29**(3), 1–7 (2004)
19. Bush, D., Finkelstein, A.: Requirements stability assessment using scenarios. In: Proceedings of 11th ICRE, pp. 23–32 (2003)
20. Yaser, S.A.M, Malik, M.I., Hsuan-Tien, L.: Learning from data. http://www.amlbook.com/support.html

Evolution of Thinking Models in Automatic Incident Processing Systems

Alexander Toschev, Max Talanov and Salvatore Distefano

Abstract In this paper we describe the evolution of the application of thinking models in automatically processing a user's incidents in natural language, starting with the model based on decision trees and ends up finishing with the human thinking model. Every model has been developed, prototyped and tested. The article contains experiments results and conclusions for every model. After evolving several theories, we found the most suitable for solving the problem of automatically processing a users incidents.

Keywords Artificial intelligence · Machine understanding · Remote infrastructure management · NLP · Reasoning · Automation · Knowledge base · Intelligent agents

1 Introduction

Today IT infrastructure outsourcing is a very popular domain for thousands of companies all over the world. This domain connect companies that provide remote support for customer's IT infrastructure. It's very expensive to have personal IT department if your firm doesn't operate in the field of IT, which is why IT infrastructure outsourcing is a large domain with a lot of companies on the market [5]. So, efficiency is very important for this business due to high competitiveness [10]. During the development of outsourcing domain a lot of information systems have been created. For example, HPOpenView [9], ServiceNow which is much more popular, it also used in CERN [1]. But all of them required human specialist to work. For example, spe-

A. Toschev (✉) · M. Talanov · S. Distefano
Kazan Federal University, Kazan, Russia
e-mail: atoschev@kpfu.ru
URL: http://www.kpfu.ru

M. Talanov
e-mail: dr.max@machine-cognition.org

S. Distefano
e-mail: salvatdi@gmail.com

G. Jezic et al. (eds.), *Agent and Multi-Agent Systems: Technology and Applications*, Smart Innovation, Systems and Technologies 58,
DOI 10.1007/978-3-319-39883-9_22

cialist registers incidents in the system, analyses them and tries to solve the issue. The system only holds workflow and sometime automatically assigns specialist.

This paper describes a system for detecting and solving incidents produced by user in natural language. The possibility of automation is based on a large amount of trivial incidents in such systems as "I do not have access to…", "Please install…" [14]. There are set of mandatory requirements:

- Understanding a request produced by a user in natural language;
- Capability to learn;
- Capability to reason (using the analogy, deduction, etc.);
- Decision making.

We demonstrate the progress and evolution of models and architectures taking in account these requirements in the following.

2 Thinking Models

To solve the problem we evaluate several techniques and models. In this section we present results of this evolution.

2.1 *Intellectual Document Processing (IDP) Model*

One of the early tasks was to parse and understand documents. This is why the first model was named Intellectual Document Processing. The system makes meta-information from an uploaded document and is able to quickly find the required information. The main idea behind this system is decision tree algorithms [2]. General workflow of the system described below:

1. System should be trained, which is why user must upload initial information;
2. System proceeds these documents and tries to extract information;
3. System annotates information with tags like: "birth date", "job title", etc.;
4. User uploads new documents and the system automatically extracts all valuable information, so you can simply navigate to this information using the tag tree.

On the Fig. 1 the architecture of the proposed solution is shown. To start use of the system training is required, so several annotated documents are provided to the Trainer component, after which the Evaluator processes them. After that Applier can work to extract information from the raw documents provided by the user.

The system has been designed with aim of supporting the HR division of an organization. So, they can quickly parse a lot of candidates proposals. Technically this system required a lot of initial data to start work. Moreover, if the system tries to processes an unsupported document format it will crash.

As the sample for testing input information, candidates application forms in the form of Word documents are used.

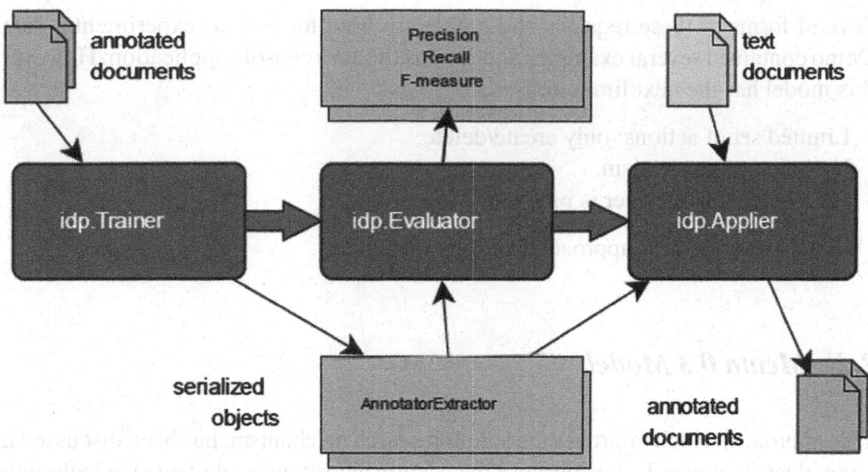

Fig. 1 IDP architecture

2.2 *Menta 0.1 Model*

The second model was developed as an intelligent request processing system for automatic software generation. For example, if you need to simply add a new field or change it, you can do so automatically, so do not need spend a lot of time on simple requests. The system can process simple requests such as "Add name to the customer". The main workflow for this system is described below:

1. Receive request and formalize it;
2. Generate action (special object, which contains information for the next step);
3. Change the target application which is represented in OWL (Web Ontology Language) [8] model using the action from previous step;
4. Generate application by OWL model;

Decision tree technology is used for the solution search (find suitable solution option for the incoming request). This model supports requests in natural language by using Stanford Parser [4] as a NLP processing tool. This approach is much more intelligent than the previous one but it still has problems like:

- If the input request contains several errors in grammar, spelling, etc. system failed to proceed such requests;
- The solution search system works only for one application architecture and can't use information for other systems. For example, system knows how to add field to the class in system A, but it can't proceed the same operation for the system B.

For example, the model works with the following requests:

- Please add Name field to the Customer;
- Please remove Lastname field from the Customer;

Several forms of these requests and others are used for tests as experimental data. Demo contained several examples and runs as the Java console application. However, this model has the next limitations:

- Limited set of actions: only create/delete;
- No learning mechanism;
- No feedback to the user is provided;

That's why the next approach has been introduced.

2.3 *Menta 0.3 Model*

This approach, which improves the solution search mechanism, has been discussed in general terms above. To this purpose the genetic algorithm is adopted also including:

- Acceptance criteria—set of rules to the solution;
- Data model in semantic network OWL representation;
- Use of logic in rules (such as induction, abduction, etc.).

The overall solution architecture presented on Fig. 2. It contains modules: Menta Client—web-service client of the system, could be the GUI application in .NET; MentaController—the main component of the system that invokes all the other components in the workflow; Communicator—component that processes the responses of the MentaClient and creates requests to MentaController; KBServer (Knowledge Base server)—the storage of ontological information used in the system; Solution Generator—the generator of the Solution for the MentaClient request. System works with specialist according to the following workflow:

1. User sets acceptance criteria by using special tool;
2. The system combines different components of the application to build by genetic algorithms application suitable to acceptance criteria;
3. The system checks final generated solution by asking user.

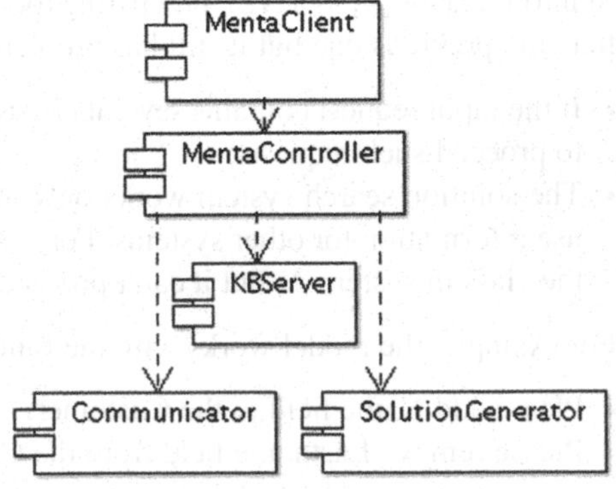

Fig. 2 Menta 0.3 architecture

Acceptance criteria set of logic rules: the system checks it by using the logic engine NARS [15]. However, after testing we found that the creation of acceptance criteria is not trivial and takes a lot of time. Moreover, on large models genetic algorithms take long time to find suitable solutions. Detailed description of this approach is available in the article [11].

For the model validation we used predefined sets of acceptance criteria, a target software application description and a set of possible modules. After setting up acceptance criteria the system generates the solution as a changeset for the existing application.

2.4 TU Model

After evaluating all models the final model had been build based on Marvin Minsky theory of six levels [7]. This model contains all general parts of the previous models. The general modules of this system are:

- Incident;
- Learning;
- Solution search;
- Solution application.

The model is much more general than the previous ones and can be used in several domains. The main concepts of Minsky's theory are: Critic, Selector, WayToThink. Critic is a probabilistic predicate. In other words it is a trigger, which activates after several events. After activation critic checks if it reacts to this event by using a set of logic rules, which are evaluated by the reasoning server PLN in [3]. After checking, it returns the Selector with the set of resource for processing incoming request. List with definitions of implemented critics follows:

- Incoming request—activates when incoming request comes from user;
- Natural language processing—activates for the incoming request and run build (Way-to-think) of the semantic model of the incoming request;
- Preliminary splitter—activates for the incoming request and run the correction appliers: grammar, spelling, annotating for the incoming request;
- Time control—activates from time to time to check request processing time;
- Emotional state control—activates when system state change and run resource allocation.

The next component is the Selector. Its job is to obtain resources from the internal knowledge base. The Selector returns another Critic or another Selector or Way-to-think. Below is the list with definitions of implemented Selectors:

- Direct instruction problem Selector;
- Problem with desired state Selector.

For example, “incoming request critic” returns “direct instruction problem” selector which returns Way-to-think “solve direct instruction problem”. The third component of the model is Way-to-think. It’s processes data modification in general. In brief, it solves incoming request. In the IT outsourcing domain waytothink can be:

- Learned knowledge—system already knows how to solve this issue;
- Adaptation—system applies existing knowledge by using the analogy such as “please install Firefox” and “please install IE”;
- Reformulation—system puts newly obtaining knowledge to database by cleaning up it to more general way such as “please install office” to “please install msi”;
- Cry4Help—system raises question to the user;
- Solution search—different ways to find the solution. The solution search algorithms use tree distance algorithm to found suitable set of how-to to solve the problem. How-to’s incapsulated in the Solution concept, which is linked to the problem. The search works between incoming problem and existing problem;
- How-to—special Way-to-think that represent solution for problem. It has been built as a general algorithm for applying solutions. Such as a set of script commands;

The second important part of the model is thinking levels: higher level incapsulates more complex behavior than including them. These levels are:

1. Instinctive
2. Learned
3. Reflective
4. Self-reflective
5. Self-conscious

In the theory it was abstract concept. In the TU model we have made domain specific and adopted to software system.

The first level—instinctive—includes instincts such is the egg-retrieval behavior of the graylag goose [6]. The level—self-conscious—includes higher goals and ideals. Regarding incident processing the first level controls auto-generated incidents such us “LOT2345 required.” If the system can’t find result on the first level it goes down to the next level. On the second level system proceed incident by natural language processing software and creates semantic networks, after that several critics become active and try to classify problem.

The 3rd level controls request state and sets goals for the system. For example, the base goal for the system is to help user. Goals have tree structure. Subgoals of the help user goal are: understand request, classify problem, find solution.

The Fourth level controls the time of the request, and if it runs out of time raises the problem on the next level.

The Fifth level initializes communication context and performs user iteration.

The 6th level controls the general system state—emotion state. It also controls system’s resources. So, if the 4th level raises the problem of request’s time it adjusts resources to solve this request as soon as possible. On this level system also performs monitoring which can indicate that replacement of hardware components or resource upgrading is required.

Another problem was how to pass data between thinking levels, current operation context and knowledge base of the system. Special concepts were introduced:

- Short term memory—exist during the current request and located in the memory;
- Long term memory—after finishing request processing system merges Short term memory to the knowledge base.

Way-to-thinks are working with short term memory and after successful incident processing the system copies newly obtained knowledge to long term memory. This way bad solution are not copied to knowledge base.

System knowledge base is build over the non-sql database engine Neo4j. Solutions, problems, Way-to-think, Critics, Selectors. That's why all system components are dynamic.

One of the main option of the system is to learn new knowledge by interaction with the user. At the beginning system has a few concepts: object and action, which increase by learning. Learning process is quite similar to request processing, but instead of searching for the connections between incoming concepts and knowledge base's concepts it creates them. For example, learning sequence: "Browser is an object. Firefox is a browser". Now the system knows two new concepts: Firefox and browser. Also it builds links between browser and Firefox.

This paper extends the Minsky's model into a new architecture shown in Fig. 3. It contains five components: MessageBus—connector between different system's modules; TUWebService—main communication component; CoreService—general processing module; ClientTarget—client component for solution applying; Data Service—data provider component, includes knowledge base. The model has been extended by specific Critics, Selectors, Way-to-thinks. Thinking levels have been

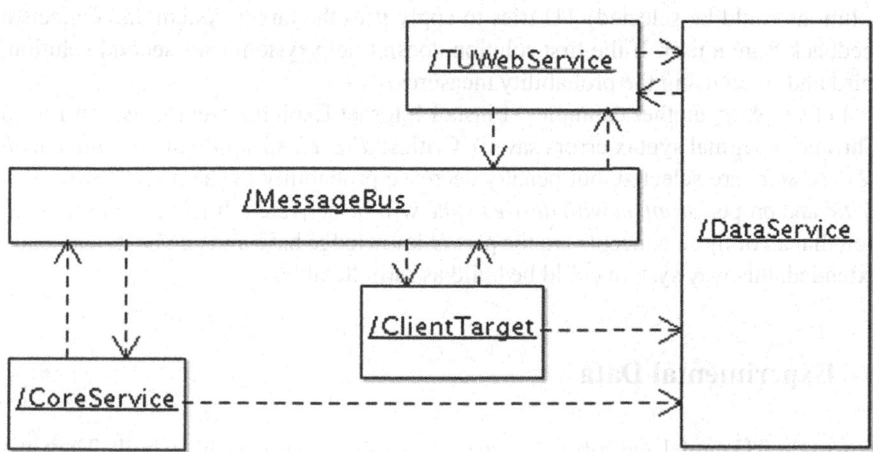

Fig. 3 TU architecture

adapted to system tasks. Since the architecture is general, the application of this model is very flexible from helpdesk to virtual assistant. More detailed description of architecture is available in [13].

3 TU. How It Works

Let's take as a example the incident "Please install Firefox". The critic "Inbound Incident Received" reacts and sets up the global goal "help user". This activates critics that are associated (a criteria of a critic matches) with this goal starts processing incident. This triggers the critic "get incident type", that activates "natural language processing" way-to-think, which in its turn extracts a semantic network from the inbound incident description. Several critics connected to the sub goal of "get incident type" are activated. All of them analyze the inbound context (short term memory) with semantic network of incident. Usually a critic has several rules, that are populated using the inbound context and evaluated via probabilistic reasoner. As the result a critic returns a probability of an action. Let's system has three incident type classification critics: direct instruction (when request is a direct instruction, like "please install something"); problem with desired state (I want office instead of Wordpad); problem without desired state (I can't open pdf file). In this case the *direct instruction* critic returns maximum probability, because request contains all required concepts for this particular critic. The approximate dependencies structure: action—please; object—Firefox. In the case of *problem with desired state* there are missing concepts: subject (which describes problem), previous state (current situation). After activation of the critic *direct instruction* returns the way-to-think which tries to search for a solution. The critic to *problem with desired state* triggers the *simulating and modeling* way-to-think that searches for a suitable action to put the system from current state to the desired state. After the solution was found (several solutions could be selected) TU tries to apply it to the target system and collects a feedback from a user. If the first solution doesn't help system tries second solution, third and so on using the probability measures.

Let's look at another example: "I install Internet Explorer previously, but I need Chrome" (original syntax errors saved). Critics: *direct instruction* and *problem with desired state* are selected, but penalty decrease probability of the *direct instruction critic* and proper *problem with desired state* will be activated. It's important to mention that all of these concepts are the part of knowledge base and can be dynamically extended, this way system could be build as rally flexible.

4 Experimental Data

To test the TU model and software a special set of requests have been built. They was extracted from real time systems of ICL-Services [12]. Initial dump contained over 1000 incidents. After processing, 43 unique incidents have been preselected. Why are they unique? Because other incidents are just variants of these 43. For example, "Please install Office" and "Please install browser" are similar—simply run install program.

TU model can be tested Among these 43 incidents 85 % have been successfully understood by the system. And the if solution is available in database—it is selected.

5 Conclusions

We described 3 models which were created to automatically process a user's request. After investigation we found that the TU model is the most suitable solution. We consider it to be a very good combination of an application and the flexibility of human's thinking. This model provides the ability to build the system in a restricted-resource environment (because it does not use neural networks), which has been proven in the research by creating software using this model.

Acknowledgments Thanks to Marvin Minsky for his incredible theory.

References

1. Alonso, R., Arneodo, G., Barring, O., Bonfillou, E., Dos Santos, M., Dore, V., Lefebure, V., Fedorko, I., Grossir, A., Hefferman, J., Lorenzo, P., Moller, M., Mira, O., Salter, W., Trevisani, F., Toteva, Z.: Migration of the cern it data centre support system to servicenow. In: 20th International Conference on Computing in High Energy and Nuclear Physics, CHEP 2013, Amsterdam; Netherlands, 14–18 Oct 2013, Code 108171. Institute of Physics Publishing, Geneva, Switzerland (2013)
2. Deng, H., Runger, G., Tuv, E.: Bias of importance measures for multi-valued attributes and solutions. In: Proceedings of the 21st International Conference on Artificial Neural Networks (ICANN). Singapore (2011)
3. Harrigan, C., Goertzel, B., Ikl, M., Belayneh, A., Yu, G.: Guiding probabilistic logical inference with nonlinear dynamical attention allocation. In: 7th International Conference on Artificial General Intelligence, AGI 2014, Quebec City, QC, Canada, 1–4 Aug 2014, Code 106897, pp. 238–241. OpenCog Foundation, Quebec City, QC, Canada (2014)
4. Kadhim, M., Afshar Alam, M., Kaur, H.: A multi-intelligent agent architecture for knowledge extraction: novel approaches for automatic production rules extraction. Int. J. Multimedia Ubiquitous. Eng. **9**(2), 95–114 (2014)
5. Kunert, P.: Blighty's it support services to decline in 2013, 2014, 2015. Channelregister **1**(1), 1–5 (2013)
6. Mazur, J.: Learning and Behavior. Prentice Hall, New Jersey (2005)
7. Minsky, M.: The emotion machine'. From pain to suffering. In: IUI-2000: International Conference on Intelligent User Interfaces, New Orleans, LA, USA, 9–12 Jan 2000, Code 57075. ACM,New York, NY, United States, New Orleans, LA, USA (2000)
8. Sheng, B., Zhang, C., Yin, X., Lu, Q., Cheng, Y., Xiao, T., Liu, H.: Common intelligent semantic matching engines of cloud manufacturing service based on owl-s. Int. J. Adv. Manuf. Technol. **1**(1), 1–16 (2015)
9. Sperling, E.: Outsourcing core competencies. Hewlett-Packard J. **47**(5), 77–80 (1996)

10. Sperling, E.: Outsourcing core competencies. Forbes (2009)
11. Talanov, M., Krekhov, A., Makhmutov, A.: Automating programming via concept mining, probabilistic reasoning over semantic knowledge base of SE domain. In: 2010 6th Central and Eastern European Software Engineering Conference, CEE-SECR 2010, Moscow, Russian Federation, 13–15 Oct 2010, Category number CFP1CER-ART, Code 85262. SECR, Moscow, Russia (2010)
12. Talanov, M., Toschev, A.: Incident analysis. http://tu-project.com/for-business/ms-ad-management/ (2013)
13. Toschev, A., Talanov, M.: Thinking lifecycle as an implementation of machine understanding in software maintenance automation domain. In: 9th KES International Conference on Agent and Multi-Agent Systems-Technologies and Applications, KES-AMSTA 2015, Sorrento, Italy, 17–19 June 2015, Code 157669, pp. 301–310. Springer Science and Business Media Deutschland GmbH, Sorrento, Italy (2015)
14. Toschev, A.: Thinking model and machine understanding in automated user request processing. In: 16th All-Russian Scientific Conference Digital Libraries: Advanced Methods and Technologies, Digital Collections, RCDL 2014, Hospitable Laboratory of Information Technologies of the Joint Institute for Nuclear Research (JINR) Dubna, Russian Federation, 13–16 Oct 2014, Code 109385, pp. 224–226. CEUR-WS, Dubna, Russia (2014)
15. Wang, P.: The assumptions on knowledge and resources in models of rationality. Int. J. Mach. Conscious. **3**(1), 193–218 (2011)

Quality Attributes in Practice: Contemporary Data

Rasul Tumyrkin, Manuel Mazzara, Mohammad Kassab, Giancarlo Succi and JooYoung Lee

Abstract It is well known that the software process in place impacts the quality of the resulting product. However, the specific way in which this effect occurs is still mostly unknown and reported through anecdotes. To gather a better understanding of such relationship, a very large survey has been conducted during the last year and has been completed by more than 100 software developers and engineers from 21 countries. We have used the percentage of satisfied customers estimated by the software developers and engineers as the main dependent variable. The results evidence some interesting patterns, like that quality attribute of which customers are more satisfied appears functionality, architectural styles may not have a significant influence on quality, agile methodologies might result in happier customers, larger companies and shorter projects seems to produce better products.

1 Introduction

Quality is the set of characteristics of an entity that describe its ability to satisfy stated and implied needs of the customer and/or of the end user; this notion has been formalized in numerous standards like the ISO 9000. For systems including soft-

R. Tumyrkin (✉) · M. Mazzara · M. Kassab · G. Succi · J. Lee
Innopolis University, Innopolis, Russia
e-mail: r.tumyrkin@innopolis.ru

M. Mazzara
e-mail: m.mazzara@innopolis.ru

M. Kassab
e-mail: m.kassab@innopolis.ru

G. Succi
e-mail: g.succi@innopolis.ru

J. Lee
e-mail: j.lee@innopolis.ru

M. Kassab
Pennsylvania State University, Malvern, PA, USA

G. Jezic et al. (eds.), *Agent and Multi-Agent Systems: Technology and Applications*, Smart Innovation, Systems and Technologies 58,
DOI 10.1007/978-3-319-39883-9_23

ware, the notion of quality has been instantiated in several standards, including the ISO 9216 [1] and IEEE 730 [2]. Software Quality is an essential and distinguishing attribute of the final product. Nevertheless, functionality usually takes the front seat during software development. This is mainly because of the nature of these quality requirements which poses a challenge when taking the choice of treating them earlier in the software development. Quality requirements are subjective, relative and they become scattered among multiple modules when they are mapped from the requirements domain to the solution space. Furthermore, Quality requirements can often interact, in the sense that attempts to achieve one can help or hinder the achievement of the other at particular software functionality. Such an interaction creates an extensive network of inter dependencies and trade-offs among quality requirements which is not easy to trace or estimate [3].

This preference for functionality over the qualities is shortsighted though. Software systems are often redesigned not because they are functionally deficient but because they are too slow, not user friendly, hard to scale or hard to maintain. In addition, quality requirements drive architectural structure more than functionality. In fact, if functionality were the only thing that mattered, there wouldn't be a need to divide the system into architectural components at all; a single monolithic blob with no internal structure would satisfy the need [4].

The key issue in implementing an improvement in industrial practices is to first identify the areas that need the most improvement. But little contemporary data exists to report on how the quality requirements and its tight coupling to architecture are perceived in industry. To remedy this deficiency and provide useful data to other researchers we conducted an exploratory survey study on quality requirements and software architecture in practice. In this article, we report on our findings from this survey. Reported data includes characteristics of projects, practices, organizations, and practitioners linked to projects qualities and their architectural structures.

While there is an endless list of qualities a software system may have to exhibit, the focus of this survey is on the pre-dominant ones as described in [5]: Availability, Interoperability, Modifiability, Performance, Security, Testability and Usability.

The rest of this paper is organized as follows: Sect. 2 describes tools and techniques to build questions and collect answers; Sect. 3 reports on the nature of respondents profiles and the businesses in which they are employed; Sect. 4 shows the actual empirical data and Sect. 5 draws some preliminary conclusions that have to be validate in future with the collection of further data and expansion of dataset.

2 Experimental Design

A web-based survey instrument was created using the web-based QuestionPro survey tool (www.QuestionPro.com). The survey consisted of 19 questions. The survey questions were designed after a careful review to specialized literature on conducting survey studies (e.g. [6–10]). A summary of our survey questions is available via the link.[1] While the respondents reported a wide range of experiences; they were asked to base their responses on only one software project that they were either currently involved with or had taken part in during the past five years.

Using the conjectures in our hypotheses as means of constructing specific questions, the survey was arranged into five sections: First section aiming at capturing general project characteristics first. Then, a series of questions were asked in the second section to determine the participants knowledge of architectural styles and whether if any were applied into the surveyed projects. In case of incorporating architectural styles into the projects; the respondents were then asked to report on the criteria they used to select these styles in the third section and the challenges they faced while incorporating them in the forth section. Since quality requirements are the major that shapes the software architecture [5]; a series of questions were then asked in the fifth section to report on the level of customers satisfaction with these qualities while the final product is in use.

We drew our survey participants from multiple sources but primarily from members of the following Linked-In professional groups, to which one or more of the authors belonged: "Software Engineering Productivity: Software Architecture", "Techpost Media","ISMG: Software Architecture" and "ISMG Architecture World". A invitation on these groups was posted under the subject "Software architecture in practice". The participation to this survey was entirely anonymous and voluntarily. Survey data was collected from May 2015 through September 2015. The survey drew 687 participants from 37 countries. Of these survey takers; 103 completed the survey to the end. The completion rate was 15 % and the average time taken to complete the survey was 10 min. We also included the results of the partially completed responses. When respondents aborted the survey, they tended to do so on or near question 15, we speculate from survey fatigue.

3 Profiles of the Respondents

In this analysis we take into consideration also the answers given by people who partially completed the questionnaire for statistical analysis not requiring pairing or correlating information, as suggested in [11]. Given this population, responses to the survey are more likely to reflect the opinions and biases of any given projects development team rather than those of other groups represented in a software development effort. In this section we will consider two aspects of respondents' profiles: *distribution of business* and *type of respondents*.

Distribution of business The distribution of businesses that survey respondents have associated themselves with entails a lot of different fields. The data indicate that respondents are well distributed across a wide range of business domains. All fourteen of the provided domains have been selected at least by few participants. Further-

[1] http://www.questionpro.com/a/summaryReport.do?surveyID=4182537.

more, the *"other"* category included responses such as social media, transportation, automotive, virtualisation, meteo and etc.

Type of respondents In order to understand the types of respondents, a number of questions regarding *"Organizational Characteristics"* have been asked. Respondents to this survey characterize themselves as programmers and developers 41 % of the time, and software engineers, 17 % of the time. One third of the respondents characterize themselves as architects and 9 % as managers (project managers, scrum managers and product owners). Other respondents include system engineers, testers, consultants—reaching a total of around 3 %. The majority (more than 36 %) of respondents represent small companies, with an annual budget of less than 5 million US dollars, within the listed business domains (all company sizes are measured both, in terms of annual budget and number of employees). It is noticeable that about one third of respondents ignore the budget of their companies (this is consistent with the fact that developers often are not exposed to financial information).

4 Analysis of the Results

In this section we will present a portion of the collected data with some projections in order to identify and highlight some aspects of Quality Attributes and satisfaction. The collected data allow several projections and analysis that cannot be reported in full in this paper due to space constraints. In Sect. 5 we will discuss these results in more detail and we will try to connect the dots. We will also anticipate how the work can be continued. In this section in particular, and more in general in this paper, we focus on Quality Attributes and satisfaction. All data are based on the responses of IT-specialists and we use the "satisfaction rate" as the main measure in evaluating quality of final product. The "satisfaction rate" is defined as the estimation made by IT-specialists of the percentage of customers satisfied with quality attributes and overall quality of the final product.

Figure 1 shows the overall satisfaction rate for each specific Quality Attribute considered in the survey. In this figure over hundreds responses of the survey have been computed. Functionality appears as the QA for which customers have an over-

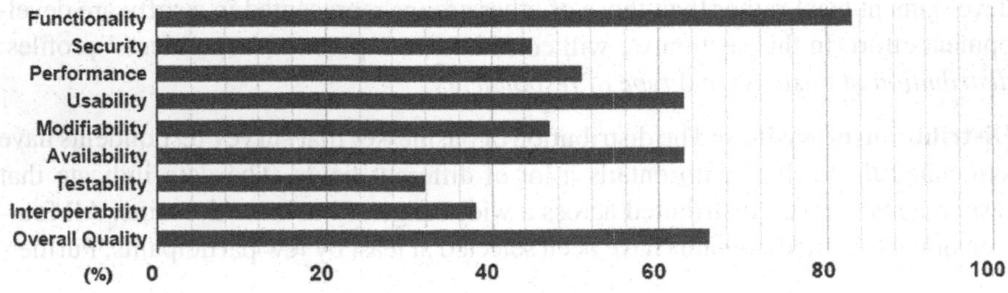

Fig. 1 The satisfaction level of the customer with this project in terms of QA

all higher satisfaction level with over 80 % of customers satisfied with functionality of the final product. Availability, usability and overall quality also shows high level of satisfaction while security, performance and modifiability suited only about half of customers. As for testability and interoperability, just one out of three respondents reported customers' satisfaction with respect to these attributes. These results can be interpreted according to a specific attention of software engineers on functionality, usability and availability resulting in higher overall quality of the final software artifacts.

Figure 2 considers Quality Attributes satisfaction rates for specific industry domains. Here only fields which had ten or more responses have been computed, therefore eight industry domains are shown: Education, Finance & Banking, Gaming & Utilities, Sales & Business Development, Telecommunications, Medical Systems & Pharmaceuticals, Government, Human Resources & Payroll. The level of customer satisfaction has been here computed individually for each domain.

Regarding *Functionality* each industry domain reports level of satisfactions of 75 % or higher with Finance & Banking and Government projects almost reaching 100 %. As for *Security*, Government projects shows an unexpectedly low 30 % not very far from games and utilities. Here HR and Finance & Banking fields reports the highest level. The highest satisfaction levels with *Performance* are shown in HR & Payroll systems and Medical systems, while Education and Sales scores the least.

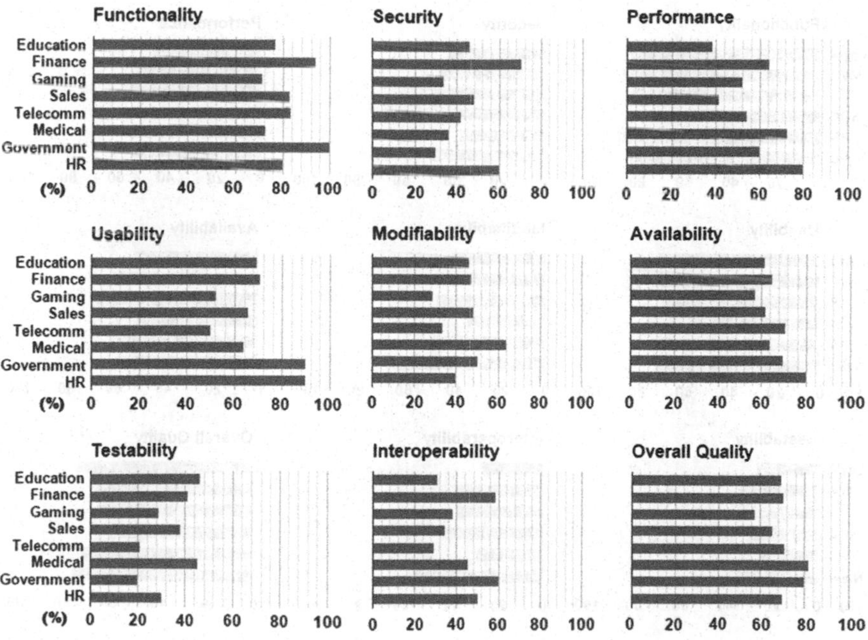

Fig. 2 Satisfaction level of the customer with QA for different industry domains

Usability satisfaction rates have Government and HR projects around 90 % and Gaming and Telecommunication with 50 %. When it comes to *Modifiability* for Gaming & Utilities also have very low 30 % levels and Medical Systems score the highest (more than 60 %). More than half customers met the expectations with respect to *Availability* in all business domain. In general, customers have not been much satisfied by *Testability*, where percentages stay between 45 and 20 %. *Interoperability* also shows low average satisfaction levels. Satisfaction concerning *Overall Quality* of the final product is generally high, in particular for Medical Systems and Government projects.

Summing up data from Fig. 2 we can conclude that satisfaction rates appear coherent with expectations of a business domain for a specific quality attribute, for example Security is a priority for Finance & Banking sector. General low rates appearing for Gaming & Utilities can be explained by the peculiar nature and needs of this domain.

Figure 3 shows the impact of architectural styles on the final product. Levels of satisfaction related to six among the most popular architectural styles have been projected. Layer, Publisher-Subscriber and Multi-tier are generally higher than MVC, SOA and Client server. However, no significant difference can be appreciated among different architectures used in projects and the fact that architectural style do not affect quality of final product can be concluded.

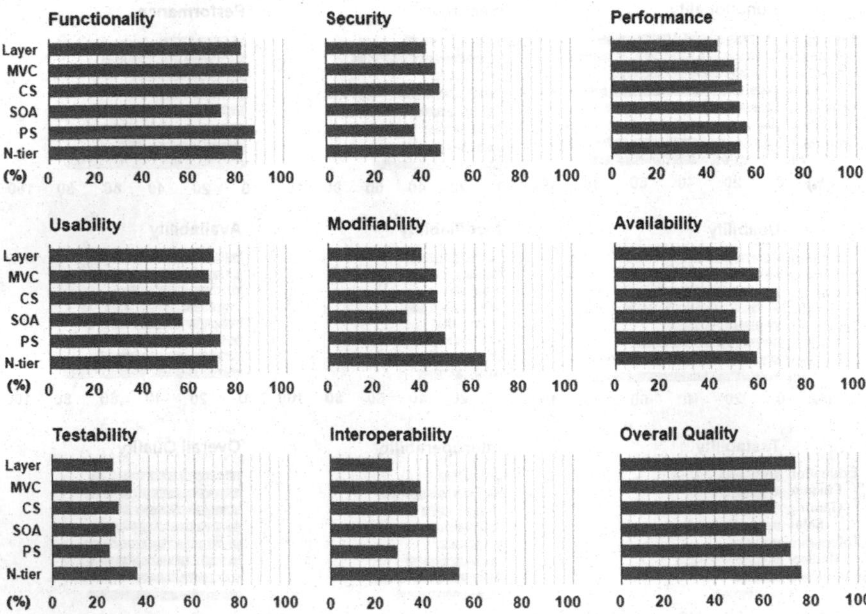

Fig. 3 Satisfaction level of the customer with QA for different architectural styles

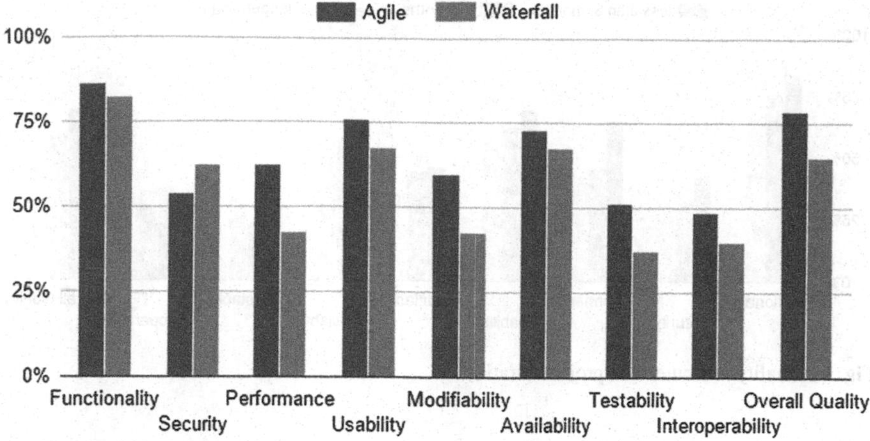

Fig. 4 Satisfaction level of the customer with QA for Agile versus Waterfall projects

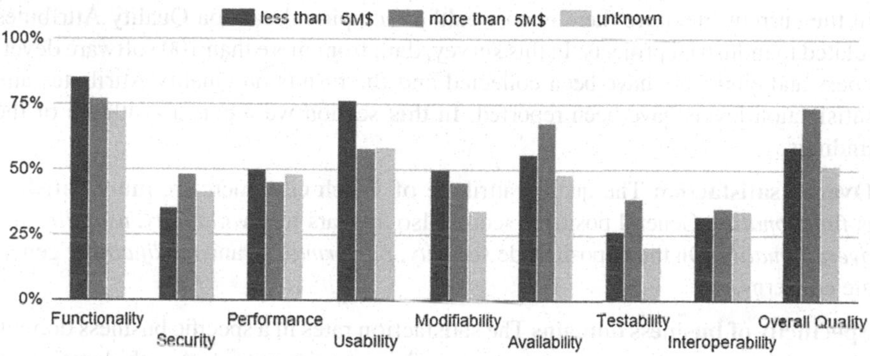

Fig. 5 Satisfaction level of the customer with QA for different annual budget

We have also analyzed quality attributes from the point of view of software development methods, and observed how projects performed following Agile methodologies shows higher rates of satisfaction than those following Waterfall (see Fig. 4).

Figure 5 we observe how the overall quality of final products developed by companies with an annual budget higher than 5M of US dollars is higher. This could be considered as the result of the fact that large companies might have more slack time for developers [12], who can then devote more time to refactoring or to improve their own development skills.

Figure 6 shows that projects with a duration inferior to six months have satisfaction rates higher than projects with a longer duration. This can be interpreted along the line that longer development activities result in more complex projects for which may be harder to maintain high quality and more generally keep customer's satisfaction high for all the process.

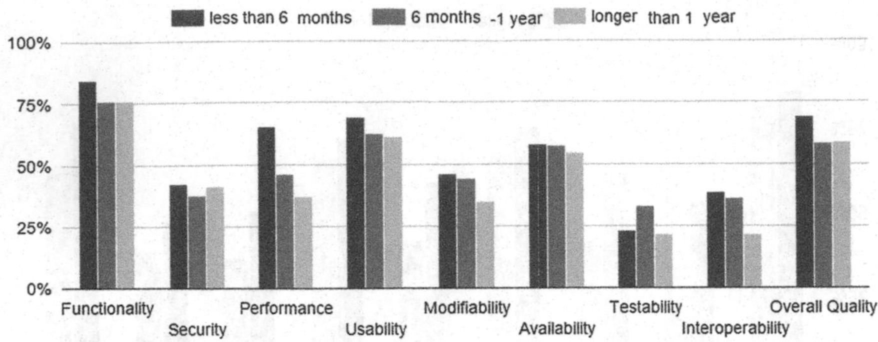

Fig. 6 Quality attributes and project duration

5 Conclusions

In the current literature there is no wealth of empirical data on Quality Attributes related to industrial projects. In this survey, data from more than 100 software developers and engineers have been collected and the results on Quality Attributes and satisfaction levels have been reported. In this section we report a synthesis of the findings.

Overall satisfaction The quality attribute of which customers are **more satisfied** is *functionality*. General positives scores also appears for *availability*, *usability* and *overall quality*. On the opposite side *security*, *performance*, and *modifiability* generate concerns.

Specificity of business domains The satisfaction rates in a specific business domain of the different quality attributes appear coherent with nature of such domain, for example usability score is very high for Government and HR while functionality scores high for telecommunication. On the other side testability is low for all this business domains. This confirms expectations in a sperimental manner.

Architectural Styles While the certain architectural styles (Layer, Publisher-Subscriber, and Multi-tier) result in higher satisfaction rates than other (MVC, SOA, and Client server), no statistically significant difference can be observed, so that there is no evident relationship between architectural style and quality of final product.

Development Methods Projects implemented using **Agile methodologies** shows higher rates of satisfaction than those following Waterfall.

Budget The overall quality of final products developed by companies with an annual budget higher than 5M of US dollars is higher than those with a lower budget.

Duration Projects lasting less than six months have satisfaction rates higher than longer projects.

Reflections Some of the results here presented confirm the intuition and the expectation, for example how **budget is affecting overall quality** and how **some Quality**

Attributes are more relevant for specific business domains. Other results shows the importance of development methods, other partly contradict the intuitions, for example project with short duration have satisfaction rates. For what reason high satisfaction cannot be achieved and maintained in longer project? While Architectural Styles seems not do significantly affect satisfaction?

This is just a preliminary work and results are synthesized by a reasonably large dataset, though it cannot be considered definitive. It is necessary to expand the number of respondents and possibly double it to reach more solid conclusions. On the other side, the nature of domains investigated is pretty broad, therefore to reach some stable conclusion will also be necessary to separately investigate the different domains.

There are two aspects in which the study has to be and can be improved. On one side, the collected data does not discriminate between IT specialist and IT managers. It would therefore be interesting to analyze the effect of separating these two classes of IT players. On the other side, once this is done it will also be possibly to distinguish between *internal* and external *quality* [1]. The collected data also opens the possibility to focus the analysis more on the dichotomy waterfall versus agile. However, in this paper we did not exploit much this data projection, leaving it as future work.

This work is intended to shed some light on the relationships between Quality Attributes and different aspects of software development, and constitutes only a starting point for the accumulation and analysis of further data. In this paper we have indeed established methods and approach to the research which will be applied and extended in future. Next steps are:

1. Expand dataset
2. Specialize on architectural patterns
3. Validate the temporary conclusion presented in this work
4. Deeper analysis on development methodologies (waterfall vs. agile)

As a matter of fact, we are already working on the first of these steps.

Acknowledgments We would like to thank Innopolis University for logistic and financial support and, in particular, the PR department and Elena Maksimenko for their contribution in reaching local IT industry and collect survey data.

References

1. ISO/IEC: ISO/IEC 9126. Software engineering—Product quality. ISO/IEC (2001)
2. IEEE standard for software quality assurance processes. IEEE Std 730-2014 (Revision of IEEE Std 730-2002), pp. 1–138, June 2014
3. Chung, L., Nixon, B., Yu, E., Mylopoulos, J.: Non-Functional Requirements in Software Engineering. International Series in Software Engineering. Springer, US (1999)
4. Shaw, M., Garlan, D.: Software Architecture: Perspectives on an Emerging Discipline. Prentice-Hall Inc, Upper Saddle River, NJ, USA (1996)

5. Bass, L., Clements, P., Kazman, R.: Software Architecture in Practice, 2nd edn. Addison-Wesley Longman Publishing Co., Inc, Boston, MA, USA (2003)
6. Hoinville, G., Jowell, R.: Survey Research Practice, 1st edn. SCPR (1982)
7. Silverman, D.: Doing Qualitative Research: A Practical Handbook. SAGE Publications (2000)
8. Marshall, C., Rossman, G.: Designing Qualitative Research. Sage Publications (2006)
9. Shaughnessy, J., Zechmeister, E., Zechmeister, J.: Research Methods in Psychology. McGraw-Hill, McGraw-Hill Higher Education (2003)
10. Groves, R.M., Fowler Jr., F.J., Couper, M.P., Lepkowski, J.M., Singer, E., Tourangeau, R.: Survey Methodology, 2nd edn. Wiley, Hoboken, N.J. (2009)
11. Kassab, M., Neill, C., Laplante, P.: State of practice in requirements engineering: contemporary data. Innovations in Systems and Software Engineering **10**(4), 235–241 (2014)
12. DeMarco, T.: Slack: Getting Past Burnout, Busywork, and the Myth of Total Efficiency. Broadway Books (2002)

Robot Dream

Alexander Tchitchigin, Max Talanov, Larisa Safina and Manuel Mazzara

Abstract In this position paper we present a novel approach to neurobiologically plausible implementation of emotional reactions and behaviors for real-time autonomous robotic systems. The working metaphor we use is the "day" and "night" phases of mammalian life. During the "day" phase a robotic system stores the inbound information and is controlled by a light-weight rule-based system in real time. In contrast to that, during the "night" phase the stored information is been transferred to the supercomputing system to update the realistic neural network: emotional and behavioral strategies.

Keywords Robotics · Spiking neural networks · Artificial emotions · Affective computing

1 Introduction

Some time ago we have asked ourselves, "Why could emotional phenomena be so important for robots and AI (artificial intelligence) systems?" As for humans: emotional mechanisms manage processes like attention, resource allocation, goal setting, etc. These mechanisms seems to be beneficial for computational systems in general and therefore for AI and robotic systems. Still these phenomena tend to be "difficult"

A. Tchitchigin (✉) · L. Safina · M. Mazzara
Innopolis University, Kazan, Russia
e-mail: a.chichigin@innopolis.ru; a.tchichigin@it.kfu.ru; sad.ronin@gmail.com

L. Safina
e-mail: l.safina@innopolis.ru

M. Mazzara
e-mail: m.mazzara@innopolis.ru

A. Tchitchigin · M. Talanov
Kazan Federal University, Kazan, Russia
e-mail: max.talanov@gmail.com

G. Jezic et al. (eds.), *Agent and Multi-Agent Systems: Technology and Applications*, Smart Innovation, Systems and Technologies 58,
DOI 10.1007/978-3-319-39883-9_24

for computational as well as AI and robotics researchers. To implement phenomena related to emotions we simulate nerobiological processes underlying emotional reactions.

Besides, without understanding and simulation of emotions the effective AI–human communication is almost impossible. Let us review one real-life example. Imagine a concierge robot in a hotel, finishing checking in a fresh lodger, when a man from room 317 rushes into concierge desk in a panic. "Water tap is broken and it starts flooding the whole floor!" Even in a case when it is optimal to finish check-in first, it is very likely that the new guest would wish a concierge to switch immediately to arisen problem. That happens due to empathy between human beings. If a robot does not understand emotions, it can not prioritize tasks appropriately, which would damage client's satisfaction and hotel's reputation.

We could identify the main contribution of this paper as an approach that allows to expand practical autonomous real-time control of a robotic platform with realistic emotional appraisal and behavior based on simulation of spiking neural network with neuromodulation.

In Sect. 2 we substantiate the need for neurobiologically plausible emotional simulation and point out the mismatch between computational resources available to current robotic systems and what is required for neuronal simulation. In Sect. 3 we introduce our concept how a robotic system execution can be separated into "day" and "night" phases in order to bridge the gap between a robotic system and supercomputer performing the simulation. In Sect. 4 we introduce the notion of "bisimulation" to answer the questions of learning and mapping from realistic neural network to rules-based control system. Section 5 provides the information about the actual topics in the field of affecting computations, notable authors and research projects in this area. Finally we sum up the ideas presented in the paper and discuss the arose questions with attention to the steps we are going to take in order to resolve them in Sect. 6.

2 The Problem

There are several cognitive architectures that implement emotion phenomena, some of the most notable are listed in Sect. 5. Rather than implementing the emotional model in a computational system, we re-implemented the neurobiological basis of emotions using simulation [30]. This was done to create a biologically plausible approach and to validate the results of our simulations from neurobiological perspective. The other way around could not provide proper evidence that the result could be regarded as emotional phenomenon. We used the model of basic mechanisms of a mammalian brain via neuromodulation and their mapping to basic affective states [14, 26–29]. We used the realistic spiking neural networks with neuromodulation reconstructing all brain structures involved into the pathways of neuromodulators of the "cube of emotions" by Hugo Lövheim [14]. Unfortunately, current robotic systems usually do not have enough memory and computational capacity to run realistic simulations of human brain activity.

For example, this is computational resources of rather advanced bipedal robotic platform AR-601:

- CPU—4th Gen Intel Core i7-4700EQ 4-Core 3.4 GHz processor;
- System Memory—1 × 204-Pin DDR3L 1333MHz SO-DIMM up to 8 GB;

However the simulation of 1 % of human brain required a cluster of 250 K-supercomputers (each contains 96 computing nodes, each node contains a 2.0 GHz 8-core SPARC64 processor and 16 GB of memory) that was done by RIKEN institute in 2013 and this simulation was slower than human brain in 1000 times [12]. According to the estimates of the Human brain project the computational capacity to simulate whole human brain should be 30 exaflop that is not feasible at the moment.

Therefore realistic simulation even for parts of mammalian brain involved into neuromodulation processes leading to emotional reactions cannot be done in real-time even at scale of rat brain (not mention human brain) on an autonomous robotic platform. To combine autonomous control with advanced realistic emotional appraisal we propose life-cycle separation into "day" and "night" phases.

3 Day Phase and Night Phase

To enable (soft) real-time behaviour from a robotic system it should run some kind of traditional rule-based control system. We do not constrain type or architecture of this system, it might be some sort of Boolean expert system, fuzzy-logic based or Bayesian-logic based system, possibly augmented with Deep-learning pre-processing cascades for visual and/or audio input channels. In our approach robot's control system should not only produce appropriate reaction for input stimulus but also record these stimuli for post-processing during "night phase".

So we propose that robotic system records and stores all input signals for further post-processing. It might store either raw inputs (video and audio streams for example) or some higher-level aggregated form like Deep-learning cascade output (if control system uses such pre-processing mechanisms). And when robot "goes to sleep", it transmits all new recorded inputs to the supercomputer over some network connection (Fig. 1). Actually robotic system might not "sleep all night long", it can stay active while supercomputer processes new information, we assume only that the robot regularly "takes a break" to transmit recorded information and later to receive results and update rules of its control system.

Finally, the supercomputer runs realistic spiking neural network that simulates brain regions involved into neuromodulation pathways. This network receives as an input actual signals recorded by robotic system in a raw or pre-processed (aggregated) form. Then the network processes inputs in the simulated time (which might be faster or slower than clock time depending on available computational power), so

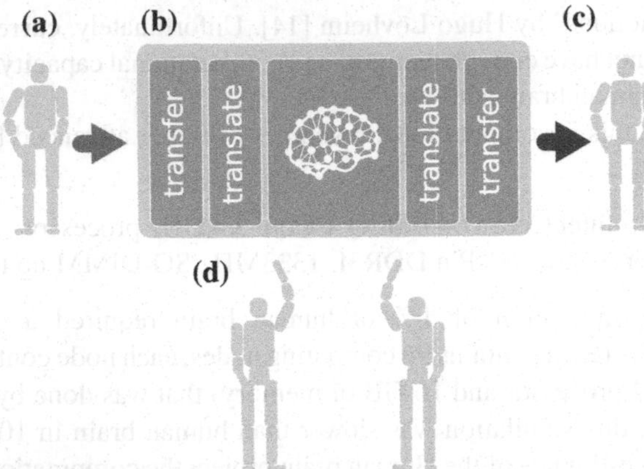

Fig. 1 Night phase and Day phase. **a** In this position a robotic system transfers the accumulated experience into spiking neural network. **b** Processing is done as follows: 1. First the accumulated experience is transferred from a robotic system to the processing center; 2. Then bisimulation starts producing a set of updated rules for robot's control system; 3. Finally update is transferred to the robotic system. **c** The updated rules of the control system are transferred to the robotic system and applied to it. **d** The robotic system continues it's job running updated control system with adjusted emotional reactions and accumulating new experience to be processed again starting from (**a**)

we can infer emotional response of the system [24, 30] from resulting neural activity and neuromodulators levels. From emotional responses of spiking neural network we generate updates to the rules of a control system which are sent back to the robot.

4 Bisimulation

With the approach of simulating separately a reaction of realistic neural network to input stimuli in order to infer an appropriate emotional response, two essential questions arise. First of all, how are we going to train our spiking neural network so it produces relevant and suitable emotional response, and after that how are we going to understand what particular rules of robot's control system we need to update to accommodate emotional mismatch if any?

To answer both questions we propose an approach that we call bisimulation by analogy with the notion in state transition systems. The idea is to simultaneously run both realistic neural network and a copy of robot's control system on the same input. This immediately gives an answer to the second question as we can see which rules of the control system gets activated on current input and what emotional response they imply so we can compare that with emotional reaction inferred from spiking neural network and make necessary adjustments.

But we can use bisimulation to address the question of spiking neural network training too. If we have a rule-based control system with reasonably good emotional reactions we can use it as a source of reward stimuli for neural network. So initially we start running both neural network and control system on sample input and whenever control system produce positive emotional value of the current stimulus, we increase dopamine level in neural network simulating positive feedback (analogously to reward reaction (e.g. to feeding) in mammals). Respectively when control system signals negative emotional value, we decrease dopamine level in neural network.

After realistic neural network learns to mimic emotional responses of control system it can evolve emotional reactions in unsupervised way as it happens in mammals. From that point we can start adjusting emotional reactions of the control system to ever improving ones of spiking neural network.

Speaking of adjusting the rules of a control system with respect to emotional feedback from neural network we highlight several aspects of decision-making that undergo influence of emotional state [1, 25]:

- decisive/less decisive
- speed of decision-making
- bias to positive or negative "thoughts"
- optimistic/conservative
- careful/risky

Let us discuss how these traits could affect a rules-based control system taking as an example NARS framework [31], based on non-axiomatic logic. NARS system was specifically designed to work under bounded amount of knowledge, computational resources and available memory. Therefore the system dynamically redistributes resources among tasks and processes that are needed to be performed according to assigned priorities and relevance. Hence NARS can spend less resources on one task and get approximate answer and much more resources on another task, which leads to obtaining a more precise solution.

Thus the most straightforward aspect to map is the speed of decision-making: the more computational and memory resources we allocate to the whole NARS system the faster it provides decisions to perform necessary actions and vice versa. This could be implemented as some kind of CPU-boost in case of excitement or fear emotional states. Furthermore we can say that NARS have built-in "conservativeness handle": parameter k in denominator of confidence formula. The higher the value of k, the more unconfident the system about its knowledge, the more effort it will spend to ascertain a solution. We can model inclination of the system to "positive" or "negative" "thoughts" adjusting relative priorities of facts with and without logical negation. Also risky behavior could be modeled by lowering the confidence threshold of accepted actions.

Similar mapping for decision-making traits can be constructed for probabilistic or fuzzy-logic based control systems, though in each particular case we have to carefully consider available parameters and adjust the mapping according to semantics and pragmatics of the control system we work with.

5 Related Work

Starting from the late 90-es the interest to emotions and emotional representations in computational systems has been growing [3, 19, 21, 22]. This rise of activity is based on understanding of the role of emotions in human intelligence and consciousness that was indicated by several neuroscientists [6, 7, 18]. Starting from the seminal book by Rosalind Picard, [20] (though we could mention earlier attempts to implement emotions in the machines [4]), we could identify two main directions in a new research field of affective computing: emotion recognition and re-implementation of emotions in a computational system. There are several cognitive architectures that are capable of the re-implementation of emotional phenomena in different extent, starting from SOAR [13] and ACT-R [11] to modern BICA [23]. The interest in implementation of emotional mechanisms is based on the role of emotional coloring in appraisal, decision making mechanisms, and emotional behavior, indicated by Damasio in [5]. Our approach takes a step further on the road for neurobiologically plausible model of emotions [30].

We have to mention other perspectives on emotional re-implementations in a computational machines: Arbib and Fellous [2, 8] created the neurobiological background for the direction to neurobiologically inspired cognitive architectures; the appraisal aspects are in focus of Marsella and Gratch research [10, 16, 17] as well as in Lowe and Ziemke works [15, 32]. As the neuropsychological basement for our cognitive architecture we used "Cube of emotions" created by Hugo Lövheim [14]. It bridges the psychological and neurobiological phenomena in one relatively easy to implement programmatically model based on three-dimensional space of three neuromodulators: noradrenaline, serotonin, dopamine, which is its main advantage from our perspective. For the implementation we used realistic neural network simulator NEST [9] recreating the neural structures of the mammalian brain. As it was mentioned earlier in this paper, the processing of the simulation took 4 h of supercomputer's processing time to calculate 1000 ms.

6 Conclusion

In our article we described an idea or approach for augmentation of autonomous robotic systems with mechanisms of emotional revision and feedback. There are many open questions: input formats for realistic neural network, emotional revision thresholds and emotional equalizing (homeostasis) and so on. On the one hand, different answers to these questions allow adaptation of the approach to a range of possible architectures of robots' control systems. On the other hand, even better solution would be a software framework implementing our approach with several pluggable adapters to accommodate the most popular choices for robots' "brains".

So our next step is to implement a proof-of-concept of proposed architectural scheme for some simple autonomous robotic system. We are not going to start with a concierge android right away. Our current idea is to develop some sort of robotic vacuum cleaner prototype that does not even clean but can bump into humans or pets and receive emotional feedback, for example because it is too noisy or because of incorrectly prioritized work tasks. To simplify human feedback mechanism some remote control device could be used for example with two buttons: for positive and negative feedback. Later we can employ some audio-analysis system to infer emotional reaction from verbal commands to robo-cleaner. Main research question here is to what extent these emotional mechanisms can affect robot's behavior? And we hope to establish a foundation for desired software framework.

References

1. Ahn, H., Picard, R.W.: Affective cognitive learning and decision making: the role of emotions. In: The 18th European Meeting on Cybernetics and Systems Research (EMCSR 2006) (2006)
2. Arbib, M., Fellous, J.M.: Emotions: from brain to robot. Trends Cogn. Sci. **8**(12), 554–559 (2004)
3. Breazeal, C.: Designing Sociable Robots. MIT Press (2004)
4. Breazeal, C.: Emotion and Sociable Humanoid Robots. Elsevier Science (2002)
5. Damasio, A.: Descartes' Error: Emotion, Reason, and the Human Brain. Penguin Books (1994)
6. Damasio, A.: The Feeling of What Happens: Body and Emotion in the Making of Consciousness, New York (1999)
7. Damasio, A.R.: Emotion in the Perspective of an Integrated Nervous System. Brain Res. Rev. (1998)
8. Fellous, J.M.: The neuromodulatory basis of emotion. Neuro-Scientist **5**, 283–294 (1999)
9. Gewaltig, M.O., Diesmann, M.: Nest (neural simulation tool). Scholarpedia **2**(4), 1430 (2007)
10. Gratch, J., Marsella, S.: Evaluating a computational model of emotion. Auton. Agents Multi-Agent Syst. **11**, 23–43 (2005)
11. Harrison, A.: jACT-R: JAVA ACT-R. In: Proceedings of the 8th Annual ACT-R Workshop (2002)
12. Himeno, R.: Largest neuronal network simulation achieved using k computer (2013). http://www.riken.jp/en/pr/press/2013/20130802_1
13. Laird, J.E.: Extending the soar cognitive architecture (2008). http://ai.eecs.umich.edu/people/laird/papers/Laird-GAIC.pdf
14. Lovheim, H.: A new three-dimensional model for emotions and monoamine neurotransmitters. Med. Hypotheses **78**(78), 341–348 (2012)
15. Lowe, R., Ziemke, T.: The role of reinforcement in affective computation triggers, actions and feelings. In: IEEE Symposium on Computational Intelligence for Creativity and Affective Computing (CICAC) (2013)
16. Marsella, S., Gratch, J.: Modeling coping behavior in virtual humans: don't worry, be happy. In: Appears in the 2nd International Joint Conference on Autonomous Agents and Multiagent Systems (2003)
17. Marsella, S., Gratch, J., Petta, P.: Computational models of emotion. In: Scherer, K., Bänziger, T., Roesch, E. (eds.) A Blueprint for a Affective Computing: A Sourcebook and Manual. Oxford University Press, Oxford (2010)
18. Murata, A., Fadiga, L., Fogassi, L., Gallese, V., Raos, V., Rizzolatti, G.: Object representation in the ventral premotor cortex (area f5) of the monkey. J Neurophysiol. **78**(4), 2226–30 (1997)

19. Picard, R.W.: Affective computing. Technical report. M.I.T Media Laboratory Perceptual Computing Section (1995)
20. Picard, R.W.: Affective Computing. Massachusets Institute of Technology (1997)
21. Picard, R.W.: What does it mean for a computer to "have" emotions? In: Trappl, R., Petta, P., Payr, S. (eds.) Emotions in Humans and Artifacts (2001)
22. Picard, R.W.: Affective computing: challenges. Int. J. Hum.-Comput. Stud. **59**, 55–64 (2003)
23. Samsonovich, A.V.: Modeling human emotional intelligence in virtual agents. Technical report. Association for the Advancement of Artificial Intelligence (2013)
24. Talanov, M., Toschev, A.: Computational emotional thinking and virtual neurotransmitters. Int. J. Synth. Emot. (IJSE) **5**(1) (2014)
25. Talanov, M., Vallverdú, J., Distefano, S., Mazzara, M., Delhibabu, R.: Neuromodulating cognitive architecture: towards biomimetic emotional AI. In: Advanced Information Networking and Applications (AINA), pp. 587–592 (2015). ISSN: 1550-445X
26. Tomkins, S.: Affect Imagery Consciousness. Vol. 1: The Positive Affects. Springer, New York (1962)
27. Tomkins, S.: Affect Imagery Consciousness. Vol. I, II, III. Springer, New York (1962, 1963, 1991)
28. Tomkins, S.: Affect Imagery Consciousness. Vol. II: The Negative Affects. Springer, New York (1963)
29. Tomkins, S.: The quest for primary motives: biography and autobiography of an idea. J. Pers. Soc. Psychol **41**, 306–335 (1981)
30. Vallverdú, J., Talanov, M., Distefano, S., Mazzara, M., Tchitchigin, A., Nurgaliev, I.: A cognitive architecture for the implementation of emotions in computing systems. In: Biologically Inspired Cognitive Architectures (2015)
31. Wang, P.: Non-Axiomatic Logic: A Model of Intelligent Reasoning. World Scientific (2013)
32. Ziemke, T., Lowe, R.: on the role of emotion in embodied cognitive architectures: from organisms to robots. Cogn. Comput. **1**, 104–117 (2009)

Part VI
Business Informatics and Gaming through Agent-Based Modelling

Model-Driven Development of Water Hammer Analysis Software for Irrigation Pipeline System

Yoshikazu Tanaka and Kazuhiko Tsuda

Abstract The MDD method was used to develop software to evaluate the hydraulic safety of irrigation pipeline systems. The difficulty in developing and maintaining the software is that the diversity of hydraulic behavior and structure of various ancillary facilities constituting the irrigation pipeline system bring complexity into the code to process the boundary conditions and to enter data in the numerical calculation solver. To solve that problem, this study used the following two methods. (1) The various pipe facilities and ancillary facilities which are the domain to analyze were defined the data structure by UML and the design pattern was applied in implementation. (2) The method of entering data about the boundary conditions into the numerical analysis solver was assisted by objects automatically coded from the metadata that were stipulated using the schema language of XML.

Keywords Irrigation pipeline system · Water hammer analysis · Model driven development · XML · Object oriented programming

1 Introduction

The irrigation pipeline is the important water conservancy facilities that extended dozens km in order to intake water from a river using pumps and distribute irrigation water to paddies, upland fields and orchards. The total length of irrigation pipelines already installed in Japan have reached 45,000 km of trunk channels

Y. Tanaka (✉)
NARO Institute for Rural Engineering, 2-1-6, Kan-nondai, Tsukuba, Ibaraki 305-8609, Japan
e-mail: yokka@affrc.go.jp

K. Tsuda
Graduate School of System and Information Engineering, University of Tsukuba, 3-29-1, Otsuka, Bunkyo, Tokyo 112-0012, Japan
e-mail: tsuda@gssm.otsuka.tsukuba.ac.jp

G. Jezic et al. (eds.), *Agent and Multi-Agent Systems: Technology and Applications*, Smart Innovation, Systems and Technologies 58,
DOI 10.1007/978-3-319-39883-9_25

serving beneficiary farms of 100 ha and larger [1]. Design standard criteria of the Japan Ministry of Agriculture, Forestry and Fisheries has stipulated that numerical analysis must be executed at the design stage to evaluate the magnitude of the water hammer pressure in order to confirm the hydraulic safety of irrigation pipeline systems constructed as national government enterprises [2]. Here, water hammer pressure means the violent change of pressure inside a pipe that occurs when a valve is closed suddenly in order to control water flow in pipeline system, this hydraulic phenomenon may damage the pipeline. We assumed that the diverse boundary conditions of the numerical analysis would make it difficult to maintain the codes and prepare data in existing water hammer numerical analysis software, so we adopted the MDD (Model-Driven Development) method in developing new one in order to overcome these problems.

At first, we developed an object oriented program for water hammer analysis to avoid maintenance problems that could occur later. We will call that program OOWHL (the Object Oriented Water Hammer Library). As some cases of development in the numerical computation field, there are the object oriented linear algebra libraries (for example, LAPACK++ [3] and IML++ [4] etc.) which is important for solving partial differential equations. However, there is no library of the boundary conditions. New contribution of this paper is introduction of a boundary library for an applied MDD [5].

Next, in order to simplify the complication of entering data, we proposed the data conduct method that uses XML as a file entering into the numerical analysis software. Thus, XML contains information about pipe facilities and ancillary facilities inside the pipeline system of which the metadata were regulated based on RELAX. We will call that ISML (the Irrigation System Markup Language). Here, RELAX is a kind of the schema language of XML [6]. In the field of business informatics, the government promotes CALS/EC (Continuous Acquisition and Life-cycle Support/Electronic Commerce) [7, 8] to manage the information of national government enterprises. Our approach for data management is to not keep the CAD data of detailed shape, but to manage the behavior and specifications of the structure as XML data. The advantage is that this makes it possible to inspect the hydraulic behavior of a pipeline system during planning and maintenance.

The results show how the class group architected based on UML was skillfully linked to the XML file described based on the schema language. The results also show how the entering data method into the numerical analysis software increases the robustness and smoothness by linking with KML which is used for geographical information in Google Earth. Thanks to the method of entering data, it makes possible to assist entering data and also to visualize the results of the numerical simulation.

2 Characteristics of the Irrigation Pipeline Water Hammer Analysis Software

This section examines the characteristics and problems of the water hammer analysis software. The software performs a one-dimensional numerical calculation about hydraulic transient phenomena in pipelines. Generally, the numerical analysis software has three main parts: the pre-processing part for entering initial conditions and boundary conditions, the computational part for performing numerical calculation, and the post-processing part for visualizing simulation results.

In the computational part, taking account for the elasticity of pipe material and water in the pipe facilities, Eqs. (1) and (2) which are the equations of motion and mass conservation are derived.

$$g\frac{\partial H}{\partial x}+\frac{\partial V}{\partial t}+V\frac{\partial V}{\partial x}+\frac{fV^2}{2D}=0 \tag{1}$$

$$\frac{a^2}{g}\frac{\partial V}{\partial x}+\frac{\partial H}{\partial t}+V\left(\frac{\partial H}{\partial x}+\sin\gamma\right)=0 \tag{2}$$

where, H is pressure head, V is average flow velocity inside the pipe, x is the distance in the pipe length direction, t is elapsed time, f is the coefficient of frictional loss, g is acceleration due to gravity, D is the diameter of the pipe, a is the propagation velocity inside the pipe, and γ is the gradient of the pipe. These equations are solved by the method of characteristics. As shown in Fig. 1, the average flow velocity inside the pipe V_{M}^{n+1} and the pressure head H_{M}^{n+1} at the next time step n+1 are obtained by overlapping two values of each the average flow velocity inside the pipe and the pressure head at the locations where the two characteristics lines of C^{+} (the travelling wave) and C^{-} (the backward wave) intersect the computational grid at the last time step n. However, as it is often the case, the characteristic lines often do not pass above the computational grid points, so the value at M is needed to calculate from the values at R and S based on Eqs. (3) and (4).

$$\begin{aligned}V_{\mathrm{M}}^{n+1}=\frac{1}{2}\Big[&V_{\mathrm{R}}^{n}+V_{\mathrm{S}}^{n}+\frac{g}{a}\left(H_{\mathrm{R}}^{n}-H_{\mathrm{S}}^{n}\right)\\&-\frac{f\Delta t}{2D}\left(V_{\mathrm{R}}^{n}\left|V_{\mathrm{R}}^{n}\right|+V_{\mathrm{S}}^{n}\left|V_{\mathrm{S}}^{n}\right|\right)-\frac{g\left(V_{\mathrm{R}}^{n}-V_{\mathrm{S}}^{n}\right)}{a}\sin\gamma\Big]\end{aligned} \tag{3}$$

$$\begin{aligned}H_{\mathrm{M}}^{n+1}=\frac{a}{2g}\Big[&H_{\mathrm{R}}^{n}+H_{\mathrm{S}}^{n}+\frac{a}{g}\left(V_{\mathrm{R}}^{n}-V_{\mathrm{S}}^{n}\right)\\&-\frac{f\Delta t}{2D}\left(V_{\mathrm{R}}^{n}\left|V_{\mathrm{R}}^{n}\right|-V_{\mathrm{S}}^{n}\left|V_{\mathrm{S}}^{n}\right|\right)-\frac{g\left(V_{\mathrm{R}}^{n}+V_{\mathrm{S}}^{n}\right)}{a}\sin\gamma\Big]\end{aligned} \tag{4}$$

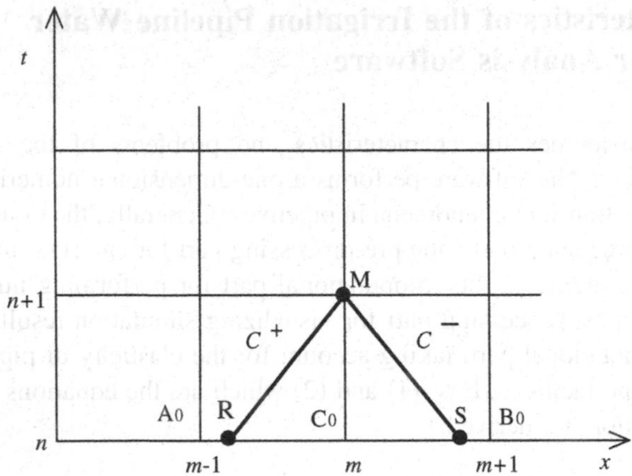

Fig. 1 Positional relation of characteristic and calculation points

where, the average flow velocity inside the pipe and the pressure head at the intermediate points R and S are obtained by interpolating in the grid A_0–C_0 and the grid B_0–C_0 at the last time step *n*. Here, the notice is that the boundary conditions are the behavior represented by the ancillary facilities at the upstream and downstream edges of the computational grid. An irrigation pipeline consists of pipe facilities and various ancillary facilities, as shown in Table 1. Those boundary conditions are not only fixed pressure and fixed average flow velocity inside the pipe, but also behave dynamically in different ways, so there are various treatments for boundary conditions. That diver sity causes the problem that many conditions branches are needed to process the boundary conditions in the solver.

Table 1 Various ancillary facilities in an irrigation pipeline system

Facility Group	Examples of specific facilities
Regulating facilities	Regulating reservoir, Farm pond
Pressure control facilities	Check stand, Pressure-reducing valve, Float valve type pressure-reducing tank
Pump facilities	Water source pump, Relay pump, Booster pump, Water suction tank, Water discharge tank
Diversion facilities	Stand type division work, Closed type division work, Hydrant
Ventilation facilities	Air stick, Air valve
Protective facilities	Water hammer buffer device, Safety valve
Control facilities	Control valve, Check valve
Others	Bent pipe, Sudden contraction pipe, Sudden enlarged pipe, End pipes, Branches

Besides, another problem is that the entering data method becomes complicated, because their mechanisms and shapes of ancillary facilities are different each other.

3 Methods Based on MDD of the Irrigation Pipeline

3.1 Method of Object Oriented Modelling

In object oriented modelling, at first, the abstract classes were defined by abstracting the behavior and attributes of the pipe and ancillary facilities. Then, the concrete ancillary facilities were defined as the derived class that achieves the specifications by overriding the contents of the method with the same name in the abstract class or by including another ancillary facilities class. Instances of derived class can act in the code using the polymorphism to represent complex structures and behaviors of various ancillary facilities. We designed these pipe and ancillary facilities using UML and implemented Java objects library OOWHL.

OOWHL is explained below. The basic unit of computation model that constitutes an irrigation pipeline is one pipe that is placed between the ancillary facilities located at upstream edges and downstream edges. The pipeline that some basic units have connected is a hydraulic unit. A long irrigation pipeline consists of multiple hydraulic units.

Pipe Facilities
To perform one-dimensional numerical analysis of the basic Eqs. (1) and (2), the pipe is represented as a path on a diagram. A path is a diagram consisting of two points, a start point and an end point. Therefore, as shown in left side of Fig. 2, the

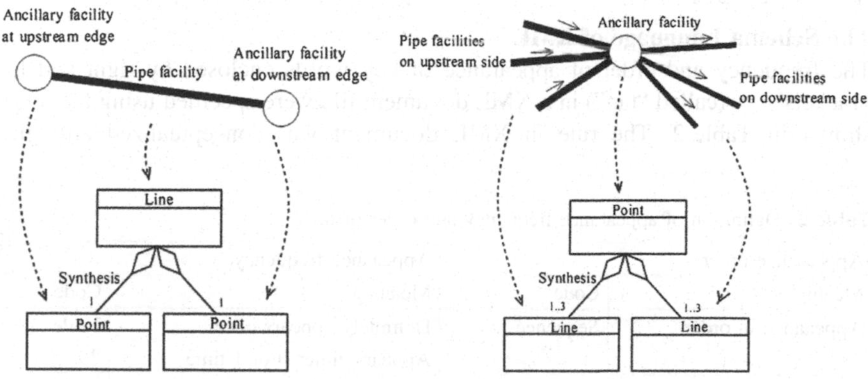

Fig. 2 The connection relationship with point classes and a line class

upstream and downstream edges are defined as Point class as explained later, and the path is defined as Line class that includes these two edges. In the Pipe class derived from Line class, an algorithm that performs numerical calculation based on the method of characteristics is implemented. The various specific pipes were derived from the Pipe class. An irrigation pipeline consists of linked instances made from those derived Pipe classes.

Ancillary Facilities
The derived classes representing ancillary facilities inherited Point class which have attributes such as three-dimensional space coordinate location information and fluid dynamic values etc. and methods coded function to execute the entry data, the computation, and the output (right side in Fig. 2). The methods defined to compute boundary conditions in Point class has been overridden in the derived classes of ancillary facilities in order to represent their different characteristic behavior.

In order to utilize the polymorphism of each instances generated those classes, the design pattern called Factory Method was applied to generate those instances.

3.2 Method of Document Orientation Modelling

This section proposes a data entry method by using XML to gather various attributes of each ancillary facility in one place. The role of XML in this proposal is to represent structurally any information in various pipe and ancillary facilities that constitute an irrigation pipeline system. The schema compiler of XML makes it possible to code the class equivalent to the information in XML, so the solver can easily get information about the initial and boundary conditions from those generated instances. Additionally, it is expected to be linked with KML that has specified geographical information in order to facilitate data entry.

The Schema Language of XML
The frequency and order of appearance of key words enclosed by right and left brackets < > (called "tag") in a XML document file were specified using the codes shown in Table 2. The rule in XML document was conceptualized arranging

Table 2 Definition of appearance frequency and order of tags

Appearance order		Appeaance frequency	
Meanig	Code	Meaning	Code
Appeaance in order	Sequence	Definitely appears once	Node
		Appears either 0 or 1 time	?
Either one appears	Choice	Appears 1 or more times	+
		Appears 0 or more times	*

Table 3 Content of tags in facility A

Tags in facility A	Meanig	Contents of lower tag
<ConnetionInfoOfA>	Connection information	ID of facility connected to upstream side and downstream side
<PositionInfoOfA>	Location information	3D location information about facility
<StructureInfoOfA>	Structure information	Typical dimensions and drawing of facility, etc.
<CharacterInfoOfA>	Characteristics information	Coefficient of flow velocity, coefficient of loss and other values representing characteristics

hierarchically the frequency and order of tags. The computer language to specialize arranging tags is the schema language. In this study, we adopted RELAX as the schema language. RELAX (Regular Language description for XML) is a simple schema language based on the hedge automata theory for structured documents.

Method of Stipulating a Pipeline in the Document

The conceptualizing ancillary facility is represented by specifying the contents of tags of the connections, locations, structures, and characteristics in Table 3. Tags express the names of information that represent each facility, and the task of defining these tags is the heuristic task first noticed when this information was needed. Consequently, the tags in Table 3 are the top-level framework established in advance, so it is possible to specialize information related to connections, locations, structures, and characteristics to conceptualize any facilities. For example, in the case of valve and bend, the bottom tags defined under the top-level tags of the connections, locations, structures, and characteristics are shown in Table 4. The bottom tags in other facilities can be also conceptualized by the same method.

Ancillary facilities with more complex structures are hierarchically conceptualized by systematically combining other ancillary facilities as components. Taking the stand type water division works shown in Fig. 3 as an example, the components are spillway, inlet, outlet, valve, gate, tanks and barrier wall, and those components also possess the information of connection, location, structure and characteristics.

The concept of the pipeline was stipulated as the schema language ISML (Irrigation System Markup Language) for the XML document that describes the information about the irrigation pipeline. The Left side in Fig. 4 shows the overall basic structure of ISML.

Table 4 Difference between bottom tags in the concepts of valve and bend

Valve <IS_Valve>			Bend <IS_Bend>		
Top tag	Button Tag	Contents	Top tag	Button Tag	Contents
<ConnectInfoOfValve?>	<upper?>	ID of pipe on upstream side	<ConnectInfoOfBend?>	Same as on left	
	<lower?>	ID of pipe on downstream side			
<PositionInfoOfValve?>	<displaymethod?>	Method of displaying location information	<PositionInfoOfBend?>	Same as on left	
	<longitude?>	Longitude			
	<latitude?>	Latitude			
	<altitude?>	Altitude			
	<x?>	x coordinate of 3D spaces			
	<y?>	y coordinate of 3D spaces			
	<z?>	z coordinate of 3D spaces			
<StructureInfoOfValve?>	<kindOfValve>	Type of value	<StructureInfoOfBend?>	<aAngle?>	Angle of a Reflection
	<ValveDiameter>	Internal diameter		<aLength?>	Length interval reflection
	<drawing*>	Blueprint		<innerDiameter>	Internal diameter
	<picture*>	Photograph		<timeOfRefrect>	Frequency of reflections in a Bend
<CharacterInfoOfValve?>	<triggerTime?>	Trigger time		<material>	Material
	<TimeORformula?>	Designation of method of representing aperture		<totalAngle>	Total angles in a Bend
	<time_totalOpenDegree*>	Aperture by time		<totalLength>	Total length in a Bend
	<formula_totalOpenDegree*>	Aperture by secondary equation		<drawing*>	Blueprint
	<hydraulicPressure>	Hydrostatic pressure		<picture*>	Photogpaph
	<initialOpenDegree?>	Aperture at initial condition	<CharacterInfoOfBend?>	<hydraulicPressure>	Hydraulic Pressure
				<coefficientOfFormDrag>	Coefficient of local loss

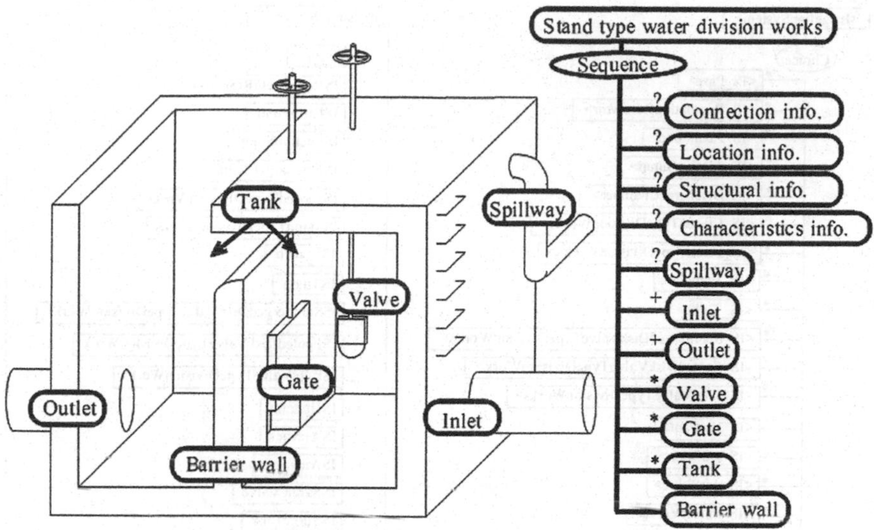

Fig. 3 Conceptual scheme showing the hierarchical structure of the stand type water division works

3.3 *Method of Obtaining Information from the XML File*

A schema compiler automatically can code classes of an object oriented program from the schema language. That feature make possible to generate object with equivalent information in the XML document. We used RELAXER [9] which has two functions as a parser and a schema compiler. RELAXER can automatically code Java classes which operate information identical to the tags defined based on stipulations described by RELAX. The classes shown in the right side of Fig. 4 were generated from the tags of ISML shown in the left side of Fig. 4. We will call that objects library ISOG (Irrigation System's Objects Generated). The fields corresponding to the names of bottom tags and the methods to get and set the values of each fields have been coded inside each classes. The fields and methods about valve have been coded by RELAXER shown in Table 5 from the tags listed on the left side of Table 4. Because names of tag and names of domain are in a one-to-one correspondence, these methods let the solver operate value about data in the XML document from equivalent object without via DOM.

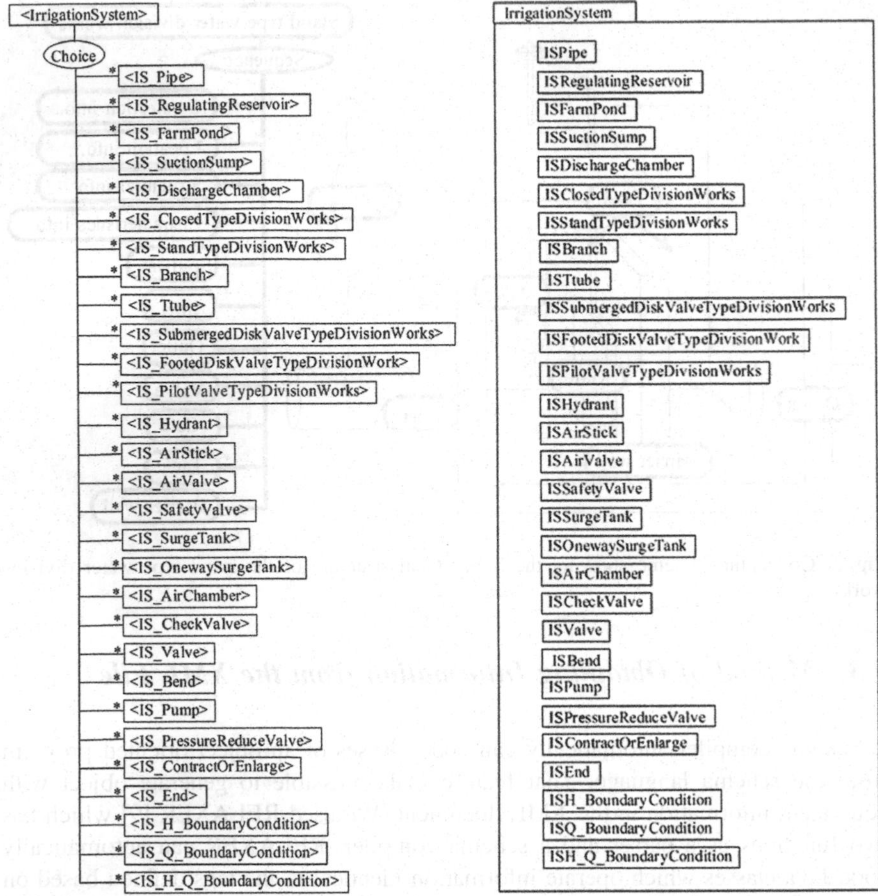

Fig. 4 Outline of the tag structure in ISML (*Left*) and ISOG formed by RELAXER (*Right*)

4 Estimations and Discussion

4.1 Estimations of the Synergy of Object Oriented and Document Oriented Modelling

First, we examine the benefit of OOWHL program that performs water hammer analysis of an irrigation pipeline as follows. The algorithm in the solver that numerically analyze water hammer phenomenon processes boundary conditions based on the poly-morphism of instances, so adding new facility or correcting an existing facility do not affect the code in the solver. That feature will facilitate maintenance, because the simultaneous multiple modifications that happen in the

Table 5 Generated fields and methods in Valve class which is equivalent to XML for valve

Class name	Fields	Methods	Contents
ConnectInfoOfValve	Lower	getId()	Gets ID of facility connected on downstream side
	Upper	getId()	Gets ID of facility connected on upstream side
PositionInfoOfValve		getDisplaymethod()	Gets location infromation display method (geodetic reference or Euclid coordinates)
	Longitude	getContent()	Gets longitude
	Latitude	getContent()	Gets latitude
	Altitude	getContent()	Gets elevation
	X	getContent()	Gets x coordinate value in Euclid coordinates
	Y	getContent()	Gets y
	Z	getContent()	Gets z
StructureInfoOfValve		getKindOfValve()	Gets type of valve
	valveDiameter	getContent()	Gets diameter of valve
CharacterInfoOfValve	TriggerTime	getContent()	Gets that valve operation status (trigger time)
		getTimeOfFormula ()	Gets valve operating time and aperture
	HydrostaticPressure	getContent()	Gets initial value of pressure head
	InitialOpenDegree	getContent()	Gets initika value of aperture of valve

case of adding or correcting the code in the software written by a structured programming language (Ex. Fortran) do not occur.

Next, the fact that information about ancillary facilities is described by the XML file is expected to offer the following benefits.

First, it makes possible to reuse data at the design stage. Taking a long irrigation pipeline as an example, at first step in the design stage, numerical analysis is executed separately for each hydraulic unit divided in order to evaluate safety. At the next step, the numerical analysis for an overall is executed after connecting with the hydraulic units to evaluate their mutual effect. If the XML document of pipe and ancillary facilities that constitute the pipeline and the file that specified the range of the hydraulic units are separately controlled, the task of connecting with hydraulic units is basically completed simply by correcting the file that specifies the range of

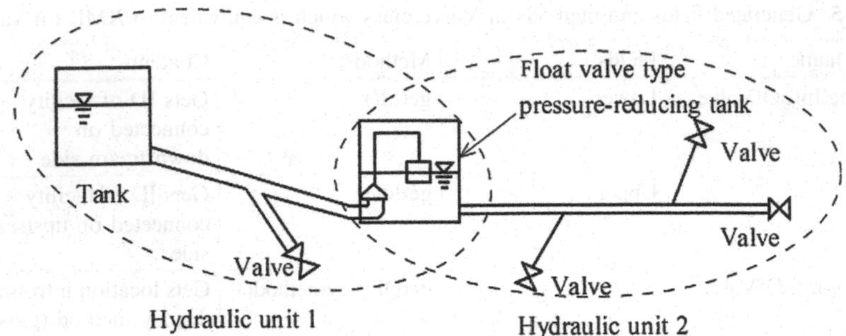

Fig. 5 Example of a simple irrigation pipeline system consisting of two hydraulic units

the hydraulic units, and it is possible to reuse most XML file for each hydraulic unit. For the simple irrigation pipeline system consisting of two hydraulic units as shown in Fig. 5 for example, as shown in Fig. 6, regarding 3 and 1001 in the XML files described ancillary facilities, it is merely necessary to integrate the valve and

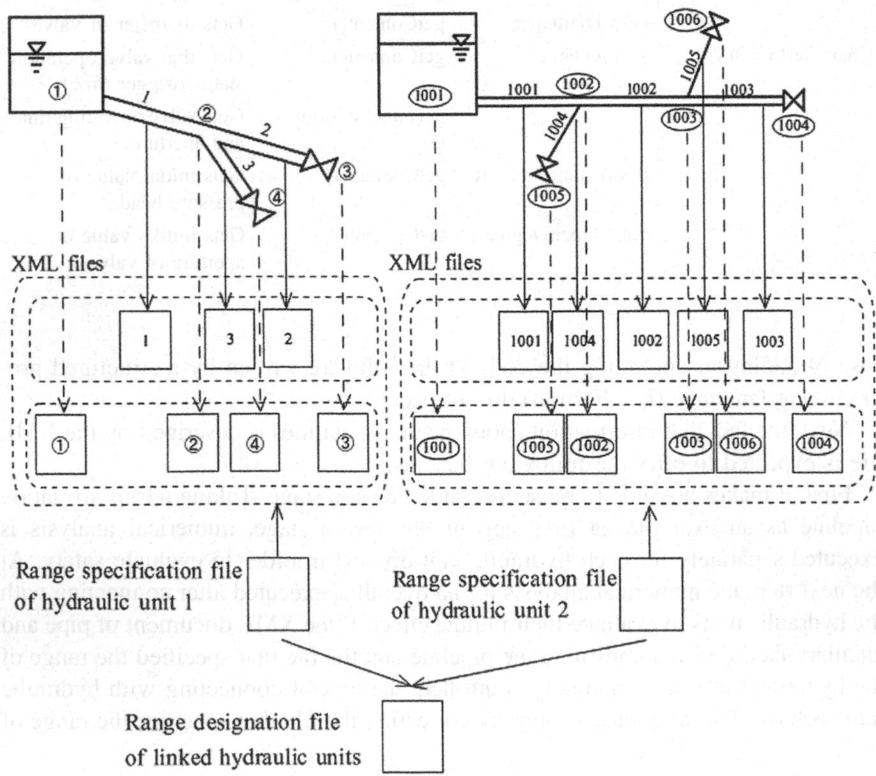

Fig. 6 Structure of the file when hydraulic units 1 and 2 are combined

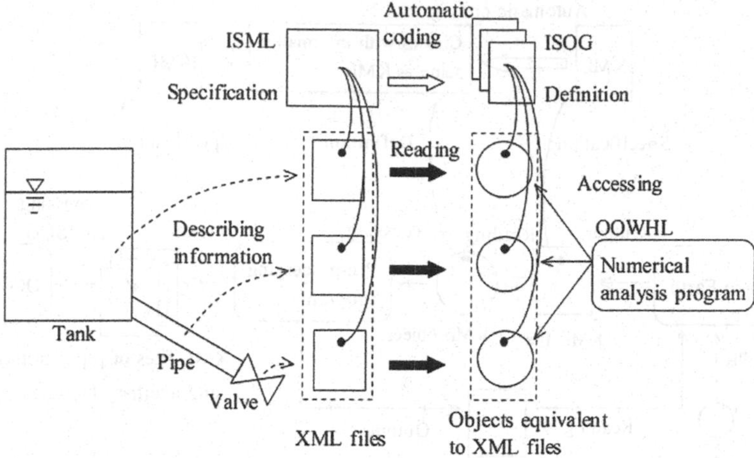

Fig. 7 Concept of entering data in the numerical analysis software

the tank on the float valve type pressure-reducing tank, and to correct only the file that specifies the ranges of the hydraulic units.

Secondly, the information in the XML file is easily incorporated in the program through objects. ISOG is class library which possesses attributes equivalent to ISML, so as shown in Fig. 7, the objects inside OOWHL can obtain information of the XML document via ISOG without forming DOM.

4.2 Discussion About the Developed Software

As shown in Fig. 7, Something put name of either of the ancillary facilities could be converted via XML files that represent information in ancillary facilities to Objects in order to numerically analyze water hammer phenomenon. That mechanism makes possible to assist users by linking to the structured document about the geographical information and using the geographical information system (GIS). In Google Earth, which is a simple GIS, shapes and representations on the map are controlled by a KML file. KML is the description language specified for displaying geographical information on the map in Google Earth. In 2008, KML became a standard specification of OGC (Open Geospatial Consortium) which is a group that promotes the standardization of GIS. Therefore, the proposed pre-processing and post-processing are methods of processing KML files in cooperation with using Google Earth.

If the place mark of a point is identified with an ancillary facility, the information about a series of ancillary facilities in pipeline is prepared by making place marks on the map of Google Earth. The pipe facilities are also prepared from the connection information of ancillary facilities. The pre-processing program can assist

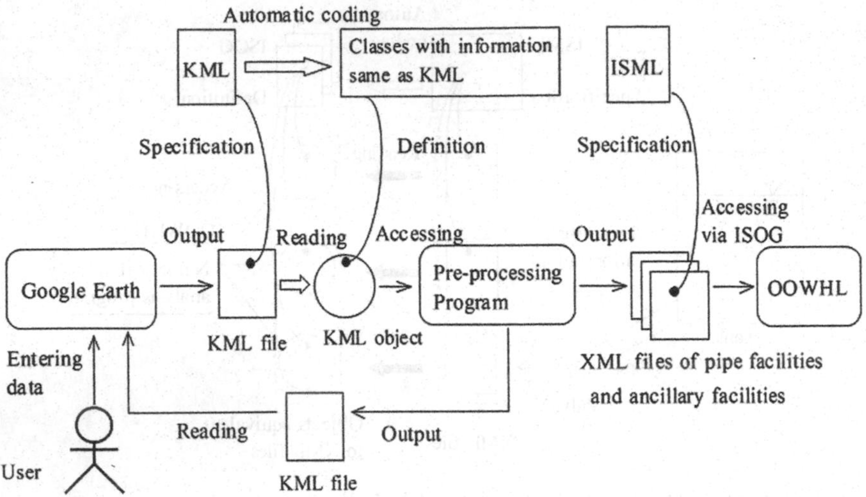

Fig. 8 Concept of the proposed work flow of pre-processing

users to describe information by letting change appropriate default values against tags attached in advance to the text area on the pop-up menu displayed above clicked place mark in Google Earth. After users have completed the task on Google Earth, the pre-processing program abstracts information in the text area in KML to prepare an XML file of pipe and ancillary facilities. The solver gets entry data about pipeline from objects equivalent to the XML files. The general work flow of this procedure is shown in Fig. 8.

The post-processing program converts the CSV file which is output from the numerical analysis program to a KML file, and superimposes the results on the map of Google Earth (Fig. 9).

The maximum water hammer pressure at ancillary facilities was displayed the shape of the polygon diagram drawn on the map. The <Description> tag under the <Placemark> tag can be described in HTML format, so a hyperlink to graph image about the time-series history of the pressure head at the ancillary facilities is applied. The post-processing program prepares graphs of the numerical analysis results and makes image files of these. The conversion of these to a KML file is done with freeware called GE-Graph and GE-Path [10]. As a specific analysis case, an example of evaluating the hydraulic safety about a trunk pipeline of irrigation system N at Prefecture T in Japan is shown. This is a pump pressure type pipeline that shown in Fig. 10. The length is about 6 km and the difference in elevation between the water levels in the pump and the charge chamber is 30 m. There are 74 ancillary facilities including 5 pumps at a pump station and two surge tanks in the

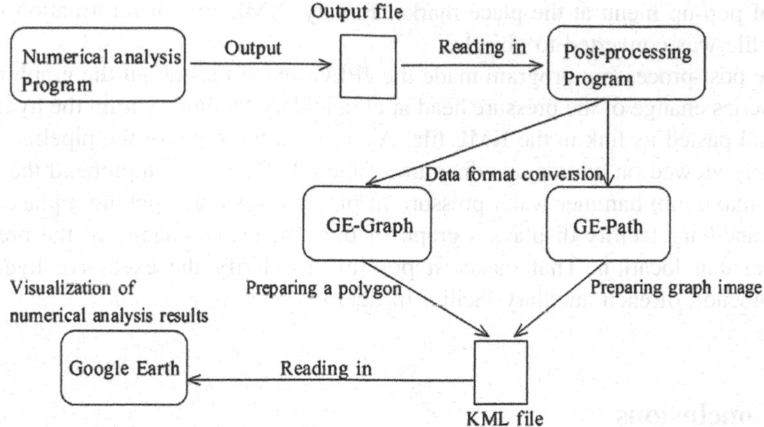

Fig. 9 Concept of the proposed work flow of post-processing

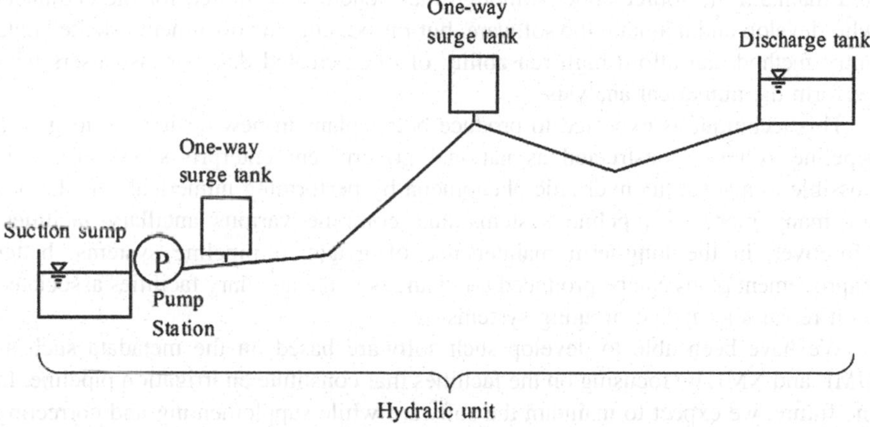

Fig. 10 Basic structure of the hydraulic unit as an analysis sample

pipeline system. The purpose of the example was to perform a numerical analysis to simulate the hydraulic transient phenomenon that occurs when two pumps are abruptly stopped by a sudden loss of power.

Specifically, the data entry work was done using Google Earth in cooperation with the pre-processing program. First, the place marks of ancillary facilities were prepared on the map of Google Earth. Next, in the pre-processing program, we corrected XML format information about the ancillary facilities pasted on the text area of pop-up menu at the place marks. Finally, XML format information in the KML file was converted to objects.

The post-processing program made the JPEG image files about the graph of the time-series change of the pressure head at all ancillary facilities within the hydraulic unit and pasted its link in the KML file. As a result, the route of the pipeline could be easily viewed on a birds-eye diagram of Google Earth to comprehend the value of the maximum hammer water pressure in pipeline system. Clicking a place mark of an ancillary facility displays a graph of the time-series change of the pressure head at that location. That makes it possible to clarify the excessive hydraulic phenomenon of each ancillary facility in detail.

5 Conclusions

Introducing object oriented modelling to design the numerical analysis of water hammer in the irrigation pipeline can afford the advantages which to easily expand and maintain the source code. Although that benefit was limited for the engineers who develop and maintain the software, but introducing the document oriented data entry method can afford high reusability of once created data for also users who perform the numerical analysis.

This technique is expected to produce better plans in new designs of irrigation pipeline systems constructed as national government enterprises, because it is possible to inspect the hydraulic phenomena by performing numerical calculations for many plans of pipeline systems that combine various ancillary facilities. Moreover, in the long-term maintenance of irrigation pipeline systems, better improvement plans can be produced for changes in the ancillary facilities associated with repair and update of aging systems.

We have been able to develop such software based on the metadata such as UML and XML by focusing on the facilities that constitute an irrigation pipeline. In the future, we expect to maintain the software while supplementing and correcting these metadata.

References

1. Ministry of Agriculture, Forestry and Fisheries. http://www.maff.go.jp/j/council/seisaku/nousin/gizyutu/h20_1/pdf/data04-1a.pdf
2. Ministry of Agriculture, Forestry and Fisheries. Rural Development Bureau, Rural Infrastructure Department, Design Division, Planning and design standards and operation of land improvement projects, commentary, and design, "Pipelines", pp. 322–378 (2009)
3. Stimming, C.: http://lapackpp.sourceforge.net/
4. IML++. http://math.nist.gov/iml++/
5. Tanaka, Y.: Proposal of development and maintenance management method based on object orientation of water hammer analysis program. IDRE J. **284**, 1–11

6. Murata, M.: XML[I]-XML schema and RELAX-. J. Inst. Electron. Inf. Commun. Eng. **84**(12), 890–894 (2001)
7. JACIC. https://www.jacic.or.jp/hyojun/densinouhin_model_1308.pdf
8. Ministry of Agriculture, Forestry and Fisheries. http://www.maff.go.jp/j/nousin/seko/nousin/seko/nouhin_youryou/doboku.html
9. Asami, T.: Web Development Using Java/XML. Peason Education (2001). (in Japanese)
10. Sgrillo, R.: http://www.sgrillo.net/googleearth/index.htm

A Health Policy Simulation Model of Ebola Haemorrhagic Fever and Zika Fever

Setsuya Kurahashi

Abstract This study proposes a simulation model of a new type of infectious disease based on Ebola haemorrhagic fever and Zika fever. SIR (Susceptible, Infected, Recovered) model has been widely used to analyse infectious diseases such as influenza, smallpox, bioterrorism, to name a few. On the other hand, Agent-based model begins to spread in recent years. The model enables to represent behaviour of each person in the computer. It also reveals the spread of an infection by simulation of the contact process among people in the model. The study designs a model based on Epstein's model in which several health policies are decided such as vaccine stocks, antiviral medicine stocks, the number of medical staff to infection control measures and so on. Furthermore, infectious simulation of Ebola haemorrhagic fever and Zika fever, which have not yet any effective vaccine, is also implemented in the model. As results of experiments using the model, it has been found that preventive vaccine, antiviral medicine stocks and the number of medical staff are crucial factors to prevent the spread. In addition, a modern city is vulnerable to Zika fever due to commuting by train.

Keywords Infectious disease · Ebola haemorrhagic fever · Zika fever

1 Introduction

Infectious diseases have been serious risk factors in human societies for centuries. Smallpox has been recorded in human history since more than B.C 1100. People have also been suffering from many other infectious diseases such as malaria, cholera, tuberculosis, typhus, AIDS, influenza, etc. Although people have tried to prevent and

S. Kurahashi (✉)
Graduate School of Business Sciences, University of Tsukuba,
3-29-1 Otsuka, Bunkyo, Tokyo, Japan
e-mail: kurahashi.setsuya.gf@u.tsukuba.ac.jp

G. Jezic et al. (eds.), *Agent and Multi-Agent Systems: Technology and Applications*, Smart Innovation, Systems and Technologies 58,
DOI 10.1007/978-3-319-39883-9_26

hopefully eradicate them, a risk of unknown infectious diseases including SARS, a new type of infectious diseases, as well as Ebola haemorrhagic fever and Zika fever have appeared on the scene.

A model of infectious disease has been studied for years. SIR (Susceptible, Infected, Recovered) model has been widely used to analyse such diseases based on a mathematical model. After an outbreak of SARS, the first SIR model of SARS was published and many researchers studied the epidemic of the disease using this model. When an outbreak of a new type of influenza is first reported, the U.S. government immediately starts an emergency action plan to estimate parameters of its SIR model. Nevertheless the SIR model has difficulty to analyse which measures are effective because the model has only one parameter to represent infectiveness. For example, it is difficult for the SIR model to evaluate the effect of temporary closing of classes because of the influenza epidemic. The agent-based approach or the individual-based approach has been adopted to conquer these problems in recent years [1–4]. The model enables to represent behaviour of each person. It also reveals the spread of an infection by simulation of the contact process among people in the model.

In this study, we developed a model to simulate smallpox and Ebola haemorrhagic fever and Zika fever based on the infectious disease studies using agent-based modelling. What we want to know is how to prevent an epidemic of infectious diseases not only using mechanisms of the epidemic but also decision making of health policy [5]. Most Importantly, we should make a decision in our modern society where people are on the move frequently world wide, so we can minimise the economic and human loss caused by the epidemic.

2 Cases of Infectious Disease

2.1 Smallpox

The smallpox virus affects the throat where it invades into the blood and hides in the body for about 12 days. Patients developed a high fever after that, but rashes do not appear until about 15 days after the infection. While not developing rashes, smallpox virus is able to infect others. After 15 days, red rashes break out on the face, arms and legs, and subsequently they spread over the entire body. When all rashes generate pus, patients suffer great pains; finally 30 % of patients succumb to the disease. For thousands of years, smallpox was a deadly disease that resulted in thousands of deaths.

2.2 *Ebola Haemorrhagic Fever*

A source of Ebola infection is allegedly by eating a bat or a monkey, but it is unknown whether the eating these animals is a source of the infection. The current epidemic, which began in Guinea in Dec. 2013, 23 people have died. The authorities of Guinea, Liberia and Sierra Leone have each launched a state committee of emergency and have taken measures to cope with the situation. The prohibition of entry over the boundary of Guinea is included in these measures.

There is a risk that a cough and a sneeze includes the virus, so the infection risk is high within 1 m in length of the cough or sneeze. The incubation period is normally 7 days, and then the person gets infected after showing the symptoms. The symptoms in the early stage are similar to influenza. They are fever, a headache, muscular pain, vomiting, diarrhoea, and a stomachache. The fatality rate is very high; 50 to 90 % There is no effective medical treatment medicine confirmed officially and several medicines are currently being tested. According to a guideline of WHO, the serum of a recovered patient is one of most effective treatments.

2.3 *Zika Fever*

Zika fever is an illness caused by Zika virus via the bite of mosquitoes. It can also be potentially spread by sex according to recent report [6, 7]. Most cases have no symptoms and present are usually mild including fever, red eyes, joint pain and a rash [8], but it is believed that the Zika fever may cause microcephaly which severely affects babies by a small head circumference.

3 Related Work

3.1 *Smallpox and Bioterrorism Simulation*

Epstein [9, 10] made a smallpox model based on 49 epidemics in Europe from 1950 to 1971. In the model, 100 families from two towns were surveyed. The family includes two parents and two children thus the population is each 400 from each town. All parents go to work in their town during the day except 10 % of adults who go to another town. All children attend school. There is a communal hospital serving the two towns in which each 5 people from each town work. This model was designed as an agent-based model, and then simulation of infectious disease was conducted using the model. As results of experiments showed that (1) in a base model in which any infectious disease measures were not taken, the epidemic spread within 82 days and 30 % of people died, (2) a trace vaccination measure was effective but it was difficult to trace all contacts to patients in an underground railway or an airport, (3) a mass vaccination measure was effective, but the number of vaccinations would be huge so it was not realistic, (4) epidemic quenching was also effective, and reactive household trace vaccination along with pre-emptive vaccination of hospital workers showed a dramatic effect.

3.2 Individual-Based Model for Infectious Diseases

Ohkusa [11] evaluated smallpox measures using an individual-based model of infectious diseases. The model supposed a town including 10,000 habitants and a public health centre. In the model, one person was infected with smallpox virus at a shopping mall. They compared between a trace vaccination measure and a mass vaccination measure. As a result of simulation, it was found that the effect of trace vaccination dropped if the early stage of infection was high and the number of medical staff is small, while the effect of mass vaccination was stable. Therefore timely and concentrate mass vaccination is required when virus starts spreading. The estimation about the number, place and time of infection is needed quickly and the preparation of an emergency medical treatment and estimation system is required for such occasions.

Summary of related work From these studies, the effectiveness of an agent-based model has been revealed, yet these are not sufficient models to consider a relationship between vaccination and antiviral medicine stocks, and the number of support medical staff and medicine from other countries. In addition, authorities need to make a decision regarding blockade, restrictions on outings including cars and railways while considering economic loss of the policy. This study takes into account these extensions.

4 A Health Policy Simulation Model of Infectious Disease

We designed a health policy simulation model of infectious disease based on Epstein's smallpox model. The model includes smallpox and Ebola haemorrhagic fever.

4.1 A Base Model of Smallpox

We assume all individuals to be susceptible which means no background of immunity. 100 families live in two towns.

The family includes two parents and two children. Therefore the population is each 400 in each town. All parents go to work in their town during the day except 10 % of adults commute to another town. All children attend school. There is a communal hospital serving two towns in which 5 people from each town work. Each round consists of an interaction through the entire agent population. The call order is randomised each random and agents are processed or activated, serially. On each round, when an agent is activated, she identifies her immediate neighbours for interaction. Each interaction results in a contact. In turn, that contact results in a transmission of the infection from the contacted agent to the active agent with probability.

The probability of contact at an interaction is 0.3 at a workplace and a school, while 1.0 at a home and a hospital. The probability of infection at a contact is 0.3 at a workplace and a school, while 1.0 at a home and a hospital. In the event the active agent contracts the disease, she turns blue to green and her own internal clock of disease progression begins. After twelve days, she will turn yellow and begins infecting others. Length of noncontagious period is 12 days, and early rash contagious period is 3 days. Unless the infected individual is vaccinated within four days of exposure, the vaccine is ineffective. At the end of day 15, smallpox rash is finally evident. Next day, individuals are assumed to hospitalize. After eight more days, during which they have a cumulative 30 % probability of mortality, surviving individuals recover and return to circulation permanently immune to further infection. Dead individuals are coloured black and placed in the morgue. Immune individuals are coloured white. Individuals are assumed to be twice as infectious during days 1 through 19 as during days 12 through 15.

4.2 A Base Model of Ebola Hemorrhagic Fever

In the event the active agent contracts the disease, she turns blue to green and her own internal clock of disease progression begins. After seven days, she will turn yellow and begins infecting others. However, her disease is not specified in this stage. After three days, she begins to have vomiting and diarrhoea and the disease is specified as Ebola. Unless the infected individual is dosed with antiviral medicine within three days of exposure, the medicine is ineffective. This is an imaginary medicine to play the policy game. At the end of day 12, individuals are assumed to hospitalize. After four more days, during which they have a cumulative 90 % probability of mortality, surviving individuals recover and return to circulation permanently immune to further infection. Dead individuals are coloured black and placed in the morgue. Immune individuals are coloured white. Other settings are the same as smallpox.

4.3 A Base Model of Zika Fever

About 80 % of cases have no symptoms which is called latent infection, but the latent patients can transmit Zika virus to other mosquitoes. The incubation period of Zika virus disease is not clear, which is likely to be 3 to 9 days. After the incubation period, the symptoms including fever, skin rashes, conjunctivitis, muscle and joint pain, malaise, and headache occur and last for 6 days. Zika virus disease is relatively mild and requires no specific treatment, so any strategies are not selected in the model.

Zika virus is transmitted to people through the bite of an infected mosquito from the *Aedes* genus. This is the same mosquito that transmits dengue. Therefore, the model of Zika fever is based on the life and habits of mosquitoes that transmit dengue. mosquitoes bite and thus spread infection at any time of day. Humans are the primary host of the virus, but it also circulates in mosquitoes. An infection can be acquired via a single bite. A mosquito that takes a blood meal from a person infected with Zika fever, becomes itself infected with the virus in the cells lining its gut. About 10 days later, the virus can be spread to other human. An adult mosquito can live for 30 days with the virus.

Mosquitoes live around each town, office, school in Zika fever model. The areas they live overlap with human areas. Therefore the Zika virus can be transmitted between mosquitoes and human. Additionally, mosquitoes also live around a rail station in another Zika fever model where people use a train to commute their offices every day. A mosquito bite human once per four days. An infection rate from a mosquito to human is set at 0.5 and from human to a mosquito is set at 1.0. Figure 1 shows Interface view of a simulation model of Zika fever.

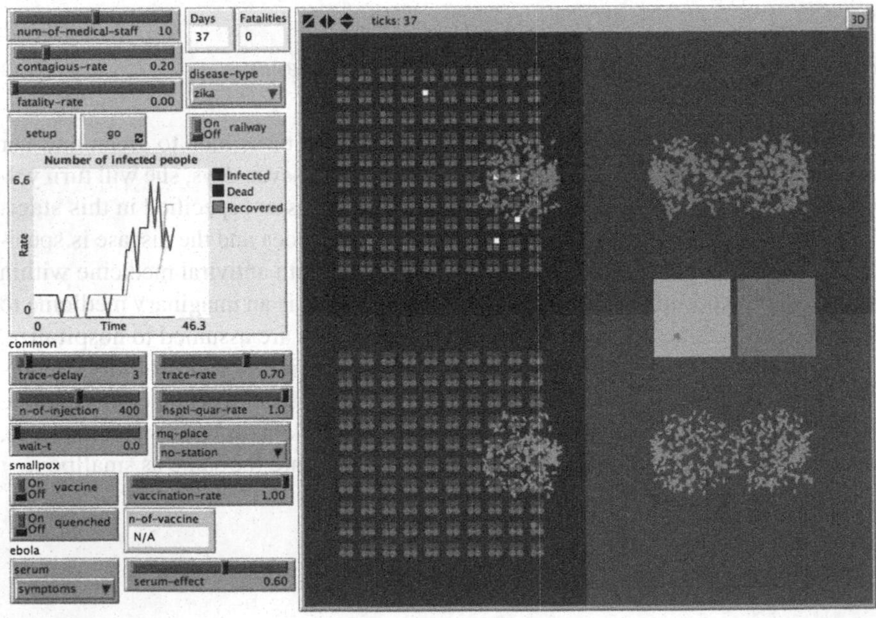

Fig. 1 Interface view of a health policy simulation model of Zika fever

4.4 Vaccination Strategies for Smallpox and Ebola Hemorrhagic Fever

The vaccination strategies we can select in the model are mass vaccination and trace vaccination. Each of them has advantages and disadvantages.

Mass vaccination As preemptive vaccination, the mass vaccination strategy adopts an indiscriminate approach. First all of the medical staff is vaccinated to prevent infection. When the first infected person is recognised, certain per cent of individuals in both towns will be vaccinated immediately. The vaccination rate and the upper limit number of vaccination per day are set on the model for the strategy.

Trace vaccination All of the medical staff is vaccinated as pre-emptive vaccination. Given a confirmed smallpox case, medical staff traces every contact of the infected person and vaccinates that group. In addition of the mass vaccination strategy, the trace rate and the delay days of contact tracing are able to be set according to the model for the trace vaccination strategy.

Trace serum or antiviral medicine dosing All of the medical staff is given serum or antiviral medicine as TAP (Target antivirus prophylaxis). Given a confirmed Ebola hemorrhagic fever case, medical staff traces every contact of the infected person and provides the medicine to that group. In addition to the mass vaccination strategy, the trace rate and the delay days of contact tracing are set according to the model.

5 Experimental Results

5.1 A Base Model of Ebola Hemorrhagic Fever

The process of infection in an Ebola hemorrhagic model is plotted in Fig. 2. The model employs non-intervention to the disease. A solid line, a dotted line and a line with marker indicate the number of infected, dead and people who have recovered respectively. When a player adopts non-intervention, it takes approximately 231 days until convergence of the outbreak and more than 716 people have died.

Figure 3 shows the results of 100 experiments. The number of fatalities under 20 people was 59 times out of 100 experiments, but the number of fatalities over 700 people in 800 inhabitants recorded 39 times. The result indicates that the infection phenomenon is not based on the normal distribution. It means we need to carefully consider all possibility and risk beyond the scope of the assumption regarding infection spread.

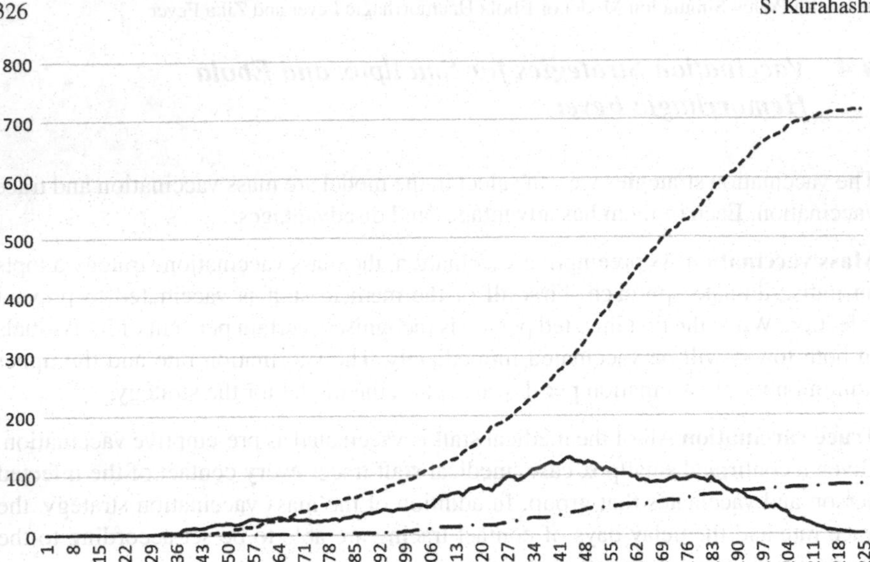

Fig. 2 The experimental result of the Ebola base model: non-intervention

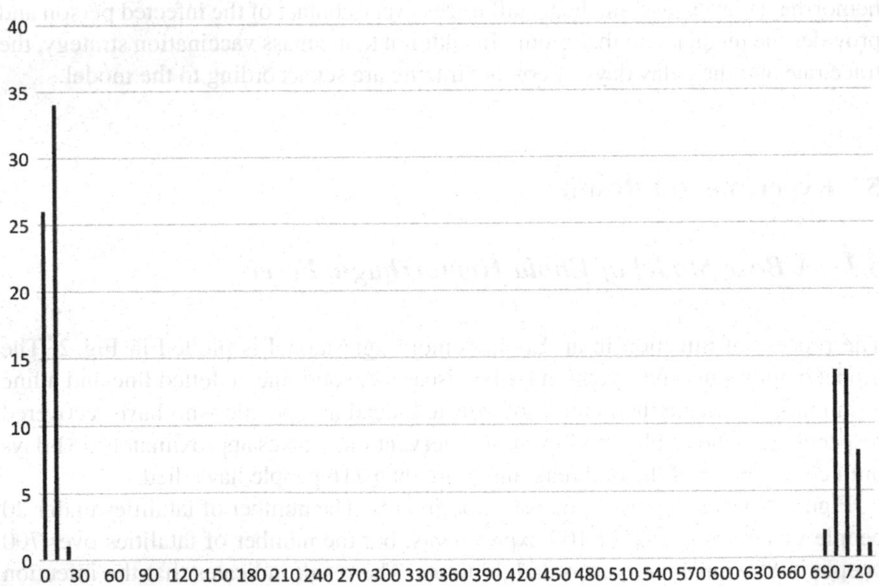

Fig. 3 The frequency chart of the Ebola base model: non-intervention

5.2 *Mass Vaccination Model*

The process of infection is plotted with the mass vaccination strategy in which individuals are vaccinated randomly after three days when given a confirmed smallpox case. The policy succeeds to prevent the outbreak because the number of vaccination per day is 600 and three fourths of inhabitants are vaccinated per day. On the other hand, the short ability of vaccination ends in failure because the number of vaccination per day is 400 and a half of inhabitants are vaccinated. The ability of vaccination per day bifurcates the results.

5.3 *Trace Vaccination Strategy*

It was found that the ability of more than 50 vaccinations per day was able to control an epidemic in most cases. In the case of using a public transportation to commute, however, it makes a substantial difference. 400 vaccinations per day could not prevent the outbreak, while 600 vaccinations per day succeeded to prevent it. Trace vaccination strategy is one of the most effective policies in a town where people commute by car, but a large number of vaccinations per day, at least a half of people, is required if most of people use public transportation systems like a railway and a bus.

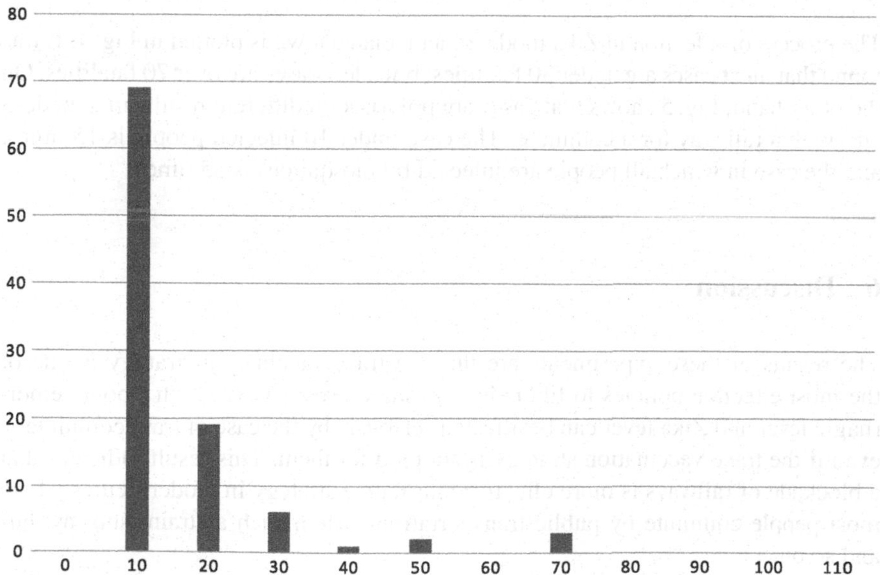

Fig. 4 The experimental result of Zika fever model without a railway

Fig. 5 The experimental result of Zika fever model with a railway

5.4 *Isolated Town Model of Zika Fever With/Without a Rail Station*

The process of infection in Zika model of an isolated town is plotted in Fig. 4. It was found that most cases are under 30 fatalities, but a few cases are over 70 fatalities. On the other hand, Fig. 5 shows that cases are polarised to different results in a modern city with a rail way for a commute. The case under 10 infected people is 15 times, and the case in which all people are infected by mosquitoes is 85 times.

6 Discussion

The results of these experiments are that (1) trace vaccination strategy is one of the most effective policies to Ebola hemorrhagic fever, but (2) both Ebola hemorrhagic fever and Zika fever can be affected seriously by the case of train commuters, even if the trace vaccination strategy is adopted for them. This result indicates that a blockade of railways is more effective alternative strategy in modern cities where most people commute by public transportation system such as train. subway, bus and so on.

7 Conclusion

This study proposes a simulation model of Ebola haemorrhagic fever and Zika fever. It also evaluates health policies to prevent an epidemic. As health policies, vaccine stocks, antiviral medicine stocks, the number of vaccinations per day by medical staff, mass vaccination, and trace vaccination are implemented in the model. As a result of experiments, it has been found that vaccination availability per day and the number of medical staff are crucial factors to prevent the spread for the mass vaccination strategy. On the other hand, small quantities of vaccination for approximately 10 % of inhabitants are vaccinated per day are able to control an epidemic in most cases. In the case of using a public transportation to commute, however, even if half of inhabitants were vaccinated per day, it would not prevent the epidemic. Two thirds of vaccinations per day are required each day to prevent the epidemic.

In addition to that, this study also conducts the model of Zika fever to understand the mechanism of spreading Zika virus via mosquitoes. As a result of experiments, it is crucial for preventing an Zika epidemic to get rid of mosquitoes from a train station where people use it to commute to their office.

References

1. Burke, D.S., et al.: Individual based computational modeling of smallpox epidemic control strategies. In: Academic Emergency Medicine, vol. 13, no. 11, pp. 1142–1149 (2006)
2. Longini, Jr., Ira, M., et al.: Containing a large bioterrorist smallpox attack: a computer simulation approach. Int. J. Infect. Dis. **11**(2), 98–108 (2007)
3. Gilbert, N.: Agent-Based Models, no. 153. Sage (2008)
4. Easley, D., Kleinberg, J.: Networks, crowds, and markets: reasoning about a highly connected world. Cambridge University Press (2010)
5. Okabe, N.: Risk and benefit of immunisation: infectious disease prevention with immunization. Iryo Shakai **21**(1), 33–40 (2011)
6. Chen, L.H., Hamer, D.H.: Zika Virus: rapid spread in the western hemisphere. Ann. Intern. Med. (2016)
7. Hennessey, M., Fischer, M., Staples, J.E.: Zika virus spreads to new areas—region of the americas, May 2015–Jan 2016. In: Centers for Disease Control and Prevention, vol. 65, no. 3, pp. 55–58 (2016)
8. WHO: Zika virus, World Health Organization, Media centre, Fact sheets, Jan 2016
9. Epstein, J.M., et al.: Toward a containment strategy for smallpox bioterror: an individual-based computational approach (2002)
10. Epstein, J.M.: Generative social science: studies in agent-based computational modeling. Princeton University Press (2006)
11. Ohkusa, Y.: An evaluation of counter measures for smallpox outbreak using an individual based model and taking into consideration the limitation of human resources of public health workers. Iryo Shakai **16**(3), 284–295 (2007)

Analyzing the Influence of Indexing Strategies on Investors' Behavior and Asset Pricing Through Agent-Based Modeling: Smart Beta and Financial Markets

Hiroshi Takahashi

Abstract This study analyzes the influence of indexing strategy on investors' behavior and financial markets through agent-based modeling. In this analysis, I focus on smart beta index, which is proposed as a new stock index and condsidered to have better characteristics than traditional market capitalization-weighted indices. As a result of intensive computational simulation studies, we have concluded that a smart beta strategy is effective even in cases where the initial number of smart beta investors is small. This study also finds a significant relationship between the number of smart beta investors and trading volume. These results are significant from both practical and academic viewpoints.

Keywords Financial markets · Stock indices · Smart beta · Agent-based model · Asset management business

1 Background

Stock indices play a significant role in the asset management business. The arguments about the role of stock indices have a close relationship to arguments in traditional financial theories [3, 9, 14]. According to the concept of market efficiencies—which is the central hypothesis in traditional financial theories—, a strategy which follows a market index is supposed to be the most optimal investment strategy [6, 14]. Along with the progress of these theories, many arguments regarding securities investment in practical business affairs have been actively discussed and various kinds of analyses focusing on stock indices have been conducted [8, 14–16].

This research investigates the influence of stock indices through agent-based modeling. In particular, this analysis focuses on smart beta, which has been proposed as a new benchmark for investments in place of market capitalization-weighted indices

H. Takahashi (✉)
Graduate School of Business Administration, Keio University, 4-1-1 Hiyoshi, Kohoku-ku, Yokohama 223-8572, Japan
e-mail: htaka@kbs.keio.ac.jp

G. Jezic et al. (eds.), *Agent and Multi-Agent Systems: Technology and Applications*, Smart Innovation, Systems and Technologies 58,
DOI 10.1007/978-3-319-39883-9_27

(price indices), which are currently employed in practical business affairs. Although smart beta has been becoming popular in the asset management business, the influence of smart beta on financial markets is not clear. This study investigates market behavior by changing the initial number of smart beta investors to understand the influence of those strategies on finanicial markets.

In this study, we examine the influence of smart beta through agent-based modeling (social simulation) [2, 18, 20, 21]. Among several types of smart beta, this study focuses on fundamental indexing, which is one of the most popular forms of smart beta [1]. A fundamental index is calculated based on the value of a company's financial statements—profit and so on—instead of the market price that is commonly used in business affairs. Compared to market capitalization-weighted indices, fundamental indexing is relatively unaffected by the decision bias of investors and could have better characteristics than traditional stock indexing. Furthermore, fundamental indexing could actually contribute to market efficiency. Therefore, the analysis of fundamental indexing is significant from both academic and practical points of view. The next section describes the model used in this analysis. Section 3 shows the results of the analysis. Section 4 summarizes this paper.

2 Model

A computer simulation of the financial market involving 1000 investors was used as the model for this research.[1] Shares and risk-free assets were the two types of assets used, along with the possible transaction methods. Several types of investors exist in the market, each undertaking transactions based on their own stock evaluations. This market was composed of three major stages: (1) generation of corporate earnings; (2) formation of investor forecasts; and (3) setting transaction prices. The market advances through repetition of these stages. The following sections describe negotiable transaction assets, modeling of investor behavior, setting of transaction prices, and the rule of natural selection.

2.1 *Negotiable Assets in the Market*

This market has risk-free and risk assets. There are risky assets in which all profits gained during each term are distributed to the shareholders. Corporate earnings (y_t) are expressed as $y_t = y_{t-1} \cdot (1 + \varepsilon_t)$ [13]. However, they are generated according to the process of $\varepsilon_t \sim N(0, \sigma_y^2)$ with share trading being undertaken after the public announcement of profits for the term. Each investor is given common asset holdings at the start of the term with no limit placed on debit and credit transactions (1000 in

[1] I built a virtual financial market on a personal computer with i7 4790 3.6 GHz, RAM32 GB. The simulation background is financial theory [5, 10].

Table 1 List of investors type

No.	Investor type
1	Fundamentalist
2	Forecasting by latest price
3	Forecasting by trend (most recent 10 days)
4	Forecasting by past average (most recent 10 days)
5	Smart beta investor

risk-free assets and 1000 in stocks). Investors adopt the buy-and-hold method for the relevant portfolio as a benchmark to conduct decision-making by using a one-term model.[2]

2.2 Modeling Investor Behavior

Each type of investor considered in this analysis is organized in Table 1.[3] Type 1–4 are active investors and type 5 is smart beta investor. This section describes the investors' behavior.

Smart Beta Investor This section describes the behavior of smart beta investors who try to follow a fundamental index. In this model, a fundamental index is calculated based on the value of company's profit. Smart beta investors automatically decide their investment ratio in stock based on the index's value. Smart beta investors decide their investment ratio in stock based on the value of the fundamental index.[4]

Active Investors Active investors in this market evaluate transaction prices based on their own market forecasts, taking into consideration both risk and return rates when making decisions. Each investor determines the investment ratio (w_t^i) based on the maximum objective function($f(w^i t)$), as shown below.[5]

$$f(w_t^i) = r_{t+1}^{int,i} \cdot w_t^i + r_f \cdot (1 - w_t^i) - \lambda(\sigma_{t-1}^{s,i})^2 \cdot (w_t^i)^2. \quad (1)$$

Here, $r_{t+1}^{int,i}$ and $\sigma_{t-1}^{s,i}$ expresses the expected rate of return and risk for stocks as estimated by each investor i. r_f indicates the risk-free rate. w_t^i represents the stock investment ratio of the investor i for term t. The investor decision-making model here is

[2] 'Buy-and-hold' is an investment method to hold shares for the medium to long term.

[3] This analysis covers the major types of investor behavior [16].

[4] When market prices coincide with fundamental value, a fundamental index has the same score with a capitalization-weighted index.

[5] The value of objective function $f(w_t^i)$ depends on the investment ratio(w_t^i). The investor decision-making model here is based on the Black/Litterman model that is used in securities investment [4, 11, 12].

based on the Black/Litterman model that is used in securities investment [4]. The expected rate of return for shares is calculated as follows.

$$r^{int,i}_{t+1} = (r^{f,i}_{t+1} c^{-1}(\sigma^i_{t-1})^{-2} + r^{im}_t(\sigma^i_{t-1})^{-2})/(1 \cdot c^{-1}(\sigma^i_{t-1})^{-2} + 1 \cdot (\sigma^i_{t-1})^{-2}). \quad (2)$$

Here, $r^{f,i}_{t+1}$ Cr^{im}_t expresses the expected rate of return, calculated from the short-term expected rate of return, plus risk and gross price ratio of stocks respectively. c is a coefficient that adjusts the dispersion level of the expected rate of return calculated from risk and gross current price ratio of stocks [4].

The short-term expected rate of return ($r^{f,i}_t$) is obtained where ($P^{f,i}_{t+1}, y^{f,i}_{t+1}$) is the equity price and profit forecast for term $t+1$ is estimated by the investor, as shown below:

$$r^{f,i}_{t+1} = ((P^{f,i}_{t+1} + y^{f,i}_{t+1})/P_t - 1) \cdot (1 + \eta^i_t). \quad (3)$$

The short-term expected rate of return includes the error term ($\eta^i_t \sim N(0, \sigma^2_n)$) reflecting that even investors using the same forecast model vary slightly in their detailed outlook. The stock price ($P^{f,i}_{t+1}$), profit forecast ($y^{f,i}_{t+1}$), and risk estimation methods are described below.

The expected rate of return obtained from stock risk and so forth is calculated from stock risk (σ^i_{t-1}), benchmark equity stake (W_{t-1}), investors' degree of risk avoidance (λ), and risk-free rate (r_f), as shown below [4].

$$r^{im}_t = 2\lambda(\sigma^s_{t-1})^2 W_{t-1} + r_f. \quad (4)$$

Stock Price Forecasting Method The fundamental value is estimated by using the discount cash flow model(DCF), which is a typical model in the field of finance. Fundamentalists estimate the forecasted stock price and forecasted profit from the profit for the term (y_t) and the discount rate (δ) as $P^{f,i}_{t+1} = y_t/\delta, y^{f,i}_{t+1} = y_t$. In this analysis, fudamentalists know that a profit of company follows the stochastic process.

Forecasting based on trends involves forecasting the following term's stock prices and profit through extrapolation of the most recent stock value fluctuation trends. The following term's stock price and profit is estimated from the most recent trends of stock price fluctuation (a_{t-1}) from time point $t-1$ as $P^{f,i}_{t+1} = P_{t-1} \cdot (1 + a_{t-1})^2, y^{f,i}_{t+1} = y_t \cdot (1 + a_{t-1})$.

Forecasting based on past averages involves estimating the following term's stock prices and profit based on the most recent average stock value.

Risk Estimation Method In this analysis, each investor estimates risk from past price fluctuations. Specifically, stock risk is estimated as $\sigma^i_{t-1} = \sigma^h_{t-1}$ (common to each investor). Here, σ^h_{t-1} represents the stock volatility that is calculated from price fluctuation from the most recent 100 terms.

2.3 Determination of Transaction Prices

Transaction prices are determined as the price where stock supply and demand converge ($\sum_{i=1}^{M}(F_t^i w_t^i)/P_t = N$) [17]. In this case, the total asset (F_t^i) of investor i is calculated from transaction price (P_t) for term t, profit (y_t) and total assets from the term $t-1$, stock investment ratio (w_{t-1}^i), and risk-free rate (r_f), as $F_t^i = F_{t-1}^i \cdot (w_{t-1}^i \cdot (P_t + y_t)/P_{t-1} + \left(1 - w_{t-1}^i\right) \cdot (1 + r_f))$.

2.4 Rules of Natural Selection in the Market

The rules of natural selection can be identified in this market. The driving force behind these rules is cumulative excess profit [7, 17]. The rules of natural selection go through the following two stages: (1) the identification of investors who alter their investment strategy, and (2) the actual alteration of investment strategy [19].

Each investor determines the existence of investment strategy alteration based on the most recent performance of each 5 term period after 25 terms have passed since the beginning of market transactions.[6] The higher the profit rate obtained most recently is, the lesser the possibility of strategy alteration becomes. The lower the profit, the higher the possibility becomes. Specifically, when an investor cannot obtain a positive excess profit for the benchmark portfolio, they are likely to alter their investment strategy with the probability below:

$$p_i = \min(1, \max(-100 \cdot r^{cum}, 0)). \tag{5}$$

Here, however, r_i^{cum} is the cumulative excess profitability for the most recent benchmark of investor i. Measurement was conducted for 5 terms, and the cumulative excess profit was converted to a period of one term.

When it comes to determining new investment strategies, an investment strategy that has had a high cumulative excess profit for the most five recent terms (forecasting type) is more likely to be selected. Where the strategy of the investor i is z_i and the cumulative excess profit for the most recent five terms is r_i^{cum}, the probability p_i that z_i is selected as a new investment strategy is given as $p_i = e^{(a \cdot r_i^{cum})} / \sum_{j=1}^{M} e^{(a \cdot r_j^{cum})}$.[7] Those investors who altered their strategies make investments based on the new strategies after the next step.

[6] In the actual market, evaluation tends to be conducted according to baseline profit and loss.

[7] Selection pressures on an investment strategy become higher as the value of coefficient a increases.

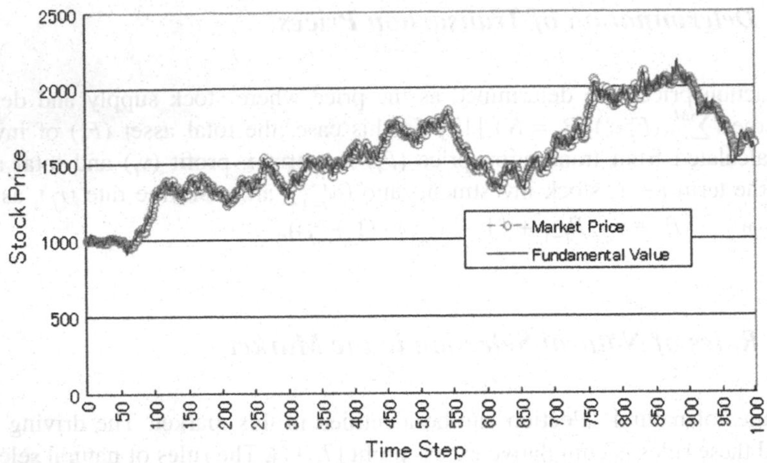

Fig. 1 Price transitions (# of smart beta = 20)

3 Result

This section analyzes the influence of a smart beta strategy on financial markets. In this study, I focus on fundametal indexing as a smart beta strategy. Firstly, this section analyzes the case where there is a small number of smart beta investors in the market; secondly, this study analyzes the cases what happens when the initial number of smart beta investors increase.

3.1 With a Small Number of Smart Beta Investors

This section analyzes the case where there is a small number of smart beta investors in the market. The number of each type of investor is as follows; fundamental index:fundamentarist:latest price:momentum trader:average trader $= 20:245:245:245:245$. Figures 1 and 2 show the typical transition of market price, volume and the number of investors with investors categorized into 5 types[8] (Table 1: FType1–5).

These results suggest that share prices closely correspond to fundamental value and the number of smart beta investors is increasing throughout the periods even though the initial number of smart beta is small. These results exhibit the validity of smart beta strategies.[9] A closer look at Fig. 2 reveals that the market situation can be divided into 3 periods, as follows: (1) the period when the number of both

[8]Investors who get excess return tend to survive in this market.

[9]This result coincides with Takahshi [22] which attempts to analyze the market with simplified conditions.

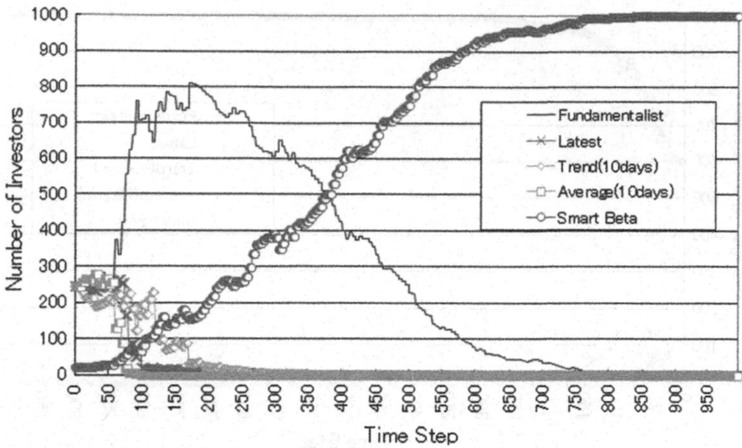

Fig. 2 Transition of the number of investors (# of smart beta = 20)

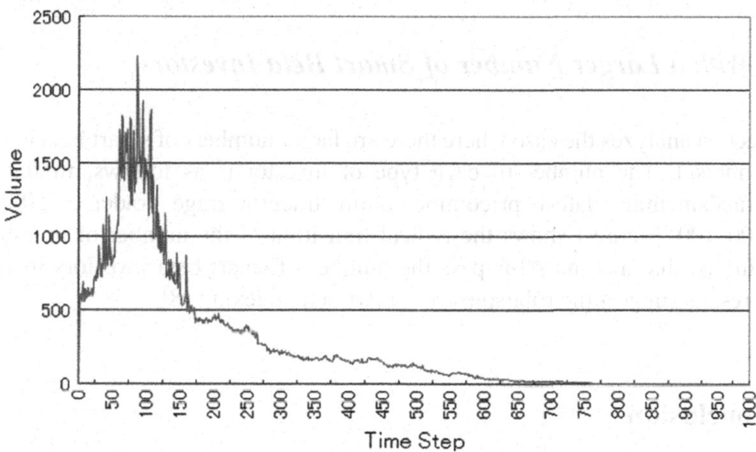

Fig. 3 Trading volume (# of smart beta = 20)

fundamentalists and smart beta investors increases; (2) the period when the number of fundamentalists decreases; (3) the period when all investors employ smart beta strategies. These transitions of market conditions shows that the effectiveness of investment strategies could depend on market conditions, including the actual mix and ratio of investor types.

Figure 3 shows the transition of volume. This result suggests that as the number of smart beta investors increases, trading volume decreases. These results suggest a relationship to arguments of paradox about market efficiencies [8].

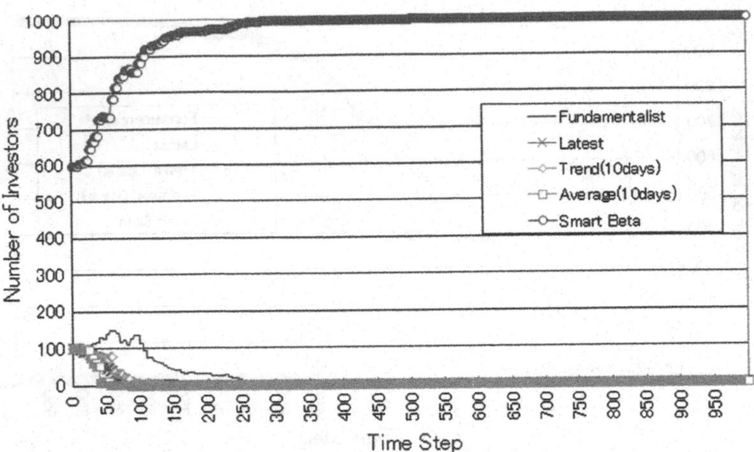

Fig. 4 Transition of the number of investors (# of smart beta = 600)

3.2 *With a Larger Number of Smart Beta Investors*

This section analyzes the case where there are larger numbers of smart beta investors in the market. The number of each type of investor is as follows; fundamental index:fundamentalist:latest price:momentum trader:average trader $= 600:100:100:100:100$. Figure 4 shows the typical transition of the number of investors. It is confirmed that as time steps pass the number of smart beta investors increases. These results suggest the robustness of smart beta indexing [8].

4 Conclusion

This research analyzes the influence of smart beta on financial markets through agent-based modeling. To address this problem, I conducted computational simulation studies and made the following finding: smart beta strategy is effective even in cases where the initial number of investors employing a smart beta strategy is small. This study also confirms a significant relationship between the number of smart beta investors and trading volume. This research has attempted to tackle the problem that the practical asset management business faces, and has produced results which may prove significant from both academic and practical points of view. These analyses also demonstrate the effectiveness of agent-based techniques for financial research. More detailed analyses relating to trading volume, market design and several other matters of interest are planned for the future.

Acknowledgments This research was supported by a grant-in-aid from Zengin Foundation for Studies on Economics and Finance.

A List of Main Parameters

This section lists the major parameters of the financial market designed for this paper. Explanations and values for each parameter are described.

M: Number of investors (1000)
N: Number of shares (1000)
F_t^i: Total asset value of investor i for term t ($F_0^i = 2000$: common)
W_t: Ratio of stock in benchmark for term t ($W_0 = 0.5$)
w_t^i: Stock investment rate of investor i for term t ($w_0^i = 0.5$: common)
y_t: Profits generated during term t ($y_0 = 0.5$)
σ_y: Standard deviation of profit fluctuation ($0.2/\sqrt{200}$)
δ: Discount rate for stock (0.1/200)
λ: Degree of investor risk aversion (1.25)
σ_n: Standard deviation of dispersion from short-term expected rate of return on shares (0.01)
c: Adjustment coefficient (0.01)

References

1. Arnott, R., Hsu, J., Moore, P.: Fundamental indexation. Financ. Anal. J. **61**(2), 83–99 (2005)
2. Axelrod, R.: The Complexity of Cooperation—Agent-Based Model of Competition and Collaboration, Princeton University Press (1997)
3. Black, F., Scholes, M.: Pricing of options and corporate liabilities. Bell J. Econ. Manag. Sci. **4**, 141–183 (1973)
4. Black, F., Litterman, R.: Global Portfolio Optimization. Financ. Anal. J. pp. 28–43 (1992). September–October
5. Brealey, R., Myers, S., Allen, F.: Principles of Corporate Finance, 8E, The McGraw-Hill (2006)
6. Fama, E.: Efficient capital markets: a review of theory and empirical work. J. Financ. **25**, 383–417 (1970)
7. Goldberg, D.: Genetic Algorithms in Search, Optimization, and Machine Learning, Addison-Wesley (1989)
8. Grossman, S.J., Stiglitz, J.E.: Information and competitive price systems. Am. Econ. Rev. **66**, 246–253 (1976)
9. Ingersoll, J.E.: Theory of Financial Decision Making, Rowman & Littlefield (1987)
10. Luenberger, D.G.: Investment Science, Oxford University Press (2000)
11. Martellini, L., Ziemann, V.: Extending black-litterman analysis beyond the mean-variance framework: an application to hedge fund style active allocation decisions. J. Portf. Manag. **33**(4), 33–45 (2007)
12. Meucci, A.: Beyond black-litterman in practice. Risk **19**, 114–119 (2006)
13. O'Brien, P.: Analysts' Forecasts as Earnings Expectations. J. Account. Econ. pp. 53–83 (1988)
14. Sharpe, W.F.: Capital asset prices: a theory of market equilibrium under condition of risk. J. Financ. **19**, 425–442 (1964)
15. Shiller, R.J.: Irrational Exuberance, Princeton University Press (2000)
16. Shleifer, A.: Inefficient Markets, Oxford University Press (2000)

17. Takahashi, H., Terano, T.: Agent-based approach to investors' behavior and asset price fluctuation in financial markets. J. Artif. Soc. Soc. Simul. **6**, 3 (2003)
18. Takahashi, H., Takahashi, S., Terano, T.: Analyzing the influences of passive investment strategies on financial markets via agent-based modeling. In: Edmonds, B., Hernandes, C., Troitzsch, K. (eds.) Social Simulation Technologies: Advances and New Discoveries (Representing the best of the European Social Simulation Association conferences), Idea Group Inc. (2007)
19. Takahashi, H.: An analysis of the influence of fundamental values' estimation accuracy on financial markets. J. Probab. Stat. pp. 17 (2010). doi:10.1155/2010/543065
20. Takahashi, H., Takahashi, S., Terano, T.: Analyzing the validity of passive investment strategies employing fundamental indices through agent-based simulation. In: Agent and Multi-Agent Systems: Technologies and Applications (LNAI6682), pp. 180–189, Springer-Verlag (2011)
21. Takahashi, H.: An Analysis of the Influence of dispersion of valuations on Financial Markets through agent-based modeling. Int. J. Inf. Technol. Decis. Mak. **11**, pp. 143–166 (2012)
22. Takahashi, H.: Analyzing the influence of market conditions on the effectiveness of smart beta. In: Agent and Multi-Agent Systems: Technologies and Applications, Smart Innovation, Systems and Technologies, **38**, pp. 417–426. Springer-Verlag (2015)

Text Analysis System for Measuring the Influence of News Articles on Intraday Price Changes in Financial Markets

Keiichi Goshima and Hiroshi Takahashi

Abstract This study constructs a text analysis system for analyzing financial markets. This system enables us to investigate the influence of news article on intraday price changes. In this study, we examine the automobile companies in Japan to analyze the relationship between news articles and stock price reactions. As a result of empirical analyses, we confirmed that stock prices reflect news information in a timely manner. These results are suggestive from both academic and practical view points. More detailed analyses are planned for the future.

Keywords Finance · Natural language processing · Market micro structure · Asset pricing · Text mining · Asset management business

1 Introduction

In the asset management business, evaluating financial assets such as stocks and futures is a major concern [2, 13, 17, 20]. In investing in financial assets, it is essential to assess securities swiftly and correctly using a variety of information sources, such as financial statements and so on [3, 12, 19]. In the practical asset management business, institutional investors make their investment decisions by utilizing various kinds of information, including textual information from media outlets such as

[1]In parallel with the rapid progress of computational technologies, various kinds of methods proposed in computer science -such as machine learning, agent-based modeling and network analysis- have been applied to financial research [1, 15, 18, 21, 22].

K. Goshima (✉)
Interdisciplinary Graduate School of Science and Engineering,
Tokyo Institute of Technology, 4259 Nagatsuda, Midori-ku, Yokohama, Japan
e-mail: goshima.k.aa@m.titech.ac.jp

H. Takahashi
Graduate School of Business Administration, Keio University, 4-1-1 Hiyoshi,
Yokohama, Kohoku-ku 223-8572, Japan
e-mail: htaka@kbs.keio.ac.jp

G. Jezic et al. (eds.), *Agent and Multi-Agent Systems: Technology and Applications*, Smart Innovation, Systems and Technologies 58,
DOI 10.1007/978-3-319-39883-9_28

newspapers, in addition to numerical data.[1] Also, there are several analyses suggesting that textual information conveys different information to financial markets when compared to numerical information [5, 8, 9, 11, 14, 16, 23, 24].

With this in mind, this research focuses on headline news as a source of information for investors, and analyses the influence of headline news on stock markets. The remainder of the paper is organized as follows: First, we detail data and text analysis system constructed in this study. Next, we show some empirical results confirmed through this system. Finally, we summarize this paper.

2 Data

The objective of this paper is to construct a text analysis system investigating the reaction of stock markets to headline news transmitted through, for example, the computer-based news services of Reuters and Bloomberg. Headline news is one of the most important information sources for a fund manager. We use the news data (2009–2010) offered from Thomson Reuters which is one of the biggest news agencies in the world. Naturally, the content of news articles includes many different fields, such as business trends, macroeconomic indicators, the foreign market trends, information on individual stocks etc. Our study focused on the Japanese news articles in relation to Japanese stock market. We also used tag information relating to Reuters news (date time, ticker symbol).

3 Text Analysis System for Financial Markets

In this paper, we construct a text analysis sytem for financial markets. One of the most salient features of this system is our ability to analyze the relationship between news articles and intraday price changes (tick data).

Figure 1 gives an outline of the system. This system analyzes whether the information disseminated by news has an impact or not on the intraday price changes of a target company. The impact of this information naturally depends on its content. We split news articles into three basic categories: positive negative and neutral.

The system consists of the following five procedures;

1. Measurement of market reactions through the event study analysis [4, 7] (Fig. 2)
2. Constructing classification algorithms from market reactions to news articles through surpervised learnings[2] [1, 6] (Fig. 3)

[2]In this study, we selected Support Vector Regression with the linear kernel as a supervised learning. But our proposed system can work with other kinds of supervised learning.

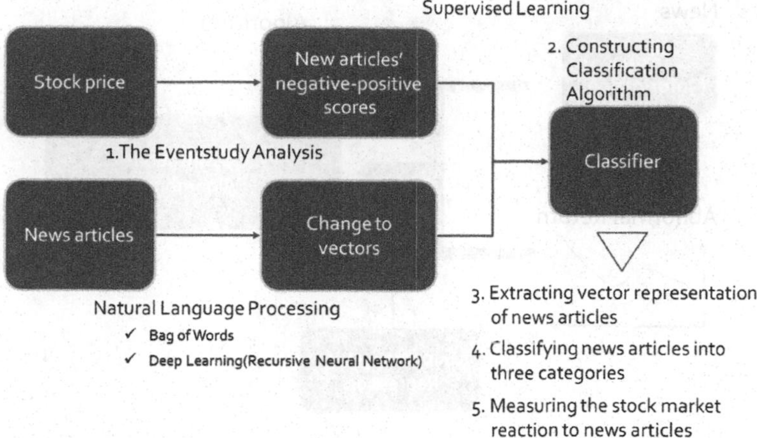

Fig. 1 Outline of procedure of text analysis system

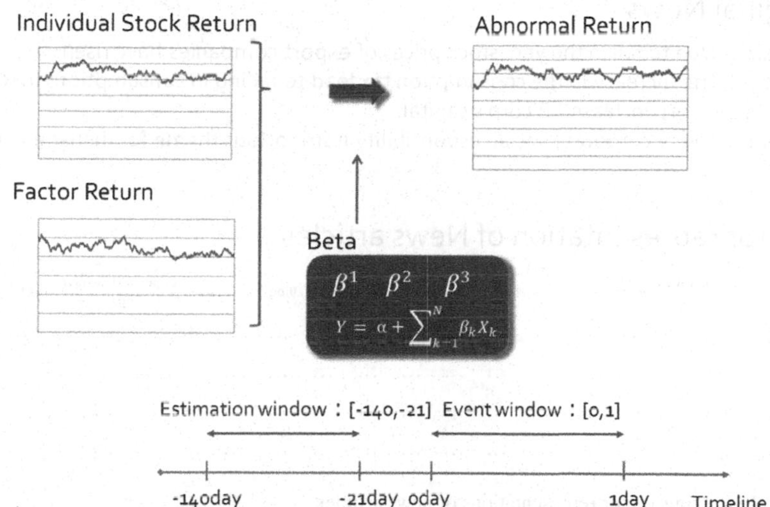

Fig. 2 Measurement of market reactions to news articles: the event study analysis

3. Extracting vector representations of news articles (Fig. 4)
4. Classification of each news article into three categories (Fig. 5)
5. Measuring market intraday-reactions to news articles using data of intraday price changes (Fig. 6).

In procedure 1 and 2, the system constructs classification algorithms[3] (positive news, neutral news, negative news). There are several alternatives for the natural

[3]In procedures 1 and 2, we analyze all Japanese news articles in relation to the Japanese stock market during the sample periods.

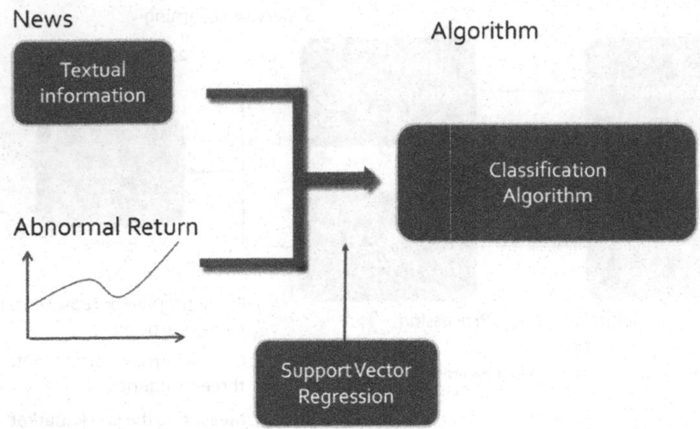

Fig. 3 Construction of classification algorithms

Original News

News.1 : Due to fall in the yen, stock prices of export companies have risen.
News.2 : The increase in the consumption tax lead to falling in consumptions in retail.
News.3 : A corp incleased B corp's capital.
News.4 : Today corporate social responsibility is important theme for management.

Vector representation of News articles

	capital.	companies	consumption	consumptions	corp's	corp	corporate	due	export	fall	falling	for	have	important	incleased	increase	lead	management.
News.1	0.00	...	...	...	...	...	...	...	...	...	...	...	...	...	...	...	...	0.00
News.2	...	...	...	...	...	...	...	...	...	...	...	...	...	...	...	...	...	...
News.3	...	...	...	...	...	...	...	...	...	...	...	...	...	...	...	...	...	...
News.4	0.00	...	...	...	...	...	...	...	...	...	...	...	...	...	...	...	...	0.00

Fig. 4 Extracting vector representations of news articles

language processing model. This study employs a 'bag-of-words' model, which is one of the most popular model to break down news articles into a representational vector.[4] In procedure 3–5, the system analyzes the reaction of the market to the news articles.

[4]Although, we don't employ a deep learning model in this study, in another article [10], we employ a recursive neural network model to analyze the relationship between news and stock indices. Detailed analysis using both a recursive neural network model and intraday price changes is one of our future plans.

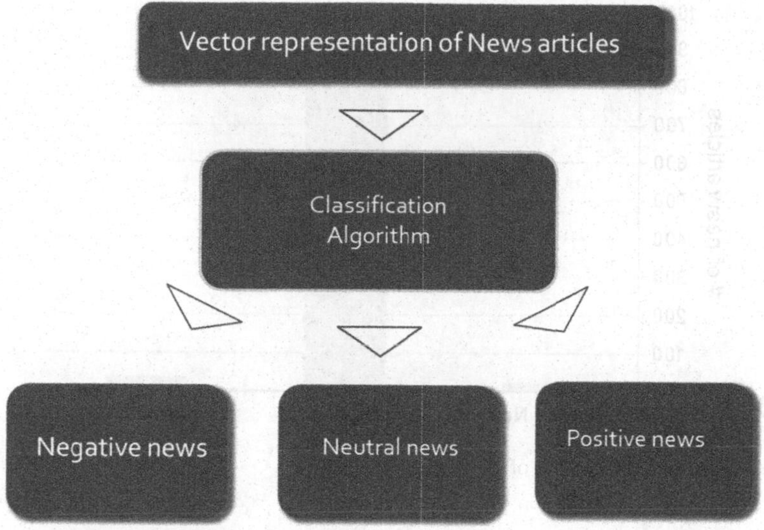

Fig. 5 Classification of news articles

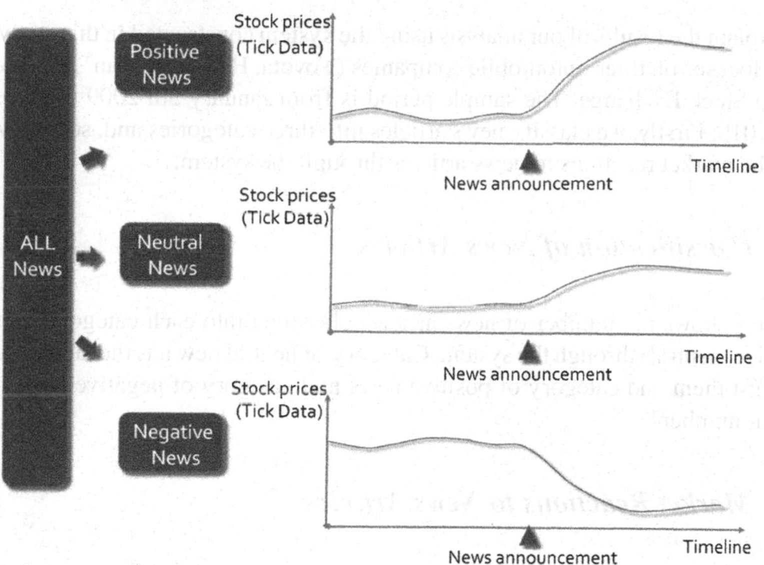

Fig. 6 Measuring market reactions to news articles with intraday price change data (tick data)

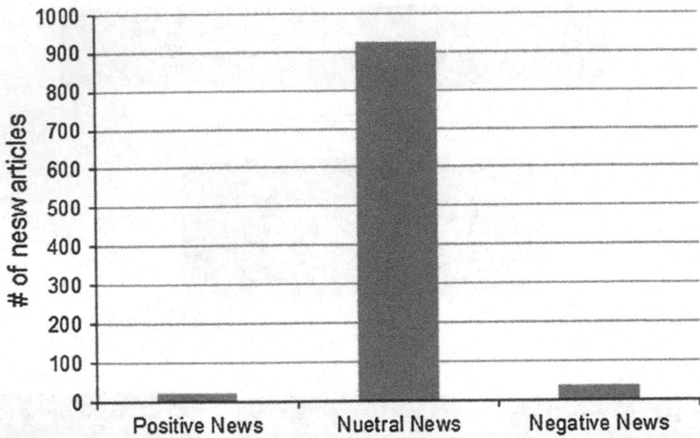

Fig. 7 Number of news articles of each category

4 Results

We explain the results of our analysis using the system constructed in this study. This study focuses on three automobile companies (Toyota, Honda, Nissan[5]) listed on the Tokyo Stock Exchange. The sample period is from January 5th 2009 to December 30th 2010. Firstly, we classify news articles into three categories and, secondly, analyze the market reactions to news articles through the system.

4.1 Classification of News Articles

Figure 7 shows the number of news articles classified into each category (positive, negative, neutral) through the system. Category of neutral new has the largest number amongst them and category of positive news and category of negative news have a similar number.[6]

4.2 Market Reactions to News Articles

Figure 8 shows the market reaction to news articles through the system.[7] The x-axis shows time step(min.) and the y-axsis shows market reactions. There is a rapid change in stock prices from -1 to $+1$ s. This result suggests that the stock market reflects news information in a timely manner.

[5]These companies are ranked in the top three companies in the automobile sector in Japan.

[6]The number of each category crucially depends on the classification procedure (Figs. 3, 5). A detailed analysis is planned for the future.

[7]Time = 0 in x-axis shows the point when news articles are released to investors.

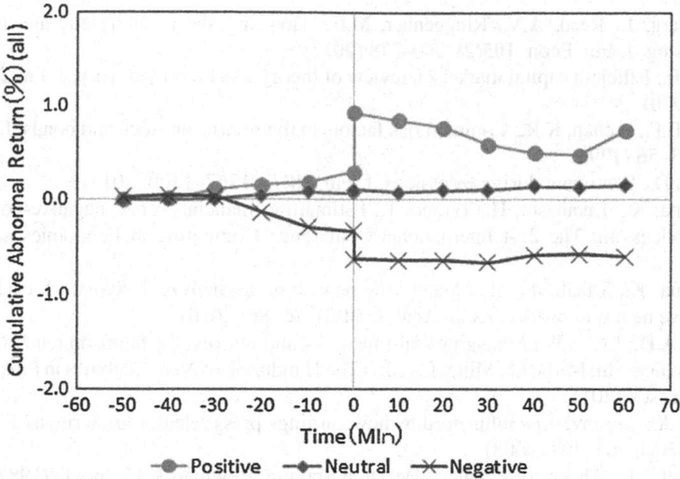

Fig. 8 Intraday price changes after the announcement of news articles (automobile companies)

5 Summary

This study constructs a text analysis system which enables us to analyze the influence of news articles—one of the most frequently used sources of information in asset management—on intraday price changes in financial markets. The system we constructed here consists of several procedures(measurement of market reactions, construction of classification algorithms, extracting vector representation of news articles, classification, measuring market reactions). In this study, we also attempted to analyze the relationship between news articles and intraday stock price fluctuations focusing on several automobile companies (Toyota, Honda, Nissan) and confirmed that stock prices do indeed react significantly. These results are suggestive from both academic and practical view points. The dataset for this study is limited. Analyzing all listed companies is one of our future projects.

Acknowledgments This research was supported by a grant-in-aid from the Telecommunications Advancement Foundation.

References

1. Bishop, C.M.: Pattern Recognition and Machine Learning. Springer (2006)
2. Black, F., Scholes, M.: The pricing of options and corporate liabilities. J. Polit. Econ. **81**(3), 637–654 (1973)
3. Brealey, R.A.: Principles of Corporate Finance 8th Edition. Irwin/McGraw-Hill (2006)
4. Campbell, J.Y., Lo, A.W., MacKinlay, A.C. : The Econometrics of Financial Markets. Princeton University Press (1997)

5. Engelberg, J., Reed, A.V., Ringgenber, M.C.: How are shorts informed? and information processing. J. Fin. Econ. **105**(2), 260–278 (2012)
6. Fama, E.: Efficient capital markets: a review of theory and empirical work. J. Fin. **25**(2), 383–417 (1970)
7. Fama, E.F., French, K.R.: Common risk factors in the returns on stock and bonds. J. Fin. Econ. **33**(1), 3–56 (1993)
8. Garcia, D.: Sentiment during recessions. J. Fin. **68**(3), 1267–1300 (2013)
9. Goshima, K., Takahashi, H., Terano, T.: Estimating financial words' negative-positive from stock prices. In: The 21st International Conference Computing in Economics and Finance (2015)
10. Goshima, K., Takahashi, H.: Quantifying news tone to analyze Tokyo stock exchange with recursive neural networks. Secur. Anal. J. **54**(3), 76–86 (2016)
11. Healy, A.D., Lo, A.W.: Managing real-time risks and returns: the thomson reuters newsscope event indices. In: Mitra, G., Mitra, L. (eds.) The Handbook of New Analytics in Finance. Wiley, West Sussex (2011)
12. Henry, A.: Are investors influenced by how earnings press releases are written? J. Bus. Commun. **45**(4), 363–407 (2008)
13. Ingersoll, J.E.: Theory of Financial Decision Making. Rowman & Littlefield (1987)
14. Kearney, C., Liu, S.: Textual sentiment in finance : a survey of methods and models. Int. Rev. Fin. Anal. **33**, 171–185 (2014)
15. Kikuchi, T., Takahashi, H., Terano, T.: The propagation of bankruptcies of financial institutions? an agent model of financing behavior and asset price fluctuations. In: The 9th International Workshop on Agent-based Approach in Economic and Social Complex Systems (2015)
16. Loughran, T., McDonald, B.: When Is a liability not a liability? textual analysis, dictionaries, and 10-Ks. J. Fin. **66**(1), 35–65 (2011)
17. Luenberger, D.G.: Investment Science. Oxford University Press, Oxford (2000)
18. May, R.M. Arinaminpathy, N.: Systemic risk: the dynamics of model banking system. J. R. Soc. Interface **7**(46), 823–838 (2010)
19. O'Brien, P.C.: Analysts' forecasts as earnings expectations. J. Acc. Econ. **10**(1), 53–83 (1988)
20. Sharpe, W.F.: Capital asset prices: a theory of market equilibrium under condition of risk. J. Fin. **19**(3), 425–442 (1964)
21. Takahashi, H., Terano, T.: Agent-based approach to investors' behavior and asset price fluctuation in financial markets. J. Artif. Soc. Soc. Simul. **6**(3) (2003)
22. Takahashi, H.: An analysis of the influence of dispersion of valuations on financial markets through agent-based modeling. Int. J. Inf. Technol. Decis. Making **11**, 143–166 (2012)
23. Tetlock, P.C.: Giving content to investor sentiment: the role of media in the stock market. J. Fin. **62**(3), 1139–1168 (2007)
24. Tetlock, P.C., Saar-Tsechansky, M., Macskassy, S.: More than words: quantifying language to measure firms's fundamentals. J. Fin. **63**(3), 1437–1467 (2008)

Author Index

G. Jezic et al. (eds.), *Agent and Multi-Agent Systems: Technology and Applications*, Smart Innovation, Systems and Technologies 58,
DOI 10.1007/978-3-319-39883-9